Allosteric Enzymes

Editor

Guy Hervé, Ph.D.

Director of Research
Department of Enzymology
Centre National de la Recherche Scientifique
Gif-sur-Yvette, France

CRC Press, Inc.
Boca Raton, Florida

Library of Congress Cataloging-in-Publication Data

Allosteric enzymes/editor, Guy Hervé.
   p.   cm.
   Includes bibliographies and index.
   ISBN 0-8493-6854-5
   1. Allosteric enzymes. I. Hervé, Guy
QP601.A555 1989   88-19473
574.19′23—dc19   CIP

INTRODUCTION

Thirty years after the first report of feedback regulation of enzyme activity, an enormous amount of literature has accumulated on allostery and regulatory enzymes. The theory aimed at explaining the complex behavior of these enzymes and especially the phenomenon of cooperativity has been extensively developed. Under the pressure of experimental data, the two original types of models, namely the concerted model and the sequential model, were the object of modifications, bringing one closer to the other, to constitute a theoretical pattern in which these original models are now only extreme cases.

Experimentation is far behind these theoretical developments, and in spite of an over-whelming amount of laboratory investigations, none of the allosteric systems studied is entirely and clearly described in terms of the relationship between structure and properties.

The aim of this book is to assemble the experimental results obtained from a dozen of the most extensively studied enzymes, to compare their characteristics, to look for emerging common features, and to examine to what extent the conclusions that can be drawn on the basis of these experimental investigations actually meet the different aspects of the theoretical developments.

Although strictly speaking hemoglobin is not an enzyme, it could hardly be forgotten in a volume dealing, in a large part, with the phenomenon of molecular cooperativity. Apart from three to four enzymatic systems whose presence in this review was obviously justified, a choice had to be made and this presentation is, of course, not exhaustive. Some enzymes deserved to be here alongside those that are present. This injustice might be repaired in a further edition.

Guy Hervé

THE EDITOR

Guy L. Hervé, Ph.D. is Director of the Institute of Enzymology at the Centre National de la Recherche Scientifique (CNRS) in Gif-sur-Yvette, France.

Dr. Hervé obtained his training in Biology, Chemistry, Radiochemistry, and Biochemistry at the University of Paris. He obtained his Ph.D. in the Service de Biologie at the Nuclear Research Center in Saclay working on the enzymatic and chemical modifications of aminoacyl-tRNA. After doing postdoctoral work in Dr. Stark's laboratory in the Biochemistry Department at Stanford University in California, he was appointed laboratory chief in the Service de Biochimie in Saclay where he started to develop a research program on the biosynthesis and on the allosteric properties of aspartate transcarbamylase, a work for which he was awarded by the French Academy of Sciences. In 1975, he was appointed laboratory chief in the CNRS Institute of Enzymology in Gif-sur-Yvette, of which he became director in 1985. Dr. Hervé has taught as guest professor in many universities in France, North Africa, and North America. His current research interest is in the molecular mechanisms of allosteric regulation and cellular enzymology.

CONTRIBUTORS

K. Ravindra Acharya, Ph.D.
Doctor
Laboratory of Molecular Biophysics
University of Oxford
Oxford, England

David Barford, D. Phil.
Laboratory of Molecular Biophysics
University of Oxford
Oxford, England

Guang-Zuan Cai, Ph.D.
Research Associate & Postdoctoral
Department of Biochemistry
School of Medicine
St. Louis University
St. Louis, Missouri

David A. Case, Ph.D.
Member
Department of Molecular Biology
Research Institute of Scripps Clinic
La Jolla, California

Staffan Eriksson, Ph.D.
Assistant Professor
Department of Biochemistry I
Medical Nobel Institute
Karolinska Institute
Stockholm, Sweden

Bruce Gelin, Ph.D.
Production Marketing Manager
Polygen Corporation
Waltham, Massachusettes

Janos Hajdu, Ph.D.
Research Fellow
Laboratory of Molecular Biophysics
University of Oxford
Oxford, England

Yoav I. Henis, Ph.D.
Associate Professor
Department of Biochemistry
George S. Wise Faculty of Life Sciences
Tel Aviv University
Tel Aviv, Israel

Guy Hervé, Ph.D.
Director of Research
Department of Enzymology
C. N. R. S.
Gif-sur-Yvette, France

Lyndal K. Hesterberg, Ph.D.
Amgen Inc.
Boulder, Colorado

Boi Hanh Huynh, Ph.D.
Associate Professor
Department of Physics
Emory University
Atlanta, Georgia

Louise Johnson, Ph.D.
Lecturer
Department of Molecular Biophysics
University of Oxford
Oxford, England

Martin Karplus, Ph.D.
Professor
Department of Chemistry
Harvard University
Cambridge, Massachusetts

Michel Laurent, Ph.D.
Laboratory of Enzymology
University of Paris-Sud
Orsay, France

Angel Wai-mun Lee, Ph.D.
Medical Staff Fellow
National Heart, Lung, and Blood Institute
National Institutes of Health
Bethesda, Maryland

James C. Lee, Ph.D.
Professor
Department of Biochemistry
St. Louis University
St. Louis, Missouri

Alexander Levitzki, Ph.D.
Professor
Department of Biological Chemistry
The Hebrew University of Jerusalem
Jerusalem, Israel

Michael A. Luther, Ph.D.
Biochemistry Department
Schering Corp.
Bloomfield, New Jersey

Paul McLaughlin, Ph.D.
Department of Structural Studies
MRC Laboratory of Molecular Biology
Cambridge, England

Nikos G. Oikonomakos
Researcher
Department of Biochemistry
National Hellenic Research Foundation
Athens, Greece

Jacques Ricard, D.Sc.
Professor
Center of Biochemistry and Molecular
  Biology
C. N. R. S.
Marseille, France

Mosé Rossi
Professor
Department of Organic & Biological
  Chemistry
University of Naples
Naples, Italy

Britt-Marie Sjöberg, Ph.D.
Professor
Department of Molecular Biology
University of Stockholm
Stockholm, Sweden

David I. Stuart, Ph.D.
Lecturer
Laboratory of Molecular Biophysics
University of Oxford
Oxford, England

Robert L. Switzer, Ph.D.
Professor
Department of Biochemistry
University of Illinois
Urbana, Illinois

Attila Szabo, Ph.D.
Research Chemist
Laboratory of Chemical Physics
National Institute of Diabetes and
  Digestive and Kidney Diseases
National Institutes of Health
Bethesda, Maryland

P. D. J. Weitzman, D. Sci.
Dean of Science
Institute of Higher Education
Cardiff, Wales

Edward P. Whitehead, Ph.D.
Scientific Officer
Biology Division, D.-G. XII
Commission of the European
  Communities
Department of Biochemical Sciences
University of Rome
Rome, Italy

Jeannine M. Yon, Ph.D.
Director of Research & Professor
Department of Enzymology
C. N. R. S. — University of Paris-Sud
Orsay, France

TABLE OF CONTENTS

Chapter 1

CONCEPTS AND MODELS OF ENZYME COOPERATIVITY

Jacques Ricard

TABLE OF CONTENTS

I. INTRODUCTION

Twenty years after the formulation of the concepts of enzyme cooperativity and allostery,[1,2] it may appear, and it is often considered in practice, that the models which allow us to tackle quantitatively how subunit interaction in a polymeric enzyme may modulate its response to a given ligand have been formulated in their definitive form. Many review articles[3-13] and even two books[14,15] have been devoted to this matter, and one may therefore wonder whether any novel concept or problem related to enzyme cooperativity is still to be discovered.

The classical models of oligomeric enzyme[1,2] cooperativity are based on the theory of multiple equilibria[16-19] which indeed postulates the existence of thermodynamic equilibria between protein and the ligand molecules that are to be bound to the protein. However, what is of main importance in the field of enzyme cooperativity is not to understand how subunit interactions and conformational constraints modulate substrate binding to an enzyme, but rather to understand how these interactions and constraints control the rate of product appearance under steady state conditions.

Two different approaches of this problem have been followed so far. The first one simply consists in postulating, implicitly or explicitly, that the reaction rate is proportional to the fractional degree of saturation of the enzyme by the substrate. One may then attempt to fit reaction velocity data to either the Monod et al.[1] or the Koshland et al.[2] model. We shall see later on that this assumption is untenable. Another approach to study oligomeric enzyme cooperativity has been to derive tractable steady-state rate equations of allosteric enzyme reactions.[12,14,20-22] These equations however are purely phenomenological, that is, they tell us nothing about how subunit coupling and conformational constraints modulate the rate of product appearance, under steady-state conditions.

The main goal of this paper is precisely to offer models that allow an understanding of how subunit interactions of an oligomeric enzyme affect its kinetic cooperativity and how this type of cooperativity may differ from classical substrate binding cooperativity. These new concepts and models of enzyme kinetic cooperativity have already been presented elsewhere.[23-27]

II. MULTIPLE EQUILIBRIA AND LIGAND BINDING COOPERATIVITY

The classical models of protein cooperativity rest on the theory of multiple equilibria developed by Wyman.[16-19] It is therefore of interest to summarize briefly some basic concepts of binding cooperativity in the frame of multiple equlibria theory.

Let there be the sequential binding process of a ligand A on multisubunit protein. One has

$$E_0 \underset{}{\overset{K_1}{\rightleftharpoons}} E_1 \underset{}{\overset{K_2}{\rightleftharpoons}} E_2 \ldots E_{n-1} \underset{}{\overset{K_n}{\rightleftharpoons}} E_n \tag{1}$$

where E_0 is the unliganded protein, E_1 the protein state that has bound one ligand molecule and so on. K_1, K_2, ... K_n represent the macroscopic binding constants. The fractional saturation of the protein by the ligand, $\bar{\nu}$, may be defined as follows:

$$\bar{\nu} = n\bar{Y} = \frac{[E_1] + 2[E_2] + \ldots + n[E_n]}{[E_0] + [E_1] + \ldots + [E_n]} \tag{2}$$

$\bar{Y}$ is obviously the scaled fractional saturation. Setting

$$\psi_1 = K_1 \qquad \psi_2 = K_1K_2, \ldots, \qquad \psi_n = K_1K_2 \ldots K_n \tag{3}$$

Equation 2 assumes the form

$$\bar{\nu} = n\bar{Y} = \frac{\sum_{i=1}^{n} i \, \psi_i \, [A]^i}{1 + \sum_{i=1}^{n} \psi_i \, [A]^i} \tag{4}$$

which is the general expression of Adair equation.[28] A plot of $\bar{\nu}$, or $\bar{Y}$, as a function of [A] should, in general, result in a complex binding curve, which is the expression of cooperativity between sites.

Let us define

$$\rho = 1 + \sum_{i=1}^{n} \psi_i \, [A]^i \tag{5}$$

$\ln\rho$ is the binding polynomial.[16,17] Moreover one has

$$d \ln\rho = d \ln\left\{1 + \sum_{i=1}^{n} \psi_i \, [A]^i\right\} = \frac{d\left\{1 + \sum_{i=1}^{n} \psi_i \, [A]^i\right\}}{1 + \sum_{i=1}^{n} \psi_i \, [A]^i} \tag{6}$$

and

$$d\left\{1 + \sum_{i=1}^{n} \psi_i \, [A]^i\right\} = \sum_{i=1}^{n} i \, \psi_i \, [A]^{i-1} \, d \, [A] \tag{7}$$

$$d \, [A] = [A] \, d \ln \, [A] \tag{8}$$

Therefore,

$$d \ln \rho = \frac{\sum_{i=1}^{n} i \, \psi_i \, [A]^i}{1 + \sum_{i=1}^{n} \psi_i \, [A]^i} \, d \ln \, [A] \tag{9}$$

and the expression of the Adair equation becomes

$$\bar{\nu} = n\bar{Y} = \frac{\partial \ln \rho}{\partial \ln \, [A]} \tag{10}$$

This is the most general and elegant expression that can be given for a binding equation.

The expression of the ψs that appear in the binding polynomial relies on the values of the macroscopic binding constants K. But one may also define microscopic binding constants K's that refer to specified binding sites. If the binding sites are all equivalent, the following general relation between the macroscopic and microscopic binding constants assumes the form

$$K_i = \left\{\binom{n}{i} / \binom{n}{i-1}\right\} K_i' \tag{11}$$

which can be reexpressed as

$$K_i = \frac{n - i + 1}{i} K_i' \tag{12}$$

when the sites are all equivalent, the ψs may be expressed in terms of microscopic binding constants, namely

$$\psi_i = K_1 K_2 \ldots K_i = \prod_{j=1}^{i} \frac{n - j + 1}{j} K_j' \tag{13}$$

Since

$$\prod_{j=1}^{i} \frac{n - j + 1}{j} = \frac{n!}{i!(n - i)!} = \binom{n}{i} \tag{14}$$

the Expression 13 assumes the form

$$\psi_i = \binom{n}{i} K_1' K_2' \ldots K_i' \tag{15}$$

and the ρ-polynomial is now

$$\rho = 1 + \sum_{i=1}^{n} \binom{n}{i} K_1' K_2' \ldots K_i' [A]^i \tag{16}$$

This polynomial may be reexpressed as

$$\rho = \prod_{i=1}^{n} ([A] + \lambda_i) \tag{17}$$

where the λ_i are the Encke's roots of the equation $\rho = 0$. These values are equivalent to macroscopic dissociation constants and become identical to these constants if

$$\lambda_1 \gg \lambda_2 \gg \ldots \tag{18}$$

or if

$$\lambda_n \gg \lambda_{n-1} \gg \ldots \tag{19}$$

On assuming that the sites are equivalent and independent, the Adair equation is now

$$\bar{v} = n\bar{Y} = \sum_{i=1}^{n} \frac{\partial \ln([A] + \lambda_i)}{\partial \ln [A]} = \sum_{i=1}^{n} \frac{\partial([A] + \lambda_i)}{\partial [A]} \frac{[A]}{[A] + \lambda_i} \tag{20}$$

which can be reexpressed as

$$\bar{v} = n\bar{Y} = \sum_{i=1}^{n} \frac{[A]}{[A] + \lambda_i} \tag{21}$$

If Relation 18 or 19 applies, the Adair Equation 21 becomes

$$\bar{\nu} = n\bar{Y} = \sum_{i=1}^{n} \frac{[A]}{[A] + 1/K_i} = \sum_{i=1}^{n} \frac{K_i [A]}{1 + K_i [A]} \tag{22}$$

and is equivalent to a binding equation that describes the saturation of different classes of independent sites.

When the binding sites are all equivalent and independent, there is only one binding microscopic constant K' and the ρ-polynomial assumes the form

$$\rho = 1 + \sum_{i=1}^{n} \binom{n}{i} K'^i [A]^i = (1 + K' [A])^n \tag{23}$$

The expression of the binding function is now

$$\bar{\nu} = n\bar{Y} = \frac{\partial \ln(1 + K' [A])^n}{\partial \ln [A]} = \frac{\partial(1 + K' [A])^n}{\partial [A]} \frac{[A]}{(1 + K' [A])^n} \tag{24}$$

which is equivalent to

$$\bar{\nu} = n\bar{Y} = \frac{n K' [A]}{1 + K' [A]} \tag{25}$$

Obviously there is no cooperativity between sites under these conditions.

The existence of equilibrium binding cooperativity may therefore be derived from the mathematical structure of the binding polynomial. With the Monod et al.[1] model, for instance, the expression of the ρ-polynomial is

$$\rho = (1 + K'_R [A])^n + L(1 + K'_T [A])^n \tag{26}$$

where L is the transconformation constant between the R- and T-states. This expression is obviously different and cannot be reduced to Expression 23. Therefore, the Monod model should generate ligand binding cooperativity. This cooperativity is not suppressed if the binding is exclusive to the R-state, for the ρ-polynomial is now

$$\rho = (1 + K'_R [A])^n + L \tag{27}$$

which is still different from Equation 23. But if the two microscopic binding constants have equal values, the ρ-polynomial is now

$$\rho = (1 + L)(1 + K' [A])^n \tag{28}$$

which may be reduced to Equation 23. Under these conditions, the cooperativity has disappeared.

III. STRUCTURAL KINETICS AND KINETIC COOPERATIVITY UNDER NONEQUILIBRIUM CONDITIONS

The definition and expression of cooperativity which has been presented above is meaningful only if the system is in equilibrium state. This is clearly not so for an enzyme reaction occurring in the living cell. In order to understand how subunit interactions affect the rate of conversion of a substrate into a product, one has to develop a new type of formalism which has been termed structural kinetics.[23-27]

A. The Principles of Structural Kinetics

The basic idea of structural kinetics is that subunit interactions may have two types of effects on a rate process. On one hand, subunit interactions may alter the rate of conformational transitions associated with that process. This alteration of activation free energy is termed protomer arrangement energy contribution. Subunit interaction, on the other hand, may often generate site distortion. The energy contribution of this intersubunit strain is termed quaternary constraint energy contribution.

Let there be a chemical process, for instance the conversion of a substrate to a product, the free energy of activation associated with that process, $\Delta G^{\neq}$, may be split into different ideal energy contributions.

- The intrinsic energy contribution, $\Delta G^{\neq*}$, corresponds to what the free energy of activation would be if subunits were not displaying any interaction.
- The protomer arrangement energy contribution, $\Sigma(^{\alpha}\Delta G^{int})$, expresses how subunit arrangement, apart from any intersubunit strain, may modulate the free energy of activation. This energy contribution is equivalent to the free energy of dissociation of a polymer into the corresponding subunits, assuming that this process does not involve any conformation change. If there exists l subunits in the A conformation and m subunits in the B conformation, simple combinatorial analysis shows that

$$\Sigma(^{\alpha}\Delta G^{int}) = \binom{l}{2}(^{\alpha}\Delta G_{AA}) + \binom{m}{2}(^{\alpha}\Delta G_{BB}) + 1m(^{\alpha}\Delta G_{AB}) \tag{29}$$

- The quaternary constraint energy contribution, $\Sigma(^{\sigma}\Delta G^{int})$, expresses how intersubunit strain controls the corresponding free energy of activation through quaternary constraints. This energy contribution may be expressed through an expression analogous to Equation 24.

One has thus

$$\Delta G^{\neq} = \Delta G^{\neq*} + \Sigma(^{\alpha}\Delta G^{int}) + \Sigma(^{\sigma}\Delta G^{int}) \tag{30}$$

The intrinsic energy contribution allows one to define an intrinsic velocity constant

$$k^* = \frac{k_B T}{h} \exp(-\Delta G^{\neq*}/RT) \tag{31}$$

where k_B and h are the Boltzmann and the Planck constants, respectively. R and T have their usual significance. Expression 31 corresponds to what the rate constant would be if subunits were isolated. The energy contributions $\Sigma(^{\alpha}\Delta G^{int})$ and $\Sigma(^{\sigma}\Delta G^{int})$ allow one to define dimensionless parameters α and σ analogous to dissociation constants

$$\alpha = \exp\{-(^{\alpha}\Delta G^{int})/RT\}$$

$$\sigma = \exp\{-(^{\sigma}\Delta G^{int})/RT\} \tag{32}$$

and the corresponding expression of the rate constant is of the type

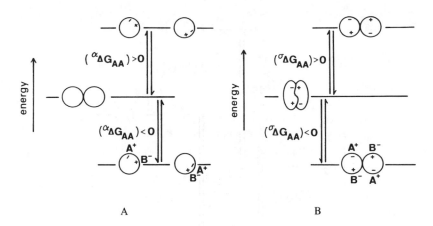

FIGURE 1. Energy diagrams which show that $(^\alpha\Delta G)$ and $(^\sigma\Delta G)$ may be positive or negative. (A) The $(^\alpha\Delta G)$ may be positive or negative; (B) The $(^\sigma\Delta G)$ may be positive or negative. (From Ricard, J. and Cornish-Bowden, A., *Eur. J. Biochem.*, 166, 255, 1987. With permission.)

$$k = k^* \, \Pi\alpha \, \Pi\sigma \qquad (33)$$

This expression is structural, in the sense that it associates a rate constant to thermodynamic parameters, α and σ, directly related to structure and structure changes of the enzyme.

It is important at this stage to express clearly what is the physical significance of these α and σ-parameters and to realize which numerical values they may take. If, for instance, a dimeric enzyme is maintained in its aggregated state by electrostatic interactions, the α-coefficients correspond to the dissociation constants of the dimer. By convention, the $(^\alpha\Delta G^{int})$ contributions are defined as the energy difference between the dissociated and the aggregated states. These energy contributions, for instance $(^\alpha\Delta G_{AA})$, may be either positive or negative (Figure 1A). This contribution is positive if the subunits ideally isolated are destabilized relative to the aggregated state. Alternatively, this contribution is negative if the ideally isolated subunits are stabilized with respect to the dimeric state (Figure 1A). This stabilization may be brought about by interaction of positive and negative charges with counterions present in the medium. These dissociation processes are indeed ideal in the sense that the enzyme is assumed to remain in the aggregated state because the energy barrier of the dissociation process is too high to be reached.

If, owing to quaternary constraints, the dimeric enzyme exists in a distorted state, the free energy of relief of intersubunit strain allows definition of the $(^\sigma\Delta G^{int})$ energy contributions and therefore the σ parameters. Here again, these energy contributions $(^\sigma\Delta G^{int})$ are defined as the energy difference between the unstrained and the strained states. One would normally expect the ideal unstrained state to be destabilized with respect to the real strained state. However, if the positive and negative charges that appear on the subunits are neutralized by counterions present in the medium, the ideal unstrained state will be stabilized relative to the real strained state (Figure 1B). These processes of strain relief, as defined by the energy $(^\sigma\Delta G^{int})$ or σ-parameters, are indeed ideal, for the corresponding activation energies are too high to be reached. From the definition of the $(^\alpha\Delta G^{int})$ and $(^\sigma\Delta G^{int})$ energies, as well as that of the α- and σ-parameters, one must have

$$0 < \alpha < 1 \qquad \text{if} \qquad (^\alpha\Delta G^{int}) > 0 \qquad (34)$$

and

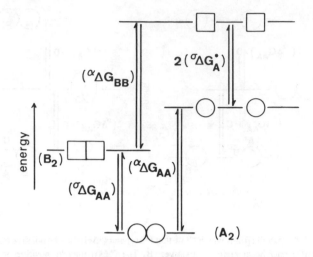

FIGURE 2.   Thermodynamic evidence that conformation B_2 is destabilized with respect to A_2.

$$\alpha > 1 \qquad \text{if} \qquad (^{\alpha}\Delta G^{int}) < 0 \tag{35}$$

Alternatively one has

$$0 < \sigma < 1 \qquad \text{if} \qquad (^{\sigma}\Delta G^{int}) > 0 \tag{36}$$

and

$$\sigma > 1 \qquad \text{if} \qquad (^{\sigma}\Delta G^{int}) < 0 \tag{37}$$

A particularly interesting situation is obtained if the enzyme, solely or largely, exists in the unstrained state A_2. The ideal, or real, strained state B_2 has then to be destabilized relative to the A_2 state. From the energy diagram shown in Figure 2, one may easily relate the intersubunit strain energy $(^{\sigma}\Delta G_{AA})$ to the intrasubunit strain energy $(^{\sigma}\Delta G_A^*)$, that is, the energy required to strain the conformation A into a conformation B. The intrasubunit strain coefficient σ_A^* is defined as

$$\sigma_A^* = \exp\{-(^{\sigma}\Delta G_A^*)/RT\} \tag{38}$$

and has a value between 0 and 1, since $(^{\sigma}\Delta G_A^*)$ is of necessity positive. The result, that the A_2 state has to be stabilized with respect to the B_2 state, implies (Figure 2) that

$$\sigma_{AA} = \frac{\alpha_{BB}}{\alpha_{AA}\,\sigma_A^{*2}} > 1 \tag{39}$$

B. Fundamental Postulates of Structural Kinetics — Coupling between Subunits

The principles which have been presented above allow derivation in structural terms of any kind of rate equation and binding isotherm. These rate equations, however, are usually extremely complex if no assumption is made as to subunit structure and coupling in the ground and transition states. The equations become beautifully simple if the three following postulates hold, at least approximately.

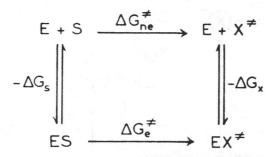

FIGURE 3. Thermodynamic box which shows that an enzyme should bind the transition state more tightly than the substrate.

1. The quaternary constraints must be relieved upon reaching either of the transition states along the reaction coordinate. This postulate is more than a simple assumption and corresponds, in fact, to the extrapolation, to polymeric enzymes, of what is known to apply to monomeric enzymes. Numerous studies have shown that conformational constraints in a monomeric enzyme have to be relieved when reaching a transition state. This appears as a consequence of simple thermodynamic principles. Let there be the simple one-substrate, one-product enzyme reaction

$$E \underset{k_{-1}}{\overset{k_1[S]}{\rightleftharpoons}} ES \underset{k_{-e}}{\overset{k_e}{\rightleftharpoons}} EP \underset{k_{-2}[P]}{\overset{k_2}{\rightleftharpoons}} E$$

Comparison of the rate constant k_e of the enzyme-catalyzed reaction to that of the corresponding uncatalyzed reaction

$$S \underset{k_{-ne}}{\overset{k_{ne}}{\rightleftharpoons}} P$$

is easily performed by using the thermodynamic box of Figure 3.
Since free energies are state functions, one has

$$\Delta G_{ne}^{\neq} - \Delta G_e^{\neq} = \Delta G_x - \Delta G_s \qquad (40)$$

Since $\Delta G_{ne}^{\neq} \gg \Delta G_e^{\neq}$, it follows that $\Delta G_x \gg \Delta G_s$. The difference between the binding energies of the transition state and of the substrate to the enzyme implies that this enzyme is more complementary to the transition state than to the substrate. This, in turn, implies that conformational constraints are, at least in part, relieved when the enzyme has bound the transition state.[29-37] In the case of polymeric enzymes, it is evident that intrasubunit strain may be relieved only if quaternary constraints have been relieved as well. The postulate of a relief of intersubunit strain in the transition states is thus an application, to polymeric enzymes, of simple physical principles that are known to apply to monomeric enzymes.

2. In the absence of quaternary constraints, the subunits exist in the minimum number of conformational states, namely two, called A and B. The substrate and the product

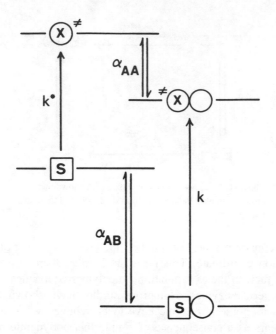

FIGURE 4.   Free energy diagram and loose coupling
of subunits. This diagram allows the derivation of a
structural expression of the catalytic rate constant k.
(From Ricard, J. and Cornish-Bowden, A., *Eur. J.
Biochem.*, 166, 255, 1987. With permission.)

stabilize the same conformation B. This postulate of the minimum number of conformations is not novel and has been used implicitly or explicitly in the Monod[1] and
Koshland[2] models.

3.   A subunit which has bound a transition state has the same conformation, whatever the
nature of this transition state. This conformation is identical to that of the unliganded
subunit, namely A. This postulate derives from the previous one (minimum number
of conformations).

Obviously, the three postulates above are certainly not always fulfilled. Nevertheless,
they allow to express *simple* models of the modulation exerted by subunit interactions on
the reaction rate.

The subunits of an oligomeric enzyme are said to be loosely coupled if the energy
contribution of quaternary constraints is nil, that is if

$$\Sigma(^{\sigma}\Delta G^{int}) = 0 \tag{41}$$

This implies that the σ-parameters all equal unity. Then subunit interactions modulate the
rate processes by altering the rate of conformational transitions involved in that process. For
the catalytic process considered thus far, one may derive the energy diagram shown in Figure
4, and this allows expression of the structural equation of the catalytic rate constant k in
terms of the intrinsic rate constant k* and of the α-parameters. One has thus

$$k = k^* \frac{\alpha_{AB}}{\alpha_{AA}} \tag{42}$$

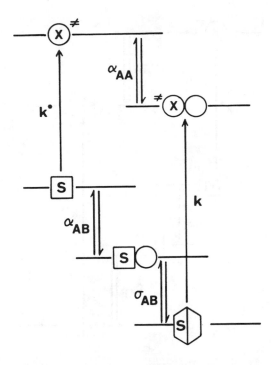

FIGURE 5. Free energy diagram and tight coupling of subunits. This diagram allows derivation of a structural expression of the catalytic rate constant k. Quaternary constraints are assumed not to be relieved in the ground states. (From Ricard, J. and Cornish-Bowden, A., *Eur. J. Biochem.*, 166, 255, 1987. With permission.)

and it appears obvious that this relation remains unchanged whatever the dimeric state is stabilized ($0 < \alpha < 1$) or destabilized ($\alpha > 1$) relative to the ideal monomeric state.

If now the subunits are coupled, one has

$$\Sigma(^\sigma\Delta G^{int}) \neq 0 \tag{43}$$

and the σ-parameters are different from unity. If subunit coupling is very tight, in such a way that conformation changes occur in a concerted way, one may have, for instance, the situation depicted in Figure 5. From this energy diagram one may derive the structural expression of the catalytic rate constant, namely

$$k = k^* \frac{\alpha_{AB}}{\alpha_{AA}} \sigma_{AB} \tag{44}$$

In the situation pictured in Figure 4, none of the strained subunits in the dimer has, owing to quaternary constraints, the A or the B conformation.

It may well occur, however, that quaternary constraints be relieved in the nonliganded and fully liganded states of the enzyme. These constraints then exist in the partially liganded states only. Thus, it becomes possible to express the extent of intersubunit strain from that of intrasubunit strain. Let there be, for instance, the situation shown in the energy diagram of Figure 6 where the nonliganded (square) subunit is strained. Structural Equation 44 applies to this situation. The energy diagram of Figure 7, however, allows expression of the extent of intersubunit strain σ_{AB} in terms of intrasubunit strain σ_A^*. One has

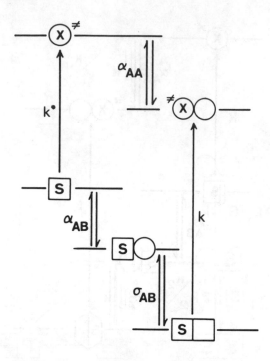

FIGURE 6.   Free energy diagram and tight coupling of subunits. This diagram allows derivation of a structural expression of the catalytic rate constant k. Quaternary constraints are assumed to be relieved in the unliganded and fully liganded ground states.

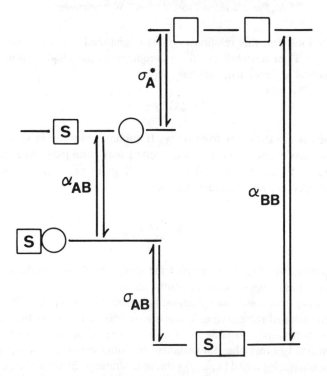

FIGURE 7.   Relationship between intersubunit strain and intrasubunit strain.

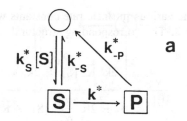

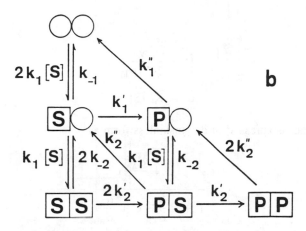

FIGURE 8. Structural kinetic model of a dimeric enzyme with loosely coupled subunits. The figure shows (a) the intrinsic process and (b) the kinetic process for the dimer. (From Ricard, J. and Cornish-Bowden, A., *Eur. J. Biochem.* 166, 255, 1987. With permission.)

$$\sigma_{AB} = \frac{\alpha_{BB}}{\alpha_{AB}} \frac{1}{\sigma_A^*} \tag{45}$$

and substituting into Equation 44 yields

$$k = k^* \frac{\alpha_{BB}}{\alpha_{AA}} \frac{1}{\sigma_A^*} \tag{46}$$

IV. STRUCTURAL RATE AND BINDING EQUATIONS

Let us consider, first, the case of a dimer with loosely coupled subunits (Figure 8). The postulates that have been formulated previously introduce constraints between rate constants in such a way that the rate equation is expressed by the ratio of two polynomials in $[S]^2$ (2:2 rate equation). The apparent catalytic constant of the intrinsic process, $\bar{k}^*$, is

$$\bar{k}^* = \frac{k^* k_{-p}^*}{k^* + k_{-p}^*} \tag{47}$$

and the apparent intrinsic affinity constant assumes the form

$$\bar{K}^* = \frac{k_s^*(k^* + k_{-p}^*)}{k_{-p}^*(k^* + k_{-s}^*)} \tag{48}$$

The significance of the various intrinsic rate constants which appear in these expressions is to be found in Figure 8. The corresponding structural rate equation is

$$\frac{v}{[E]_0} = \frac{2\bar{k}^* \bar{K}^* [S] + 2\bar{k}^* \bar{K}^{*2} \frac{\alpha_{AA}}{\alpha_{AB}} [S]^2}{1 + 2\bar{K}^* \frac{\alpha_{AA}}{\alpha_{AB}} [S] + \bar{K}^{*2} \frac{\alpha_{AA}}{\alpha_{BB}} [S]^2} \tag{49}$$

Setting

$$\bar{V} = \frac{v}{2\bar{k}^* [E]_0}$$

$$c = \bar{K}^* [S] \tag{50}$$

Equation 49 may be reexpressed in dimensionless form as

$$\bar{V} = \frac{c + \frac{\alpha_{AA}}{\alpha_{AB}} c^2}{1 + 2 \frac{\alpha_{AA}}{\alpha_{AB}} c + \frac{\alpha_{AA}}{\alpha_{BB}} c^2} \tag{51}$$

when $\alpha_{AA} = \alpha_{AB} = \alpha_{BB}$ Equations 49 and 51 become Michaelian and reduce to

$$\frac{v}{[E]_0} = \frac{2\bar{k}^* \bar{K}^* [S]}{1 + \bar{K}^* [S]} \tag{52}$$

and to

$$\bar{V} = \frac{c}{1 + c} \tag{53}$$

respectively.

The corresponding equilibrium binding equation of the substrate to the enzyme is

$$\bar{\nu} = 2\bar{Y} = \frac{2K^* \frac{\alpha_{AA}}{\alpha_{AB}} [S] + 2K^{*2} \frac{\alpha_{AA}}{\alpha_{BB}} [S]^2}{1 + 2K^* \frac{\alpha_{AA}}{\alpha_{AB}} [S] + K^{*2} \frac{\alpha_{AA}}{\alpha_{BB}} [S]^2} \tag{54}$$

and making use of the same dimensionless variable $c = \bar{K}^* [S]$, one has

$$\bar{Y} = \frac{\frac{\alpha_{AA}}{\alpha_{AB}} c + \frac{\alpha_{AA}}{\alpha_{BB}} c^2}{1 + 2 \frac{\alpha_{AA}}{\alpha_{AB}} c + \frac{\alpha_{AA}}{\alpha_{BB}} c^2} \tag{55}$$

Obviously Equations 49 and 54 as well as Equations 51 and 55 are different, and this implies that the effects of subunits are different on the reaction rate and on the substrate binding function. This matter will be discussed later on.

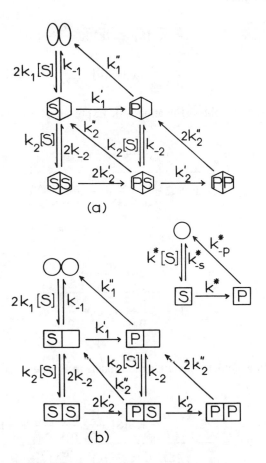

FIGURE 9. Structural kinetic model of a dimeric enzyme with tightly coupled subunits. (a) Quaternary constraints are not relieved in the unliganded and fully liganded states. (b) Quaternary constraints are relieved in the unliganded and fully liganded states. (From Ricard, J. and Cornish-Bowden, A., *Eur. J. Biochem.*, 166, 255, 1987. With permission.)

Let us consider now the case where the subunits are tightly coupled (Figure 9). If the quaternary constraints are not relieved in the nonliganded and fully liganded states, the relevant structural rate equation assumes the form

$$\frac{v}{[E]_0} = \frac{2\bar{k}^* \sigma_{AA} \bar{K}^* [S] + 2\bar{k}^* \sigma_{AA} \dfrac{\alpha_{AA}}{\alpha_{AB}} \bar{K}^{*2} [S]^2}{1 + 2\dfrac{\alpha_{AA}}{\alpha_{AB}} \dfrac{\sigma_{AA}}{\sigma_{AB}} \bar{K}^* [S] + \dfrac{\alpha_{AA}}{\alpha_{BB}} \dfrac{\sigma_{AA}}{\sigma_{BB}} \bar{K}^{*2} [S]^2} \tag{56}$$

This equation is still 2:2 in substrate concentration and reduces to Equation 49 if $\sigma_{AA} = \sigma_{AB} = \sigma_{BB} = 1$. If now the subunits are all in conformations A or B, intersubunit strain must be relieved in the nonliganded and fully liganded states. One must have $\sigma_{AA} = \sigma_{BB} = 1$ and Equation 45 applies. Equation 56 then becomes

$$\frac{v}{[E]_0} = \frac{2\bar{k}^* \bar{K}^* [S] + 2\bar{k}^* \bar{K}^{*2} \frac{\alpha_{AA}}{\alpha_{AB}} [S]^2}{1 + 2 \frac{\alpha_{AA}}{\alpha_{BB}} \sigma_A^* \bar{K}^* [S] + \frac{\alpha_{AA}}{\alpha_{BB}} \bar{K}^{*2} [S]^2} \tag{57}$$

By using dimensionless variables (Equation 50), Expressions 56 and 57 assume the forms

$$\bar{V} = \frac{\sigma_{AA} c + \frac{\alpha_{AA}}{\alpha_{AB}} \sigma_{AA} c^2}{1 + 2 \frac{\alpha_{AA}}{\alpha_{AB}} \frac{\sigma_{AA}}{\sigma_{AB}} c + \frac{\alpha_{AA}}{\alpha_{BB}} \frac{\sigma_{AA}}{\sigma_{BB}} c^2} \tag{58}$$

and

$$\bar{V} = \frac{c + \frac{\alpha_{AA}}{\alpha_{AB}} c^2}{1 + 2 \frac{\alpha_{AA}}{\alpha_{BB}} \sigma_A^* c + \frac{\alpha_{AA}}{\alpha_{BB}} c^2} \tag{59}$$

respectively.

The substrate binding isotherm pertaining to models of Figure 9a and 9b may be expressed as

$$\bar{v} = 2\bar{Y} = \frac{2 \frac{\alpha_{AA}}{\alpha_{AB}} \frac{\sigma_{AA}}{\sigma_{AB}} K^* [S] + 2 \frac{\alpha_{AA}}{\alpha_{BB}} \frac{\sigma_{AA}}{\sigma_{BB}} K^{*2} [S]^2}{1 + 2 \frac{\alpha_{AA}}{\alpha_{AB}} \frac{\sigma_{AA}}{\sigma_{AB}} K^* [S] + \frac{\alpha_{AA}}{\alpha_{BB}} \frac{\sigma_{AA}}{\sigma_{BB}} K^{*2} [S]^2} \tag{60}$$

and

$$\bar{v} = 2\bar{Y} = \frac{2 \frac{\alpha_{AA}}{\alpha_{BB}} \sigma_A^* K^* [S] + 2 \frac{\alpha_{AA}}{\alpha_{BB}} K^{*2} [S]^2}{1 + 2 \frac{\alpha_{AA}}{\alpha_{BB}} \sigma_A^* K^* [S] + \frac{\alpha_{AA}}{\alpha_{BB}} K^{*2} [S]^2} \tag{61}$$

respectively. In dimensionless form, these two equations become

$$\bar{Y} = \frac{\frac{\alpha_{AA}}{\alpha_{AB}} \frac{\sigma_{AA}}{\sigma_{AB}} c + \frac{\alpha_{AA}}{\alpha_{BB}} \frac{\sigma_{AA}}{\sigma_{BB}} c^2}{1 + 2 \frac{\alpha_{AA}}{\alpha_{AB}} \frac{\sigma_{AA}}{\sigma_{AB}} c + \frac{\alpha_{AA}}{\alpha_{BB}} \frac{\sigma_{AA}}{\sigma_{BB}} c^2} \tag{62}$$

and

$$\bar{Y} = \frac{\frac{\alpha_{AA}}{\alpha_{BB}} \sigma_A^* c + \frac{\alpha_{AA}}{\alpha_{BB}} c^2}{1 + 2 \frac{\alpha_{AA}}{\alpha_{BB}} \sigma_A^* c + \frac{\alpha_{AA}}{\alpha_{BB}} c^2} \tag{63}$$

V. SUBSTRATE BINDING AND KINETIC COOPERATIVITY — AMPLIFICATION, ATTENUATION, AND INVERSION EFFECTS

Cooperativity is usually defined, under equilibrium conditions, by the extreme Hill coefficient, that is, the extreme of the function

$$\overline{h} = \frac{d \ln\{\overline{Y}/(1 - \overline{Y})\}}{d \ln [S]} = \frac{[S]}{\overline{Y}(1 - \overline{Y})} \frac{d\overline{Y}}{d [S]} \tag{64}$$

Under steady-state conditions, the kinetic cooperativity is defined, in the same way, by the extreme of the function

$$\tilde{h} = \frac{d \ln\{(v/Vm)/(1 - v/Vm)\}}{d \ln [S]} = \frac{[S]}{(v/Vm)(1 - v/Vm)} \frac{d(v/Vm)}{d [S]} \tag{65}$$

If the subunits of a dimeric enzyme are loosely coupled, one may show that the extreme of $\overline{h}$, $\overline{h}_{1/2}$, is obtained at half-saturation by the substrate and is expressed as

$$\overline{h}_{1/2} = \frac{2\sqrt{\rho}}{1 + \sqrt{\rho}} \tag{66}$$

where

$$\rho = \frac{\alpha^2_{AB}}{\alpha_{AA} \, \alpha_{BB}} \tag{67}$$

when $\rho > 1$, $\tilde{h}_{1/2} > 1$ and substrate binding cooperativity is positive. When $0 < \rho < 1$, $0 < \tilde{h}_{1/2} < 1$ and substrate binding cooperativity is negative. When $\rho = 1$, $\tilde{h}_2^1 = 1$ and there is no substrate binding cooperativity. The thermodynamic significance of the ρ-parameter is obvious from the energy diagram of Figure 10. If the energy of the hybrid AB state is lower than half of the energy difference between the B_2 and the A_2 states, $\rho > 1$ and the substrate binding cooperativity is positive. If the energy of the hybrid AB state is higher than half of this difference, $\rho < 1$ and the substrate binding cooperativity is negative.

The extreme kinetic Hill coefficient for a dimeric enzyme with loosely coupled subunits is

$$\tilde{h}_{ext} = \frac{2}{1 + \sqrt{2 - \rho}} \tag{68}$$

The ρ-parameter is still the one defined in Equation 67. When $0 < \rho < 1$, $0 < \tilde{h}_{ext} < 1$ and the kinetic cooperativity is negative. Kinetic cooperativity is positive ($\tilde{h}_{ext} > 1$) if $2 > \rho > 1$. If $\rho > 2$, the reaction is inhibited by an excess substrate and the very concept of the Hill coefficient is meaningless. Therefore, kinetic and substrate binding cooperativities have the same sign (positive or negative). The extent of substrate binding and kinetic cooperativities, however, may be quite different. Comparison of Equations 66 and 68 shows that if cooperativity is positive, kinetic cooperativity is amplified with respect to substrate binding cooperativity. Alternatively, if cooperativity is negative, kinetic cooperativity is attenuated with respect to substrate binding cooperativity. In Figure 11, the variation of $\tilde{h}_{1/2}$ and $\tilde{h}_{ext}$ are shown as a function of the thermodynamic parameter ρ.

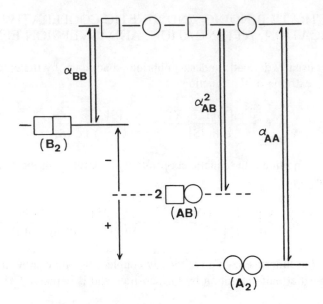

FIGURE 10. Thermodynamic significance of the ρ-parameter. If the energy level of the hybrid AB state is below the midpoint between the energies of A_2 and B_2, substrate binding cooperativity is positive. If this energy is above that midpoint, cooperativity is negative. (From Ricard, J. and Cornish-Bowden, A., *Eur. J. Biochem.*, 166, 255—272, 1987. With permission.)

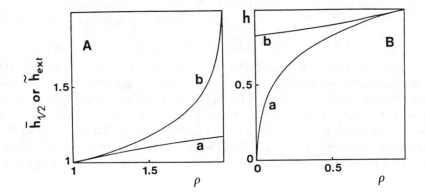

FIGURE 11. Attenuation and amplification of the kinetic cooperativity with respect to substrate binding cooperativity; A, positive cooperativity; B, negative cooperativity. In either panel curve (a) represents substrate binding cooperativity ($\tilde{h}_{1/2}$) and curve (b) kinetic cooperativity ($\tilde{h}_{ext}$). The subunits are assumed to be loosely coupled (Figure 8). (From Ricard, J. and Noat, G., *J. Theor. Biol.*, 117, 633, 1985. With permission.)

It is often considered that positive kinetic cooperativity is synonymous with sigmoidicity of the rate curve. From a mathematical viewpoint, this is not necessarily true. Positive kinetic cooperativity occurs when $\tilde{h}_{ext} > 1$ and sigmoidicity when

$$\frac{\partial^2 \overline{V}}{\partial c^2} = 0 \qquad (69)$$

for a real positive value of c. These two conditions are indeed different. For a dimeric enzyme with loosely coupled subunits, one may show that Condition 69 never holds for a positive value of c. Therefore, loose coupling of subunits can generate positive kinetic cooperativity, but not sigmoidicity of the rate curve.

If subunits are tightly coupled according to the model of Figure 9a, the expression of the extreme binding Hill coefficient is

$$\bar{h}_{1/2} = \frac{2\sqrt{\rho'}}{1 + \sqrt{\rho'}} \tag{70}$$

where

$$\rho' = \frac{\alpha_{AB}^2 \, \sigma_{AB}^2}{\alpha_{AA} \, \sigma_{AA} \, \alpha_{BB} \, \sigma_{BB}} \tag{71}$$

The thermodynamic significance of ρ' will not be discussed here for it is very similar to that of the ρ-parameter (Figure 10). As previously, when $0 < \rho' < 1$, $0 < \tilde{h}_{ext} < 1$, and the cooperativity is negative. Alternatively, when $\rho > 1$, $\tilde{h}_{1/2} > 1$ and the cooperativity is positive. The corresponding extreme kinetic Hill coefficient, under the conditions described by the model of Figure 9a is

$$\tilde{h}_{ext} = \frac{2}{1 + \sqrt{\dfrac{\sigma_{AA}}{\sigma_{AB}} \left(2 - \dfrac{\sigma_{AA}}{\sigma_{AB}} \rho'\right)}} \tag{72}$$

Comparison with Equation 70 shows that whereas substrate binding cooperativity is solely defined by the ρ'-parameter, the kinetic cooperativity, under nonequilibrium conditions, is defined by both ρ' and σ_{AA}/σ_{AB}. Since ρ' and the ratio σ_{AA}/σ_{AB} may take independent values, one may expect substrate binding and kinetic cooperativities to be different. If substrate binding cooperativity is positive ($\rho' > 1$) or negative ($0 < \rho' < 1$) one may predict the sign of the kinetic cooperativity by studying the sign of the function.

$$y = \rho' \left(\frac{\sigma_{AA}}{\sigma_{AB}}\right)^2 - 2\frac{\sigma_{AA}}{\sigma_{AB}} + 1 \tag{73}$$

If $y > 0$, kinetic cooperativity is positive; if $y = 0$, there is no cooperativity, and if $y < 0$, kinetic cooperativity is negative. Simple inspection of Equation 73 shows that if $\rho' > 1$, y can only be positive. Therefore, when substrate binding cooperativity is positive, kinetic cooperativity can only positive as well. If $0 < \rho' < 1$, y is negative if

$$\frac{1 - \sqrt{1 - \rho'}}{\rho'} < \frac{\sigma_{AA}}{\sigma_{AB}} < \frac{1 + \sqrt{1 - \rho'}}{\rho'} \tag{74}$$

or positive if

$$\frac{1 - \sqrt{1 - \rho'}}{\rho'} < \frac{1 + \sqrt{1 - \rho'}}{\rho'} < \frac{\sigma_{AA}}{\sigma_{AB}} \tag{75}$$

or if

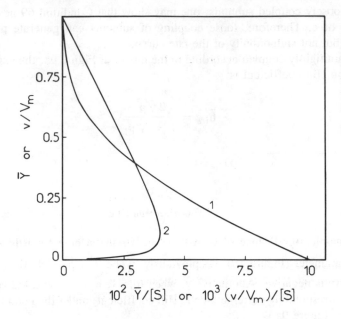

FIGURE 12.   Example of inversion of kinetic cooperativity with respect to substrate binding cooperativity for the model of Figure 9 (a). Curve 1 shows a negative substrate binding cooperativity, and curve 2 the corresponding kinetic cooperativity. The numerical values of the parameters are found in Reference 25. (From Ricard, J. and Cornish-Bowden, A., *Eur. J. Biochem.*, 166, 255—272, 1987. With permission.)

$$\frac{\sigma_{AA}}{\sigma_{AB}} < \frac{1 - \sqrt{1 - \rho'}}{\rho'} < \frac{1 + \sqrt{1 - \rho'}}{\rho'} \tag{76}$$

Then when substrate binding cooperativity is negative, kinetic cooperativity may be either positive or negative. Therefore, owing to the tight coupling of subunits, negative substrate binding cooperativity may generate positive kinetic cooperativity. The inversion of negative substrate binding cooperativity into a positive kinetic cooperativity is shown in Figure 12. If $\rho' = 1$, there is no substrate binding cooperativity, but if $\sigma_{AA}/\sigma_{AB} \neq 1$, kinetic cooperativity is positive. This is illustrated in Figure 13. Therefore, the existence of kinetic cooperativity is not necessarily accompanied by substrate binding cooperativity. What may generate kinetic cooperativity is the existence of a difference of intersubunit strain in the nonliganded and half-liganded states.

As previously mentioned, it may well occur that intersubunit strain is relieved in the nonliganded and fully liganded states. This is the situation depicted in Figure 9b. Under these conditions the substrate binding Hill coefficient is

$$\bar{h}_{1/2} = \frac{2\sqrt{\rho''}}{1 + \sqrt{\rho''}} \tag{77}$$

where

$$\rho'' = \frac{\alpha_{BB}}{\alpha_{AA}} \frac{1}{\sigma_A^{*2}} \tag{78}$$

Equations 78 and 39 show that ρ'' is of necessity greater than unity. Therefore, the substrate

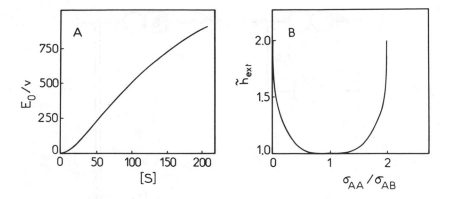

FIGURE 13. Positive kinetic cooperativity not accompanied by any substrate binding cooperativity for the model of Figure 9 (a). (A) Example of sigmoidal kinetic behavior not
accompanied by any substrate binding cooperativity; (B) variation of the kinetic cooperativity
as a function of the strain in the nonliganded and half-liganded states. The numerical values
of the parameters are found in Reference 25. (From Ricard, J. and Cornish-Bowden, A.,
Eur. J. Biochem., 166, 255—272, 1987. With permission.)

binding cooperativity can only be positive. The corresponding extreme kinetic Hill coefficient
may be expressed as

$$\tilde{h}_{ext} = \frac{2}{1 + \sqrt{\frac{\alpha_{AB}}{\alpha_{BB}} \sigma_A^* \left(2 - \frac{\alpha_{AB}}{\alpha_{BB}} \sigma_A^* \rho''\right)}}$$ (79)

Whereas the substrate binding cooperativity is solely determined by the ρ''-parameter, the
kinetic cooperativity is defined by both ρ'' and the ratio $\alpha_{AB}\sigma_A^*/\alpha_{BB}$. As previously discussed,
the nature of kinetic cooperativity may be predicted by studying the sign of the function.

$$z = \rho'' \left(\frac{\alpha_{AB} \sigma_A^*}{\alpha_{BB}}\right)^2 - 2 \frac{\alpha_{AB} \sigma_A^*}{\alpha_{BB}} + 1$$ (80)

Since ρ'' is of necessity positive, z must be positive as well. Therefore, when the conformational constraints are relieved in the nonliganded and fully liganded states, both the
substrate binding and the kinetic cooperativities are positive.

With tight coupling of subunits, sigmoidicity of the rate curve occurs if

$$\frac{1}{2} > \frac{\sigma_{AA}}{\sigma_{AB}}$$ (81)

for the model of Figure 9a, or if

$$\frac{1}{2} > \frac{\alpha_{AB} \sigma_A^*}{\alpha_{BB}}$$ (82)

for the model of Figure 9b. This implies that tight coupling of subunits may generate
sigmoidicity of the rate curve when the extent of intersubunit strain in the half-liganded state
is more than twice that occurring in the nonliganded state. The smaller the ratio σ_{AA}/σ_{AB}
or $\alpha_{AB} \sigma_A^*/\alpha_{BB}$ which appear in Equations 72 or 79 the larger is the sigmoidicity of the rate

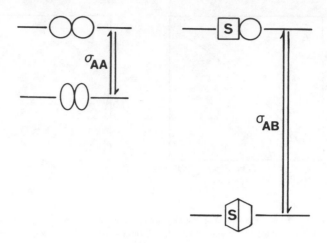

FIGURE 14.   Thermodynamic significance of the ratio σ_{AA}/σ_{AB} which controls the sigmoidicity.

curve. This is illustrated in Figure 14. From these results, it appears obvious that, even with tightly coupled subunits, strong positive cooperativity may well not be associated with any sigmoidicity.

VI. CATALYTIC EFFICIENCY OF COOPERATIVE ENZYMES

The structural formalism which has been presented, allows comparison of the extent of catalytic activity of the polymeric enzyme with respect to that of an ideally isolated subunit. This matter is of interest since, according to the classical Monod model, cooperativity is exerted at the expense of the affinity of a ligand for the protein. The larger the affinity of the ligand for the protein, the weaker is the sigmoidicity of the binding curve. Similarly, one may wonder whether kinetic cooperativity has to be paid for at the expense of catalytic efficiency. This question may be answered by comparing the actual scaled steady-state velocity of the cooperative enzyme with that of an ideal naked subunit.

If subunits are loosely coupled, the difference between the scaled velocities of the dimeric enzyme and the ideal naked subunit is

$$\overline{V} - \overline{V}^* = \frac{\left(1 - \dfrac{\alpha_{AA}}{\alpha_{AB}}\right)c^2 + \alpha_{AA}\left(\dfrac{1}{\alpha_{AB}} - \dfrac{1}{\alpha_{BB}}\right)c^3}{\left(1 + 2\dfrac{\alpha_{AA}}{\alpha_{AB}}c + \dfrac{\alpha_{AA}}{\alpha_{BB}}c^2\right)(1 + c)} \tag{83}$$

If $\alpha_{AA} < \alpha_{AB} < \alpha_{BB}$, cooperativity is positive but whatever the substrate concentration, the difference $\overline{V} - \overline{V}^*$ is always positive. This implies that loose subunit coupling may enhance the steady-state velocity of the dimer with respect to the ideal naked subunit.

If subunits are tightly coupled and if the strain is not relieved in the ground states, (Figure 9a), the expression of $\overline{V} - \overline{V}^*$ assumes the form

$$\overline{V} - \overline{V}^* = \frac{(\sigma_{AA} - 1)c + \sigma_{AA}\dfrac{\alpha_{AA}}{\alpha_{AB}}\left(1 - 2\dfrac{1}{\sigma_{AB}}\right)c^2 + \sigma_{AA}c^2 + \sigma_{AA}\alpha_{AA}\left(\dfrac{1}{\alpha_{AB}} - \dfrac{1}{\alpha_{BB}}\dfrac{1}{\sigma_{BB}}\right)c^3}{\left(1 + 2\dfrac{\alpha_{AA}}{\alpha_{AB}}\dfrac{\sigma_{AA}}{\sigma_{AB}}c + \dfrac{\alpha_{AA}}{\alpha_{BB}}\dfrac{\sigma_{AA}}{\sigma_{BB}}c^2\right)(1 + c)} \tag{84}$$

If

$$\sigma_{AA} > 1 \qquad \sigma_{AB} > 2 \qquad \sigma_{BB} > \frac{\alpha_{AB}}{\alpha_{BB}} \tag{85}$$

subunit arrangement and intersubunit strain will result in an enhancement of the reaction velocity with respect to the ideal naked subunit. However, Condition 85 is incompatible with Condition 81. Therefore, subunit arrangement and quaternary constraints cannot generate, together, rate enhancement, with respect to that of the ideal naked subunit and sigmoidicity of the rate curve. Kinetic sigmoidicity is, therefore, antagonistic to enhanced reaction velocity and is therefore paid for at the expense of catalytic efficiency.

If now the intersubunit strain is relieved in the nonliganded and fully liganded states (Figure 9b), the expression of $\overline{V} - \overline{V}^*$ assumes the form

$$\overline{V} - \overline{V}^* = \frac{\alpha_{AA}\left(\dfrac{1}{\alpha_{AB}} - 2\,\dfrac{1}{\alpha_{BB}}\,\sigma_A^*\right)c^2 + c^2 + \alpha_{AA}\left(\dfrac{1}{\alpha_{AB}} - \dfrac{1}{\alpha_{BB}}\right)c^3}{\left(1 + 2\,\dfrac{\alpha_{AA}}{\alpha_{BB}}\,\sigma_A^*\,c + \dfrac{\alpha_{AA}}{\alpha_{BB}}\,c^2\right)(1 + c)} \tag{86}$$

and this expression will be positive if[38]

$$\frac{1}{2} > \frac{\alpha_{AB}\,\sigma_A^*}{\alpha_{BB}} \tag{87}$$
$$\alpha_{BB} > \alpha_{AB}$$

The first of these conditions is identical to Condition 82. Therefore, enhancing the reaction velocity of the dimer, with respect to that of the ideal naked subunit, of necessity generates sigmoidicity of the rate curve. Then kinetic cooperativity and sigmoidicity are not paid for at the expense of catalytic efficiency. One may therefore speculate that, in the course of evolution, this type of protein design has been selected in order to generate an enzyme which displays both kinetic sigmoidicity and a high catalytic efficiency.

VII. CONCLUSION

Cooperativity between sites of an oligomeric enzyme may occur under equilibrium or nonequilibrium conditions. The classical models of Monod and Koshland describe multiple equilibria between the protein and the ligand, but these models do not apply to enzyme reactions which are studied under nonequilibrium conditions.

A novel formalism has been developed which allows expressing how subunit interactions and conformational constraints control the rate of product appearance as well as substrate binding under equilibrium conditions. Applying this formalism to enzyme reactions requires three postulates: quaternary constraints are assumed to be relieved in the transition states; under conditions where intersubunit strain does not exist, or is relieved, subunits adopt only two states, termed A and B; the subunits which have bound a transition state have, whatever the nature of this transition state, the same conformation, A.

Structural rate equations, as well as substrate binding equations, have been derived for different models of loose and tight coupling of subunits of dimeric enzymes. The rate and equilibrium binding equations, pertaining to the same type of subunit coupling, are basically different and predict different types of behavior.

With this formalism, kinetic cooperativity may be compared to substrate binding cooperativity. When subunits are loosely coupled, kinetic and substrate binding cooperativities have the same sign, positive or negative. Kinetic cooperativity, however, may be amplified or attenuated with respect to the corresponding substrate binding cooperativity. If subunits are tightly coupled and if the intersubunit strain is not relieved in the ground states, there may exist an inversion of kinetic cooperativity with respect to substrate binding cooperativity. Moreover, positive kinetic cooperativity may occur in the absence of any substrate binding cooperativity. Although loose coupling of subunits may generate positive kinetic cooperativity, this cooperativity is not associated with any sigmoidicity of the rate curve. Sigmoidicity is generated by intersubunit strain. This occurs when the extent of intersubunit strain in the half-liganded state is more than twice the one occurring in the nonliganded state.

If subunits are tightly coupled, but conformational constraints are relieved in both the nonliganded and full liganded states, substrate binding and kinetic cooperativities can only be positive. This conforms to the situation described in the concerted models of Monod and Koshland.

Contrary to a common belief, subunit coupling may enhance catalytic efficiency of an oligomeric enzyme relative to that of an ideal naked subunit. This enhancement of catalytic efficiency, however, is antagonistic to the existence of sigmoidicity for an oligomeric enzyme, with tightly coupled subunits. This means that, for such an enzyme, sigmoidicity is paid for at the expense of catalytic efficiency. This is the converse, which is to be expected, if the subunits are tightly coupled but if quaternary constraints are relieved in the nonliganded and fully liganded states. Enhancement of the catalytic efficiency of the oligomeric enzyme results in the appearance or in the strengthening of sigmoidicity. One may, therefore, speculate that this type of subunit interaction has been selected, in the course of evolution, to allow a sigmoid response and a high catalytic efficiency of the enzyme.

REFERENCES

1. **Monod, J., Wyman, J., and Changeux, J. P.,** On the nature of allosteric transitions: a plausible model, *J. Mol. Biol.,* 12, 88, 1985.
2. **Koshland, D. E., Nemethy, G., and Filmer, D.,** Comparison of experimental binding data and theoretical models in proteins containing subunits, *Biochemistry,* 5, 365, 1966.
3. **Kirschner, K.,** Temperature-jump relaxation kinetics with an allosteric enzyme: glyceraldehyde-3-phosphate dehydrogenase, in *Regulation of Enzyme Activity and Allosteric Interactions,* Kvamme, E. and Phil, A., Eds., Academic Press, New York, 1968, 39.
4. **Koshland, D. E.,** Conformational aspects of enzyme regulation, *Curr. Top. Cell. Regul.,* 1, 1, 1969.
5. **Koshland, D. E.,** The molecular basis for enzyme regulation, in *Enzymes,* Vol. 1, 3rd ed., Boyer, P. D., Ed., Academic Press, New York, 1970, 341.
6. **Koshland, D. E. and Neet, K. E.,** The catalytic and regulatory properties of enzymes, *Annu. Rev. Biochem.,* 37, 359, 1968.
7. **Levitzki, A. and Koshland, D. E.,** The role of negative cooperativity and half-of-the-sites reactivity in enzyme regulation, *Curr. Top. Cell. Regul.,* 10, 2, 1976.
8. **Sanwal, B. D.,** Allosteric controls of amphibolic pathways in bacteria, *Bacteriol. Rev.,* 34, 20, 1970.
9. **Sanwal, B. D., Kapoo, M., and Duckworth, H. W.,** The regulation of branched and converging pathways, *Curr. Top. Cell. Regul.,* 3, 1, 1971.
10. **Stadtman, E. R.,** Mechanism of enzyme regulation in metabolism, in *Enzymes,* Vol. 1, 3rd ed., Boyer, P. D., Ed., Academic Press, New York, 1970, 397.
11. **Stadtman, E. R. and Ginsburg, A.,** The glutamine synthetase of *Escherichia coli:* structure and control, in *Enzymes,* Vol. 10, 3rd ed., Boyer, P. D., Ed., Academic Press, New York, 1974, 755.
12. **Whitehead, E. P.,** The regulation of enzyme activity and allosteric transition, *Prog. Biophys.,* 21, 321, 1970.
13. **Wyman, J.,** On allosteric models, *Curr. Top. Cell. Regul.,* 6, 209, 1972.
14. **Kurganov, B. I.,** *Allosteric Enzymes. Kinetic Behavior,* John Wiley & Sons, New York, 1982.
15. **Levitzki, A.,** *Quantitative Aspects of Allosteric Mechanisms,* Springer-Verlag, Berlin, 1978.
16. **Wyman, J.,** Heme proteins, *Adv. Protein Chem.,* 4, 407, 1948.

17. **Wyman, J.,** Linked functions and reciprocal effects in hemoglobin: a second look, *Adv. Protein Chem.,* 19, 223, 1964.
18. **Wyman, J.,** Allosteric linkage, *J. Am. Chem. Soc.,* 89, 2203, 1967.
19. **Wyman, J.,** Regulation in macromolecules as illustrated by haemoglobin, *Q. Rev. Biophys.,* 1, 35, 1968.
20. **Wong, J. T. F. and Endrenyi, L.,** Interpretation of nonhyperbolic behavior in enzymic systems. I. Differentiation of model mechanisms, *Can. J. Biochem.,* 49, 568, 1971.
21. **Dalziel, K.,** A kinetic interpretation of the allosteric model of Monod, Wyman and Changeux, *FEBS Lett.,* 1, 346, 1968.
22. **Neet, K. E.,** Cooperativity in enzyme function: equilibrium and kinetic aspects, in *Methods in Enzymology,* Vol. 64, Part B, Purich, D. L., Ed., Academic Press, New York, 1980, 139.
23. **Ricard, J., Mouttet, C., and Nari, J.,** Subunit interactions in enzyme catalysis: kinetic models for one-substrate polymeric enzymes, *Eur. J. Biochem.,* 41, 479, 1974.
24. **Ricard, J. and Noat, G.,** Subunit interactions in enzyme transition states. Antagonism between substrate binding and reaction rate, *J. Theor. Biol.,* 111, 737, 1984.
25. **Ricard, J. and Noat, G.,** Subunit coupling and kinetic cooperativity of polymeric enzymes. Amplification, attenuation and inversion effects, *J. Theor. Biol.,* 117, 633, 1985.
26. **Ricard, J.,** *Organized Multienzyme Systems,* Welch, G. R., Ed., Academic Press, New York, 1985, 177.
27. **Ricard, J. and Noat, G.,** Catalytic efficiency, kinetic cooperativity of oligomeric enzymes and evolution, *J. Theor. Biol.,* 123, 431, 1986.
28. **Adair, G. S.,** The hemoglobin system. VI. The oxygen dissociation curve of hemoglobin, *J. Biol. Chem.,* 63, 529, 1925.
29. **Lumry, R.,** Some aspects of the thermodynamics and mechanism of enzymic catalysis, in *Enzymes,* Vol. 1, Boyer, P. D., Lardy, H., and Myrbäck, K., Eds., Academic Press, New York, 1959, 157.
30. **Jencks, W. P.,** *Catalysis in Chemistry and Enzymology,* McGraw-Hill, New York, 1969.
31. **Jencks, W. P.,** Binding energy, specificity and enzymic catalysis: the Circe effect, *Adv. Enzymol.,* 43, 219, 1975.
32. **Wolfenden, R.,** Transition state analogues for enzyme catalysis, *Nature (London),* 223, 704, 1969.
33. **Lienhard, G. E., Secemski, I. I., Koehler, K. A., and Lindquist, R. N.,** Enzymatic catalysis and the transition state theory of reaction rates: transition state analogs, *Cold Spring Harbor Symp. Quant. Biol.,* 36, 45, 1972.
34. **Secemski, I. I., Lehrer, S. S., and Lienhard, G. E.,** A transition state analog for lysozyme, *J. Biol. Chem.,* 247, 4740, 1972.
35. **Phillips, D.C.,** The hen egg-white lysozyme molecule, *Proc. Natl. Acad. Sci. U.S.A.,* 57, 484, 1967.
36. **Ford, L. O., Johnson, L. N., Machin, P. A., Phillips, D. C., and Tsian, R.,** Crystal structure of a lysozyme-tetrasaccharide lactone complex, *J. Mol. Biol.,* 88, 349, 1974.
37. **Chipman, D. M., Grisario, V., and Sharon, N.,** The binding of oligosaccharides containing N-acetyl-glucosamine and N-acetylmuramic acid to lysozyme, *J. Biol. Chem.,* 242, 4388, 1967.
38. **Ricard, J. and Cornish-Bowden, A.,** Review. Co-operative and allosteric enzymes: 20 years on, *Eur. J. Biochem.,* 166, 255, 1987.

Chapter 2

STRUCTURE AND FUNCTION OF HEMOGLOBIN: THE COOPERATIVE MECHANISM

Martin Karplus, David A. Case, Bruce Gelin, Boi Hanh Huynh, Angel Wai-mun Lee, and Attila Szabo

TABLE OF CONTENTS

I. INTRODUCTION

Hemoglobin, because of its vital role in oxygen transport and its status as a model for cooperative proteins, has been the subject of a wide range of physical and chemical studies for many years.[1-4] The goal of these investigations is to obtain a detailed understanding of the reversible binding of oxygen to the heme group and of the mechanism leading to the subsequent alterations of the tertiary and quaternary structure of the globin chains. A complete description of cooperativity in hemoglobin will involve its structural, thermodynamic, and kinetic aspects. Of primary importance is an understanding of the structural changes that occur on ligation and an evaluation of their relation to the thermodynamics. This is essential for an analysis of cooperativity in the normal hemoglobin molecule and for an interpretation of the relationship between structural changes and abnormal properties in mutant and modified hemoglobins.

Cooperativity plays a role in the function of many proteins. In most cases, the mechanism of cooperativity is based on a multisubunit structure and involves interactions between the different oligomers. Why look at hemoglobin rather than an alternative system? The answer is very simple; more is known about the hemoglobin molecule than about any other protein molecule with a cooperative mechanism. There are more structural data, more equilibrium studies, and more kinetic experiments than for any other system. Also, the process of reversible binding of oxygen is in many ways simpler than most enzyme reaction mechanisms. In addition, the iron atom in the heme group can be used to monitor what is happening by various spectroscopic techniques (e.g., optical, ESR, and NMR spectra). Thus, it is possible to argue that if one is going to try to understand the mechanism of cooperativity in any protein, hemoglobin is the system to study.

Section II of this review presents a brief introduction to the thermodynamic and structural aspects of cooperativity in hemoglobin. The origin of the underlying structural changes and their functional correlates are described in Section III. A detailed statistical mechanical model for cooperativity is outlined in Section IV. Section V presents a concluding discussion.

II. OVERVIEW OF STRUCTURAL AND THERMODYNAMIC ASPECTS OF COOPERATIVITY

Hemoglobin cooperativity can be defined in terms of the deviation of the oxygen binding curve from a simple absorption isotherm of the form

$$\langle y_{O_2} \rangle = \frac{K p_{O_2}}{1 + K p_{O_2}} \tag{1}$$

where K is the binding constant and p_{O_2} is the partial pressure of oxygen. The saturation function, $\langle y_{O_2} \rangle$, is the average fractional saturation with oxygen of a population of hemoglobin molecules. The value of $\langle y_{O_2} \rangle$ goes from zero to one, zero corresponding to no oxygen bound and one corresponding to four oxygens bound, one to each subunit of the hemoglobin tetramer. Equation 1 applies to the noncooperative oxygen binding of myoglobin and would be applicable to hemoglobin if binding by the four chains were independent. For hemoglobin, the variation of $\langle y_{O_2} \rangle$ with the partial pressure of oxygen above a hemoglobin solution is shown schematically in Figure 1 as the solid line. This is to be contrasted with the behavior that would result if each of the oxygen molecules were bound independently (dashed). The interest in hemoglobin, in the simplest sense, derives from the deviation of the true oxygen saturation curve from the independent binding adsorption isotherm (Equation 1). The form of the curve shows that the first oxygen molecule has the lowest affinity, and

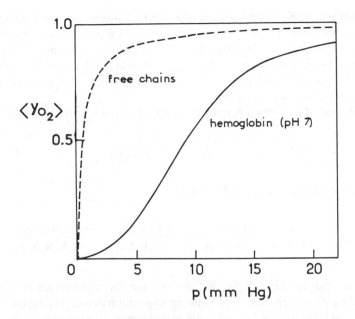

FIGURE 1. Oxygen saturation curve for hemoglobin at pH 7 (solid line) and free chains (dashed line).

that as more oxygens bind, the affinity increases. Thus, there is some interaction, which is referred to as "cooperativity", between the binding of one oxygen molecule to one chain and another oxygen molecule to another chain of the tetramer.

The biological function of the cooperative binding is that it results in the more efficient transport of oxygen. Compared with simple adsorption, the actual binding curve yields a larger difference in the fractional saturation for a given difference between the partial pressure of oxygen at the lungs and at the tissues. There are additional aspects of the cooperative mechanism that also play an important role. One of these is that the binding curve is pH dependent; that is, the binding curve in Figure 1 is shifted significantly to the right at lower pH values. Because the metabolic products make the blood more acid at the tissues, relative to when it is in the lungs, this pH dependence of the binding curve improves the oxygen transport. Hemoglobin also transports CO_2 from the tissues to the lungs and the pH dependence of the oxygen equlibrium is coupled to the CO_2 binding. In addition to the above, there are a variety of other factors involved in the oxygen binding. For example, phosphates and various other ions (effectors) alter the binding curve. All of these taken together suggest that the hemoglobin molecule is very good at doing what it does. Nature has, through evolution, developed a highly sophisticated system for carrying out the various transport operations that are required. This is, of course, what makes hemoglobin interesting. One would like to understand the cooperative mechanism and its correlates in as much detail as possible at the molecular level.

A way of modifying the simple adsorption isotherm (Equation 1) to obtain a cooperative curve is to write

$$\langle y_{O_2} \rangle = \frac{K p_{O_2}^n}{1 + K p_{O_2}^n} \tag{2}$$

The exponent "n" of the partial pressure, p_{O_2}, can take account phenomenologically of the fact that the curve is cooperative; specifically, values of $n > 1$ yield cooperative curves.

Equation 2 does not imply anything about the detailed mechanism of binding (e.g., that several oxygens are bound at the same time) but, as was found many years ago by A. V. Hill (Equation 2 is called the "Hill equation"), it does yield, with n = 2.5 to 3, a good fit to the experimental saturation curve for $<y_{O_2}>$ between 0.4 and 0.6. The exact values of n and K depend on the solution conditions, such as pH and other ion concentrations, in accord with the description given above.

An alternative way of presenting hemoglobin data is based on the so-called Adair equations

$$Hb(O_2)_i + O_2 \xrightarrow{K_{i+1}} Hb(O_2)_{i+1} \tag{3}$$

that yield an expression for $<y_{O_2}>$ of the form

$$\langle y_{O_2} \rangle = \frac{K_1 p_{O_2} + 2K_1 K_2 p_{O_2}^2 + 3K_1 K_2 K_3 p_{O_2}^3 + 4K_1 K_2 K_3 K_4 p_{O_2}^4}{4(1 + K_1 p_{O_2} + K_1 K_2 p_{O_2}^2 + K_1 K_2 K_3 p_{O_2}^3 + K_1 K_2 K_3 K_4 p_{O_2}^4)} \tag{4}$$

In this equation, the constants K_1, K_2, K_3, and K_4 are the equilibrium constants for the successive binding of oxygen molecules by the hemoglobin tetramer (see Equation 3). Values of the constants are determined by fitting the experimental binding curve. The cooperative effect appears in Equation 4 through the fact that the constants deviate from the values they would have if only statistical factors were involved (e.g., that the first oxygen can bind to four heme groups and the last oxygen can come off from four heme groups). Another way of expressing this is in terms of an apparent free energy of interaction, defined by

$$\Delta F_{app} = RT \ln \left(\frac{16K_4}{K_1} \right) \tag{5}$$

where 16 is the ratio of K_1/K_4 expected from statistical factors alone. Also, one often considers a quantity called $P_{1/2}$, which by definition is the oxygen partial pressure required for half-saturation. In fact, most binding experiments on hemoglobin are limited to finding $P_{1/2}$ and n values. These can be obtained from the middle range of the binding curve, where it is easiest to make measurements. Thus, particularly the early discussions of hemoglobin were concerned primarily with n and $P_{1/2}$ and their changes as a function of the solution variables. However, more recently, beginning with the pioneering experiments of Roughton and Lyster,[5] extensive measurements have been made of the entire binding curve from very low to very high saturation (i.e., $<y_{O_2}>$ = 0.001 to 0.999). Such data are of particular importance for testing theoretical models for hemoglobin cooperativity.

Most early models of cooperativity that went beyond the Hill equation were still essentially phenomenological in character; that is, they assumed without any structural basis that there exists some interaction which makes the Adair constants deviate from their statistical values. There are two types of widely used phenomenological models, one of which was introduced by Pauling in 1935,[6] and the other by Monod, Wyman, and Changeux in 1965.[7] In the Pauling model, elaborated by Koshland, Nemethy, and Filmer,[8] it is assumed that there is an interaction between pairs of subunits which depends on whether or not they are oxygenated. For the case in which each chain interacts with the three others ("tetrahedral" model), the equation for $<y_{O_2}>$ is

$$\langle y_{O_2} \rangle = \frac{K p_{O_2} + 3\alpha K^2 p_{O_2}^2 + 3\alpha^3 K^3 p_{O_2}^3 + \alpha^6 K^4 p_{O_2}^4}{1 + 4K p_O + 6\alpha K^2 p_{O_2}^2 + 4\alpha^3 K^3 p_{O_2}^3 + \alpha^6 K^4 p_{O_2}^4} \tag{6}$$

where K is the independent chain binding constant and α is the interaction constant, which stabilizes pairs of oxygenated chains. The interaction can be assumed to be the result of the change in tertiary structure that occurs in the chains on ligand binding. The alternative phenomenological model of Monod, Wyman, and Changeux[7] assumes that there are two different quaternary structures for the hemoglobin tetramer and that the chain binding constant for oxygen is different for the two structures. There results an expression of the form

$$\langle y_{O_2} \rangle = \frac{LcKp_{O_2}(1 + cKp_{O_2})^3 + Kp_{O_2}(1 + Kp_{O_2})^3}{L(1 + cKp_{O_2})^4 + (1 + Kp_{O_2})^4} \tag{7}$$

where the "allosteric constant", L, gives the relative stability of the two quaternary structures in the absence of ligand, and c is the ratio of the oxygen binding constants of the two structures. Both models yield cooperative binding curves and by a suitable choice of their parameters (K and α, or K, L, and c) a good fit to the experimental results for hemoglobin can be obtained. It is, thus, not possible to distinguish between the phenomenological models by the use of binding data alone. The essential problem is that hemoglobin is highly co-operative so that the intermediates with two ligands bound are only present at very low concentrations. However, it is just for these doubly liganded intermediates that the two models give different results, in general.

To go from such phenomenological models to a description which is based on the specific chemistry of the atoms making up the hemoglobin molecule, it is necessary to have structural and spectroscopic information concerning the hemoglobin tetramer as a function of ligation. Great progress has been made in this area in recent years. Of primary importance are the crystallographic studies of Perutz and his co-workers.[2] Their structural studies, made over many years, have culminated in refined high-resolution X-ray structures of two forms of hemoglobin, one which is completely deoxygenated (Hb)[9] and the other completely oxygenated (Hb(O_2)_4).[10] These provide detailed information on the atomic positions in the two forms and can be assumed to represent what happens in solution for deoxygenated and oxygenated hemoglobin, respectively.

It has been known for a long time that if a crystal of deoxyhemoglobin is oxygenated, it shatters; this suggested that there is a large-scale structural change on oxygenation. From the X-ray results, even at low resolution (5.5 Å),[11] it was evident that on oxygenation the four chains move relative to each other. This is diagrammed in Figure 2 in terms of the change of distances between the various hemes. The β-β distance changes by the largest amount and other distances also change significantly. Clearly, these distances alone do not give a complete picture of what goes on, but they do indicate that there is a significant alteration on oxygenation in the relative positions of the four hemoglobin chains. However, since the hemes are between 25 and 30 Å apart, and the interaction is on the order of several kilocalories, an interaction mediated by the protein chains is almost certain to be required. What one would like to know, then, is how such a protein mediated mechanism operates.

From the structural and related studies, it is evident that normal human adult hemoglobin is a tetramer composed of four chains that are identical in pairs (see Figure 3), both in terms of amino acid sequence and structure. The two α-chains have 141 amino acids each, and the β-chains have 146 amino acids each. The two pairs of chains are very similar in structure, both being highly helical proteins (approximately 85% of the residues are in α-helices); there are seven helices in the α-chains and eight helices in the β-chains, with the helices connected by turns or loops. The α- and β-chains making up the tetramer are not connected by covalent bonds; they can be dissociated under certain conditions (e.g., high salt concentration). Each of the chains contains a heme group; each of the heme groups contains an iron atom in the ferrous state that can reversibly bind one oxygen molecule.

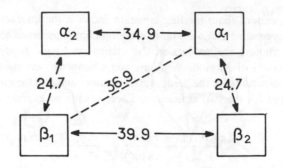

Deoxyhemoglobin

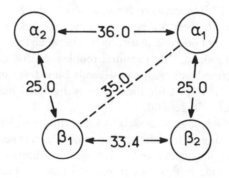

Oxyhemoglobin

FIGURE 2.   Distances (in Å) between heme groups in
deoxy- and oxyhemoglobin.

From the X-ray results for Hb and HbO$_2$, it is clear that there exist two quaternary structures
for the tetramer. The two quaternary structures are related by a relative translation and
rotation of the two αβ-dimers ($\alpha_1\beta_1$ and $\alpha_2\beta_2$) of which the hemoglobin tetramer is composed.[2,12] Recent X-ray results for intermediates (e.g., $[\alpha(FeCO)\beta(Mn)]_2)$[13] confirm that
only two quaternary structures (deoxy and oxy or T and R in the nomenclature of Monod,
Wyman, and Changeux) are required to describe the quaternary states of the system. It was
originally assumed that each subunit has only two tertiary structures (unliganded and liganded
or t and r), essentially independent of the quaternary structure of the tetramer. From more
recent structural studies[9,10] as well as theoretical calculations,[14,15] it is now clear that there
are significant differences in the unliganded and liganded tertiary structure of the individual
chains, depending on the quaternary structure of the tetramer in which they occur.

In what follows, we shall try to review our present understanding of the cooperative
mechanism. This couples the X-ray and other structural data for hemoglobin with a statistical
mechanical model for hemoglobin function. In Section III, we outline the structural results
and complementary theoretical studies that provide information on the origin of the structural
changes that occur in ligand binding and their energetic correlates. Section IV incorporates
some of the results of this analysis into a model for the thermodynamics of hemoglobin that
describes the ligation process and how it is perturbed by a variety of effectors such as pH,
2,3-diphosphoglycerate, and other ions. A concluding discussion is given in Section V.

Tyr HC2

Val E11

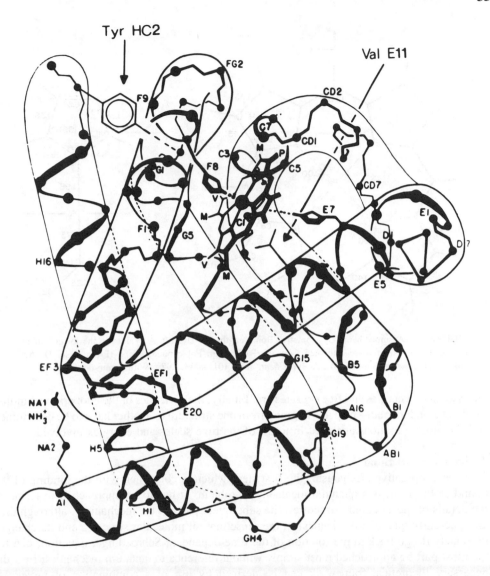

FIGURE 3. Sketch of the polypeptide chain of the β-subunit. (From Perutz, M. and Ten Eyck, L. F., *Cold Spring Harbor Symp. Quant. Biol.*, 36, 295, 1972. With permission.)

III. STRUCTURAL AND ENERGETIC ASPECTS OF THE COOPERATIVE MECHANISM

A chemical approach to hemoglobin must provide an explanation of cooperativity in terms of specific electronic and atomic interactions and their structural and energetic consequences. To look at hemoglobin from this point of view, it is convenient to divide the ligand binding process into a series of steps. The first step is the binding of the ligand to the iron in the heme group of hemoglobin; this requires an understanding of the electronic structure of the bond and of the electronic transitions involved in the bonding. Next are the structural changes induced in the heme group and the surrounding protein by the ligand binding. The changes in the neighborhood of the heme group are expected to induce other tertiary structural alterations, which in turn are coupled to the quaternary structure of the tetramer. It is necessary to know what electronic changes occur, what atom displacements are produced, and what

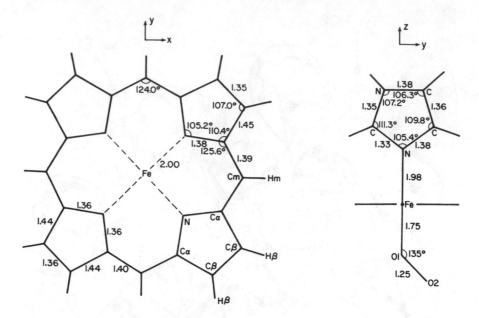

FIGURE 4.   Assumed heme-O_2 geometry. Bond lengths are given in Å. Values in the lower half of the porphyrin skeleton are estimated from the Pariser-Parr-Pople bond orders. (From Case, D. A., Huynh, B. H., and Karplus, M., *J. Am. Chem. Soc.*, 101, 4433, 1979. With permission.)

the associated energies and free energies are. Finally, the dynamics of the cooperative binding need to be understood. The paths in going from one structure to another have to be determined and the atomic motions analyzed in terms of the time scales and energies involved.

A. Iron-Ligand Bond

In the cooperative mechanism, the first step, which is concerned with the binding of the ligand to the iron, is a quantum-mechanical problem. It has to be approached in terms of the details of the electronic structure. The subsequent parts of the mechanism, although they are inherently quantum mechanical in the sense that all molecular structure and its changes ultimately do go back to the solution of the time-dependent Schroedinger equation, can for the most part be approached most simply without reference to quantum mechanics; i.e., the necessary theoretical framework can be derived by use of a combination of molecular mechanics, statistical mechanics, and dynamics.

To begin with the iron-ligand bond, an idealized picture of the heme group is shown in Figure 4. The structure is based on model compound data for liganded hemes and represents an average of a number of results. Also shown is the geometry in the perpendicular direction, with an imidazole and an oxygen as ligands. The geometry in Figure 4 is the one used in the calculations whose results are described below.

Quantum-mechanical calculations on systems of this type cover a wide range. Because of the large size of the system (there are many atoms and electrons involved), all calculations of the electronic structure are approximate. The approximations made to simplify the problem are of two types. First, simplified model systems have been introduced; e.g., each pyrrole group has been replaced by an NH_2 group and the imidazole by an NH_3 group. The other type of approximation is in the method of calculation; a wide range of techniques from simple extended Hückel treatments through limited basis-set *ab initio* calculations have been made.

The point of mentioning the approximations is to remind the reader that in trying to draw conclusions from such calculations, it is necessary to realize that there may be errors in them. If a whole gamut of calculations that differ in the approximations used agree concerning certain properties, the results are likely to be valid. If there are disagreements, then it is very hard to tell which calculation is correct, because different properties of the system may be more sensitive to one approximation than to another.

To best describe the iron-oxygen bond, it is useful to introduce limiting models for its electronic structure.[20] These are conveniently based on suitable reference configurations. The reference states that are more appropriate for the ground state of oxyhemoglobin arise from the interaction of an iron-porphyrin moiety and an O_2 molecule, with both species in either $S = 0$ or $S = 1$ valence states. In the ($S = 0$, $S' = 0$) model, the iron is ferrous low spin (t_{2g}^6), whereas the oxygen molecule resides in a spin-paired singlet configuration analogous to that of the $^1\Delta_g$ molecular state; the spin pairing in oxygen occurs because the π_g orbital in the FeO_2 plane has a lower energy than its out-of-plane partner. This reference state corresponds to the original Pauling model[16,17] and predicts oxyhemoglobin to be diamagnetic. For the CO adduct, no spin pairing is needed, since the free ligand is already in a closed-shell configuration. Calculations[20] indicate that the ($S = 0, S' = 0$) reference state is an excellent model for heme-CO, but that there are significant deviations in heme-O_2.

A second reference configuration has both iron and oxygen in $S = 1$ states. The iron is in an excited ferrous state in which one electron has been removed from the d_{xz} orbital (which is antisymmetric with respect to the FeO_2 plane in the coordinate system of Figure 4) and placed into the d_{z^2} orbital (which points along the Fe-O bond); the oxygen retains its $^3\Sigma_g$ ground-state configuration. The resulting complex is again diamagnetic owing to the pairing of the two $S = 1$ states. This description corresponds to the idealized ozone model of Goddard and Olafson.[18,19] Although the resulting orbital populations are not identical with the calculated ones (see below), there do exist many similarities between ozone and heme-O_2. Thus, it is useful to discuss heme-O_2 in terms of deviations from the idealized ozone model.

There is a third reference state that is often thought to represent many experimental features of oxyhemoglobin. This is the Weiss model[21] in which an electron is transferred from iron to dioxygen, and the complex consists of a low-spin ferric ion ($t_{2g}^5, {}^2E$), coordinated to a superoxide anion ($^2\pi_g$). The diamagnetism of heme-O_2 is presumed to be due to spin pairing of the two doublet configurations. The calculations show no evidence for such charge transfer in the ground state, although there exists a triplet-singlet pair of excited states of the Weiss type at about 1 eV.

These ideas can be made more quantitative by a schematic representation of the orbital populations. The electronic configuration for the covalent singlet-singlet coupled state ($S = 0$, $S' = 0$) is

$$Fe[(d_{x^2-y^2})^2 \ (d_{yz})^2 \ (d_{xz})^2 \ (d_{z^2})^0 \ (d_{xy})^0]$$

$$- \ O^1[(2p_{x'})^1 \ (2p_{y'})^2 \ (2p_{z'})^1]$$

$$- \ O^2[(2p_{x'})^1 \ (2p_{y'})^2 \ (2p_{z'})^1]$$

and for the covalent triplet-triplet coupled state ($S = 1, S' = 1$) is

$$Fe[(d_{x^2-y^2})^2 \ (d_{yz})^2 \ (d_{xz})^1 \ (d_{z^2})^1 \ (d_{xy})^0]$$

$$- \ O^1[(2p_{x'})^{1.5} \ (2p_{y'})^{1.5} \ (2p_{z'})^1]$$

$$- \ O^2[(2p_{x'})^{1.5} \ (2p_{y'})^{1.5} \ (2p_{z'})^1]$$

where O^1 represents the center oxygen and O^2 represents the terminal oxygen. The $x'y'z'$ axes correspond to a local coordinate system oriented with respect to the oxygen molecule;

the z′ axis is along the O-O bond, and the x′ axis is perpendicular to the Fe-O-O plane. An equal-weight average of these two states gives a configuration of the form

$$Fe[(d_{x^2-y^2})^2 \, (d_{yz})^2 \, (d_{xz})^{1.5} \, (d_{z^2})^{0.5} \, (d_{xy})^0]$$

$$- \, O^1[(2p_{x'})^{1.25} \, (2p_{y'})^{1.75} \, (2p_{z'})^1]$$

$$- \, O^2[(2p_{x'})^{1.25} \, (2p_{y'})^{1.75} \, (2p_{z'})^1]$$

Pariser-Parr-Pople calculations[20] for the system shown in Figure 4 yield orbital populations for the FeO_2 unit that correspond to the electronic configuration

$$Fe[(d_{x^2-y^2})^{2.0} \, (d_{yz})^{1.74} \, (d_{xz})^{1.36} \, (d_{z^2})^{0.53} \, (d_{xy})^{0.56}]$$

$$- \, O^1[(2p_{x'})^{1.27} \, (2p_{y'})^{1.88} \, (2p_{z'})^{1.12}]$$

$$- \, O^2[(2p_{x'})^{1.14} \, (2p_{y'})^{1.99} \, (2p_{z'})^{1.14}]$$

which is very near the equal-weight average of the (S = 0,S′ = 0) and (S = 1,S′ = 1) results. The minor differences in electronic population between the two stem mainly from the fact that, in the simplified model, only three atoms (Fe-O-O) are considered and orbitals of an isolated oxygen molecule are used as basis functions. Therefore, no d_{xy} population is obtained in the model description, and the two oxygen atoms in the FeO_2 unit retain the symmetry of the isolated ligand. The results of this model can also be compared with the valence bond calculation of Seno et al.[22]

From the calculations,[20] the iron atom in heme-O_2 has a net charge of $+0.45$ and a total 3d population of 6.19, which is consistent with a ferrous ion assignment. Although the formal oxidation state is $+2$, electron donation from the five nitrogen ligands partly neutralizes the metal. The present results suggest that σ-donation from the imidazole is also necessary. The π-system of imidazole appears to be only slightly perturbed by complexation with iron; this is in accord with *ab initio*[23] and extended Hückel results.[24] Complexation of Fe^{2+} to the porphyrin dianion involves σ-donation of 0.4 electron per nitrogen and π-back-donation of 0.1 electron per nitrogen. An additional 0.35 electron is donated to the iron by the imidazole σ-system. The ferrous ion is nearly neutralized by its five nitrogen ligands, and the binding of the oxygen molecule is mainly covalent, with little charge transfer.

The σ-donation to the iron (primarily from the 2s orbital of O^1) is almost exactly balanced by π-back-donation (into the $2p_x$ orbitals of both O^1 and O^2). The back-donation is into an antibonding orbital, and it appears that this is primarily responsible for the weakening of the O-O bond. The most significant difference between the populations in Fe-O_2 and ozone is in the $2p_y$ population of O^1; this is larger in hemoglobin because the in-plane Fe-O^1 π-bond is more strongly polarized toward O^1 than is the corresponding bond in ozone. Aside from this difference, the populations in heme-O_2 and ozone are very close.

Pariser-Parr-Pople and X-α calculations[21] indicate that, in the Fe-O_2 bond of the heme-imidazole model, the oxygen molecule is nearly neutral overall. Extended Hückel calculations[24] yield a net charge transfer to the oxygen, but all other calculations, including several *ab initio* treatments for iron-O_2 models,[18,19,23] yield a neutral oxygen molecule. For cobalt instead of iron, an extra electron is present and partial electron transfer to the oxygen does occur. This may play some role in the differences between the properties of Fe and Co hemoglobin.

There has been much discussion of various experimental measurements on the bonded Fe-O_2 system and whether they can provide information about the charge distribution. One such property is the ligand stretching frequency. The oxygen stretching frequencies, observed in myoglobin (1103 cm^{-1}), in hemoglobin (1107 cm^{-1}), and in model compounds (~1160 cm^{-1}), have been cited as evidence for the Fe^{+3}-O_2^- structure, since they fall in the range

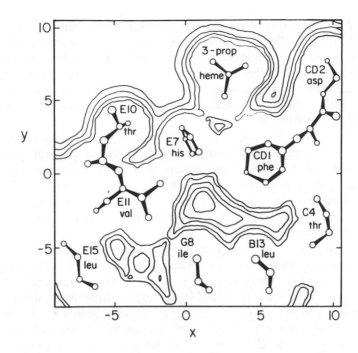

FIGURE 5. Myoglobin contour map of the (x,y) plane at z = 3.2 Å. Contours are at 90, 45, 10, 0, and −3 kcal/mol relative to a ligand at infinity. The highest contours are closest to atoms, whose projections onto the plane of the figure are denoted by circles. (From Case, D. A. and Karplus, M., *J. Mol. Biol.*, 132, 349, 1979. With permission.)

found for superoxide complexes (1100 to 1150 cm^{-1}), whereas the stretching frequencies for the O_2 molecule are 1556 cm^{-1} for the $^3\Sigma_g$ state and 1484 cm^{-1} for the $^1\Delta_g$ state. A result that appears to have been neglected is that the force constant of the O-O bond in ozone is 6.16 mdyn/Å, a value which is in the superoxide range (5.7 to 6.2 mdyn/Å). This demonstrates that a neutral O_2 unit can have the observed frequency. Clearly, what is important is not the net charge but the distribution of bond orders. The total Pariser-Parr-Pople bond order for the oxygen molecule in heme-O_2 is 1.60, which is much closer to O_2^- (1.5), and O_3 (1.65) than to values for O_2(2.0 for $^3\Sigma_g^-$ and $^1\Delta_g$).

The calculated bond orders for the CO bond in the free molecule and in heme-CO show that there is a reduction of about 0.13 in the total bond order upon incorporation into iron porphyrin (2.58 to 2.45). This can explain qualitatively the observed reduction in the stretching frequency on ligation; i.e., ν_{CO} is changed from 2143 cm^{-1} in the gas to 1970 cm^{-1} in model compounds and to about 1950 cm^{-1} in myoglobin and hemoglobin.

To gain further insight into the heme oxygen bond, it is of interest to look at some differences between the conclusions from the theoretical calculations, synthetic model compound studies, and the results found in myoglobin or hemoglobin. In the model compounds, the oxygen has the bent orientation shown in Figure 4 but is disordered; that is, there appear to be four equivalent positions for the projection of the O-O vector on the heme plane. Calculations[24] suggest that there is a significant barrier between the various positions, but that all positions have similar energies. By contrast, the oxymyoglobin structure determined by Phillips[25] indicates that one of the four positions is selected by the protein. Figure 5 shows a potential energy map calculated with empirical energy functions for the oxygen in myoglobin.[26,27] There is a low-potential energy region (the so-called heme pocket) for the ligand present in the fourth quadrant with the energy minimum near (x =

1 Å, y = −2 Å). This corresponds to the expected position of the distal O atom of an O_2 ligand forming a bent Fe-O-O bond; the iron atom would be near the top left-hand edge of the pocket and not in its center as is sometimes pictured. The predicted O_2 orientation is in agreement with the X-ray structure of oxymyoglobin,[25] which was published after the calculation was completed. It can be seen that a number of hydrophobic residues are involved in forming the pocket; the most important ones in the plane are His-E7 (the distal histidine), Val-E11, Ile-G8, Leu-B13, and Phe-CD1.

Does the orientation of the ligand found in the protein play a functional role? It turns out that the observed orientation has the oxygen and the proximal imidazole plane eclipsed as in Figure 4. Since the imidazole of the histidine is involved in the cooperative mechanism (i.e., it has a repulsive interaction with the heme group in the liganded T structure, see below), the alignment of the oxygen and imidazole may permit the oxygen to buttress the heme group and help to prevent its distortion when the imidazole is pushing against it after ligation has taken place.

Another question that arises in comparisons between the model compounds and the protein concerns the binding of carbon monoxide. Calculations for model systems suggest that CO is oriented perpendicular to the heme group in the linear arrangement characteristic of metal carbonyls. As one goes from CO to NO and O_2, one expects from simple arguments (extension of Walsh's rules)[28] that NO and O_2 should have a bent iron-ligand bond, whereas CO should be linear. That is what is found for the model compounds. In hemoglobin and myoglobin, however, the oxygen of the CO is off the axis through the iron and perpendicular to the heme plane. The carbon may or may not be on the axis; the available X-ray data are not accurate enough to tell.[29] Figure 5 suggests that the protein is pushing the oxygen off the axis; i.e., for a rigid protein and a linear CO, the oxygen would clearly be in an unstable position. However, there appears to be a delicate balance of forces. The question that has to be answered is: How much energy does it take to distort the protein relative to the energy required to displace the CO from its position in model compounds? Calculations suggest that the protein forces are sufficient to reproduce the distortion; the protein sidechains are calculated to be able to support an energy of about 7 kcal/mol[26] whereas an *ab initio* calculation for the liganded heme group indicates that only 2 kcal/mol are required to displace the CO by the observed amount.[23] However, it is possible also that the heme is distorted in such a way that the linear FeCO is off axis and remains perpendicular to the plane of the four pyrrole nitrogens; this conformation is supported by a recent interpretation of resonance Raman data.[29a]

B. Effect of Ligation on the Heme Group

For the cooperative mechanism, the essential element in the formation of the iron-oxygen bond is the effect of ligation on the heme group and the covalently linked imidazole of His F8. Figure 6 shows a simplified cross section of a "domed" heme group prior to ligand binding and the "undoming" that is induced by the ligand, in accord with model compounds results[30] and calculations.[15] As the ligand binds, the heme group approaches a much more nearly planar geometry.

The source for the heme doming in the five-coordinate systems was previously ascribed entirely to the larger effective radius of the high-spin ferrous iron with an electron in the in-plane d_{xy} orbital, which results in a repulsive interaction with the (N_p) pyrrole nitrogen σ-orbitals. It now appears that this is only one factor involved. Experiments and calculations have shown that the size of the central hole in the porphyrin (i.e., the pyrrole nitrogen N_p-N_p distance), which has the dominant role in the repulsion between the electron in iron d_{xy} orbital and the pyrrole nitrogen lone pair, is considerably more variable than had been thought, making it possible for the pyrrole nitrogens to move somewhat farther apart and reduce the iron-N_p repulsion. *Ab initio* quantum-mechanical calculations[18,19,23] and a semi-

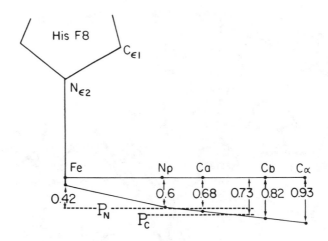

FIGURE 6. Changes in heme conformation and significant distances on ligation (relative to fixed axial histidine). Unliganded nonplanar heme is from 2-Me1mFeTPP. P_N and P_C are the mean porphyrin nitrogen and carbon planes; N_p, C_a, and C_b are the pyrrole nitrogen and carbons in standard order; C_a is the first carbon of a substituent on C_b. (From Gelin, B., Lee, A., and Karplus, M., *J. Mol. Biol.*, 171, 520, 1983. With permission.)

empirical calculation[31] suggest that an additional important factor is the steric interaction between the axial nitrogens of the imidazole (N_ϵ) and the pyrrole nitrogens (N_p). This is repulsive for an in-plane iron and is clearly decreased by an out-of-plane iron position, which is not destabilizing in the absence of a sixth ligand.

The changes in geometry expected for an isolated heme group on ligand binding have been well documented from structural studies of model compounds.[30,32] A small decrease in the lengths of all five iron-nitrogen bonds has been shown to occur; the axial Fe-$N_{\epsilon 2}$ distance decreases by 0.05 to 0.18 Å (from 2.12-2.16 Å to 1.98-2.05 Å) while the equatorial Fe-N_p bonds shorten by 0.06 to 0.08 Å (2.065-2.086 Å to 1.98-2.0 Å). Thus, in going from the five-coordinate high-spin to the six-coordinate low-spin state, the combination of an empty d_{xy} orbital and the pull on the iron due to the second axial ligand leads to a balance of forces in which an in-plane geometry for the iron is generally favored in model systems. The bond shortening and the presence of the sixth ligand in model compounds produces an "undoming" of the heme group. Thus, the porphyrin, in the absence of interactions with the globin, is expected to be essentially planar with the iron in or near the heme plane. Figure 6 shows an idealized picture of the results of ligation. It is evident that the transition from a domed to a planar geometry can produce large changes in the positions of the porphyrin atoms; if all the pyrrole atoms are planar centers, the peripheral and substituent carbons (e.g., of the vinyl group) move by nearly 1 Å while the $N_{\epsilon 2}$-P_C distance changes by about 0.75 Å. This leads to the possibility of amplification in the heme group of the effect of ligand binding to the iron; i.e., a 0.1-Å change in bond lengths is amplified to a 1-Å displacement.

C. Heme-Globin Interactions: The Allosteric Core

To investigate the possible effects of ligand binding on the globin,[14,15] we determine first the type of perturbation that is expected to be generated by the structural changes in the heme imidazole group just described and then examine the structural and energetic consequences of such a perturbation. In deoxyhemoglobin, the atoms of the globin surrounding the heme (Figure 7) are arranged to accommodate the domed structure. With respect to its

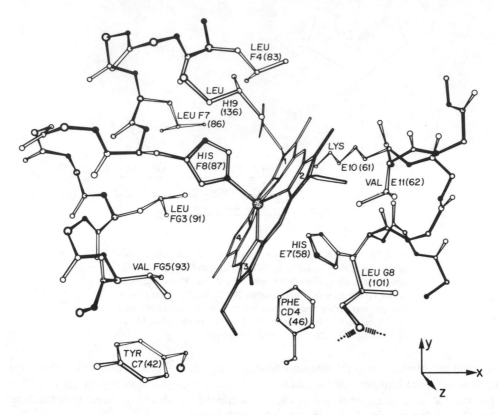

FIGURE 7. Heme group and its environment in the unliganded α-chain.[35] Filled bonds denote polypeptide backbone. Only selected side chains are shown: the heme 4-propionate is omitted for clarity. (From Gelin, B., Lee, A., and Karplus, M., *J. Mol. Biol.*, 171, 542, 1983. With permission.)

nonbonded interactions with the rigid globin, the domed heme is located in a broad potential well for translation perpendicular to its mean plane (Figure 8).[14,15] The calculated heme position, 0.15 Å on the near side of the potential minimum, corresponds to a strain energy of less than 1 kcal/mol. Because the net force on the heme is smaller on a per atom basis than for the globin as a whole, the apparent strain is not significant. In terms of the "distributed energy" model,[33] it could be argued that the nonbonded heme-globin interaction represents a small fraction of the total strain energy stored in displacements throughout the globin. This seems unlikely from independent studies of protein energetics. In a detailed examination of tyrosine side chain rotations in the bovine pancreatic trypsin inhibitor,[34] it has been shown that much larger rigid protein nonbonded interactions can be greatly reduced by small atomic displacements whose energy cost is negligible; further, most of the residual strain energy present after allowing the protein to relax was found to remain in the local nonbonded terms.

The above conclusion concerning the lack of strain on the unliganded heme group in the globin is in agreement with a number of experimental results. The X-ray structure of deoxyhemoglobin[9,35] confirms that the iron is not significantly farther from the mean plane of the heme than in five-coordinate model compounds.[32,36] Correspondingly, synchrotron radiation studies of deoxyhemoglobin[37,37a] imply that the Fe-N distances are not altered when the unliganded tetramer is shifted from the deoxy to the oxy quaternary structure. Also, resonance Raman studies[38] suggest that the five-coordinate heme is essentially unchanged upon incorporation into the globin.

An essential consequence of the bond shortening and "undoming" is the interaction

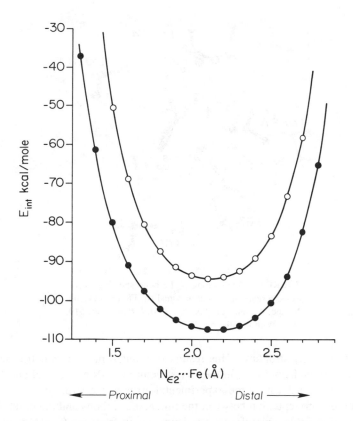

FIGURE 8. Potential due to heme-globin nonbonded interactions for translation of heme in unliganded α-chain; the zero of energy is at infinite separation and the Fe-N$_\epsilon$ bond energy is not included. (From Gelin, B., Lee, A., and Karplus, M., *J. Mol. Biol.*, 171, 523, 1983. With permission.)

between the axial histidine and certain porphyrin atoms (see Figure 7). From the X-ray coordinates of the deoxy tetramers,[9,35] His F8(87) is asymmetrically oriented with respect to the heme. In addition to its intrinsic asymmetry (i.e., the N$_{\epsilon 2}$-C$_{\epsilon 1}$ bond is shorter than the C$_{\delta 2}$-N$_{\epsilon 2}$ bond), the imidazole ring has a projection on the heme that forms an angle of 13° with the Fe-1N bond, the Fe-N$_{\epsilon 2}$ bond is not perpendicular to the heme plane (the angles N$_{\epsilon 2}$-Fe-1N and N$_{\epsilon 2}$-Fe-3N are 100.4° and 113°, respectively), and the imidazole ring is not symmetric with respect to the Fe-N$_{\epsilon 2}$ bond (the angles C$_{\delta 2}$-N$_{\epsilon 2}$-Fe and C$_{\epsilon 1}$-N$_{\epsilon 2}$-Fe are 129° and 120°, respectively). Energy refinement of the X-ray structure changed the individual angles only slightly and the basic asymmetry remained.[14,15] Further, displacements of the imidazole ring that reduce this asymmetry increase the nonbonded interaction energy between the histidine and the rest of the globin; the major contributions to this increase come from the F-helix and a "hydrophobic cage" that surrounds the histidine (see Figure 9).

Two recent crystal structures provide more detailed information on the position of the key His F8 residue. They are the 1.7 Å resolution structure of human deoxyhemoglobin[9] and the 1.8 Å resolution structure of human oxyhemoglobin.[10] Comparison of the deoxy- and oxyhemoglobin structures[9] confirms the asymmetric (tilted) position of His F8 noted[14,15] in the lower resolution deoxy structure[35] and the nearly symmetric position in the oxy structure; i.e., the key distances between C$_\epsilon$ (F8), C$_\delta$ (F8), and the closest pyrrole nitrogens (1N and 3N, respectively) are 3.26 Å (3.12 Å) and 3.78 Å (3.74 Å), respectively, in the deoxy α-chain (the β-chain values are given in parenthesis). Fermi et al.[9] note that the C$_\epsilon$-1N distance decreases by 0.1 to 0.3 Å, while the C$_\delta$-3N distance decreases by 0.8 Å, in

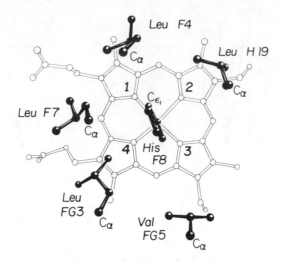

FIGURE 9.   View along N_ϵ-Fe bond showing hydrophobic residue cage about proximal His F8. (From Gelin,
B., Lee, A., and Karplus, M., *J. Mol. Biol.*, 171, 519,
1983. With permission.)

the transition to the oxy structure. This results in a much more symmetric position of the histidine relative to the heme; i.e., in the oxy structure C_ϵ-1N and C_δ-3N are 2.9 Å (3.0 Å) and 3.2 Å (3.1 Å), equal within the experimental error of the structure.

Although there is no repulsion between the imidazole carbons and the domed heme group in the unliganded structure, significant repulsion sets in from the $C_{\epsilon1}$-1N contact when the heme group becomes more nearly planar and the histidine moves toward it on ligand binding; as is evident from Figure 6 and the structural data just cited, the $C_{\epsilon1}$-1N distance is much smaller after ligation. The quaternary constraints in the deoxy structure on the F-helix, of which His F8 is a part, appear to prevent the imidazole from taking a more symmetric position. This prevents the local strain from being dissipated and contributes to the reduced ligand affinity. As mentioned above, the eclipsed O_2, imidazole position may be a contributing factor in this strain mechanism. The repulsive contacts result in tilting and distortion of the heme, whose displacement initiates other tertiary structural changes of the globin. These determine a reaction path for transmitting the effect of ligand binding from the heme group to the surface of its subunit.

To investigate the nature of the relaxation of the globin due to such repulsive contacts, a model calculation[14,15] has been performed in which a *deoxy*hemoglobin α-chain was perturbed by introducing a heme group that simulates ligation. Specifically, a planar heme with an Fe-$N_{\epsilon2}$ distance of 1.85 Å and a 20° tilt about an axis through the nitrogens of pyrrole 2 and 4 with pyrrole 1(3) displaced in the distal (proximal) direction was introduced, and the protein was allowed to relax in the presence of this perturbation. In the calculation, such an exaggerated perturbation was used to overcome barriers to structural change. The minimization method[15] would have resulted in the system remaining in a local minimum if a smaller, more realistic perturbation had been employed. Although the forces in the perturbed system are initially very high, they are rapidly dissipated by the distortion of the globin resulting from energy minimization. The distribution of atom shifts in the energy-minimized structure is highly skewed, with 90% of the globin atoms moving less than the overall rms shift and 5% more than twice its value. This implies, as is confirmed by the specific atom displacements,[14,15] that the heme perturbation is propagated through the globin along a well-defined pathway.

In the heme pocket, the most affected residue is Val FG5(93), which is pushed away by pyrrole 3 with an average atom shift of nearly 1 Å. This valine, as suggested by Anderson,[39] is a key residue in the propagation of tertiary structural change. Other residues in the heme pocket that undergo significant displacements are those adjacent to and interacting with Val FG5(93); they are Arg FG4(92), Leu FG3(91), and Asp G1(94). Phe G5(98), one turn of the G-helix from Asp G1(94), also moves slightly. These residue shifts can be summarized as a movement of the FG corner (see Figure 3). Comparison of the X-ray structures for deoxyhemoglobin and acrylamide restrained methemoglobins[39] shows corresponding displacements. On the distal side of the heme, less movement occurs in the model calculation, partly because it was done without a ligand. Lys E10(61) is pushed away by the heme, but the side chain is sufficiently flexible to damp the motion before it reaches the E-helix. Leu G8(101) moves slightly forward due to its attraction by pyrrole 3.

An essential contact between Val FG5(93) and the heme group involves pyrrole 3 and its vinyl group. The absence of the vinyl group in deuteroporphyrin may provide an explanation of its reduced cooperativity. Also, Asakura and Sono[40] have demonstrated that substitution of the vinyl by a formyl group on pyrrole 3 greatly reduces cooperativity while the corresponding substitution on pyrrole 2 has no effect. The absence of cooperativity in hemoglobin Köln, which has the β-chain Val FG5 (98) replaced by Met, is of interest though it has been suggested that the heme is lost from the same chain.[3] It would be of importance to have structural data to permit a more detailed analysis of these results.

In addition to its essential role in the displacement of the FG corner, Val FG5(93) appears to be involved in a feedback loop that stabilizes the proximal histidine and, thus, contributes to the tilting of the heme. The side chain of Leu FG3(91) supports the imidazole ring and prevents it from tilting backward or rotating to diminish its interaction with pyrrole 1. When Val FG5(93) is displaced by the heme, it pushes Leu FG3(91) toward His F8(87), thereby completing the loop. Further, as is shown by calculations of the nonbonded interactions between other globin residues and the imidazole ring of the histidine, leucines F4(83), F7(86), and H19(136) form a cage with Leu FG3(91) to help fix His F8(87) in its special position (see Figure 9). Leu FG3(91) and Leu H19(136) also act to anchor the tilt axis of the heme, while Leu F4(83) and Leu F7(86) make short contacts with pyrrole 1 after ligation and help to tilt the heme.

That there is strain on a *liganded* heme when it is in the deoxy quaternary structure is demonstrated by a number of experimental studies.[41,42] The most striking evidence is perhaps that obtained from nitrosylhemoglobin, in which the $Fe-N_{\epsilon 2}$ bond in the α-chains appears to be broken or at least greatly stretched when the quaternary structure is shifted from the oxy to the deoxy form by the addition of inositol hexaphosphate.[43,44]

The region of the α-subunit that is involved in the $\alpha_1\beta_2$ interface, identified by Perutz and Ten Eyck[11,45] as an essential component of the cooperative mechanism, is significantly altered in the model calculations.[14,15] In addition to the displacement of the FG region, residues of the C-helix are affected. As Val FG5(93) is repelled by pyrrole 3, it rotates the aromatic ring of Tyr C7(42) by 20° about the ring axis and displaces the hydroxyl oxygen, whose hydrogen bond with Asp G1(99) β could thus be weakened. Further, Thr C3(38) and Thr C6(41) come in contact with Tyr C7(42) and are slightly shifted.

The importance of Tyr HC2(140) in breaking the interchain salt bridges has been stressed by Perutz and Ten Eyck.[11,45] The tyrosine itself moves only slightly in the model calculation, but the motion of Val FG5(93) stretches and bends the hydrogen bond between its carbonyl oxygen and the hydroxyl group of Tyr HC2(140). This is in accord with the conclusion[14,15] that the tertiary changes resulting from ligation do not eject the tyrosine in a tetramer that has the deoxy quaternary structure, but do weaken the salt bridges by decreasing the stabilizing effect of the tyrosine-valine hydrogen bond.

To proceed further in the analysis of the cooperative mechanism, it is useful to compare the tertiary changes that occur on ligand binding by a subunit in the deoxy tetramer with the tertiary structural difference between an unliganded subunit in the deoxy tetramer and a liganded subunit in the oxy tetramer. Such a comparison makes it possible to investigate how the ligand-induced perturbation in a single subunit affects and is affected by the quaternary structure of the tetramer. Since there are high resolution X-ray results for the unliganded deoxy structure and the liganded oxy structure, detailed information concerning their differences is available. The original comparison of human deoxy and horse methemoglobin made by Perutz in his classic paper[45] has been supplemented by the more detailed study of Baldwin and Chothia,[46] which includes results for human carbonmonoxyhemoglobin. Their analysis has been confirmed by the higher resolution deoxy- and oxyhemoglobin structures now available.[9,10] In accord with the tertiary structural changes already described for ligation in the deoxy tetramer,[14] these authors focused on the heme and its surroundings. They found that in the deoxy-to-oxy transition of the tetramer the heme group, His F8, the F- and FG-helices of both chains, and part of the E-helix of the β-chain are the regions of the subunits that undergo the largest tertiary alterations; the chain termini also have significant displacements. Other parts of the chains have much smaller changes; e.g., in the α-subunit, the deoxy-oxy differences are within experimental error for the rest of the molecule. In the transition from the deoxy to the oxy structure, Baldwin and Chothia[46] noted that "the F helix is translated across the heme plane (in the direction from N1 to N3) by ~1 Å and is tilted with respect to it (i.e., the F9 end is 0.8 Å closer to the heme plane); this movement of the F helix takes the proximal histidine from its asymmetric position in deoxy to a more symmetric position in liganded hemoglobin. The tilting of the F helix is associated with a movement of the iron atom towards the heme plane." The tilt of the F-helix relative to the heme is approximately 10°, and it is about the axis found in the α-subunit heme-tilt calculation.[14,15] As already mentioned, the observed final position of the histidine in the oxy structure is such that the critical imidazole-heme distances are nearly equal. The FG corner moves little *relative* to the heme, with Leu FG3(91) and Val FG5(93) showing conformational changes that permit them to remain in Van der Waals contact with the heme. Slight displacements were noted also for Phe G5(98), Leu G8(101), Phe CD4(45), and His 45(CD3), as well as portions of the H-helix.

The important point to note is that the tertiary structural changes that occur when the subunits become liganded *and* when there is also a quaternary structural transition[46] are related to those arising from ligation in the deoxy tetramer, *as far as the relative motions of the heme, His F8, part of the F-helix, and the FG corner are concerned.* Thus, it appears appropriate to focus on this region, the "allosteric core",[15] as an essential element of the hemoglobin cooperative mechanism. It is convenient also to separate the internal structural alteration from the overall displacement of the allosteric core with respect to the rest of the globin chain. The *internal* structural changes of the allosteric core produced by ligand binding are coupled to the quaternary transition; i.e., as shown by calculations[14,15] and observed by Anderson[39] and Pulsinelli,[91] the changes on ligation in the deoxy tetramer have elements in common with, but are smaller than, those found when the transition from the unliganded deoxy to liganded oxy tetramer takes place (see below). However, the main factor differentiating ligation in the deoxy and oxy tetramers concerns the overall displacement of the core. Upon ligation in the deoxy tetramer, the allosteric core is prevented from moving relative to the rest of the globin chain. The quaternary constraints are such that only displacements internal to the core can occur easily; that is, because the F-helix is held in place by intersubunit contacts involving the FG corner, His F8 can only move toward the heme, which apparently distorts and tilts; this in turn leads to the localized distortions of the FG corner and nearby residues described above. By contrast, when the system is liganded and the deoxy-to-oxy transition occurs, the allosteric core is able to move as a whole. Both the heme and the F-helix slide sideways in the direction of the $N_1 \rightarrow N_3$ vector into the interior of the globin with the F-helix moving about 1 Å further than the heme. It is this extra displacement of the helix, combined with a helix tilt, that is required for the structural changes in allosteric core associated with ligation, as described above. In the α-chain, the displacement of the allosteric core in going from the deoxy to the oxy tetramer is such that the heme plane ends up shifted but nearly parallel to its original position in the unliganded deoxy tetramer; in the β-subunit, a heme tilt remains after relaxation. The tertiary structural changes in the deoxy-to-oxy transition are such that the allosteric cores of the α- and β- chains of the $\alpha\beta$ dimer ($\alpha_1\beta_1$) and $\alpha_2\beta_2$) move toward each other; as pointed out by Baldwin and Chothia,[46] the (FGα_1)-(FGβ_1) corner distance is shorter by approximately 2.5 Å in the oxy than the deoxy tetramer.

The recent X-ray study[13] of $[\alpha(FeCO)\beta(Mn)]_2$ provided the first refined structure of a liganded α-subunit in a tetramer with the deoxy quaternary structure. Although the results are only at 3.0 Å resolution, they supplement the earlier work of Anderson[39] and Pulsinelli.[91] The structure of Arnone et al.[13] serves as a partial check for the theoretical model for the liganded α-chain in the deoxy quaternary structure. The results, as far as they go, are in accord with the model of the allosteric core as the essential element in coupling tertiary and quaternary structural change. The quaternary constraints in the deoxy tetramer prevent the large overall displacement of the allosteric core that occurs in the transition to the oxy tetramer. Arnone et al.[13] find that the changes within the allosteric core are significantly smaller than the differences between the deoxy- and oxyhemoglobin structures.[46] As to the direction of the small alterations within the allosteric core, there is a displacement of His F8 toward the heme, but the change in the relative orientation of His F8 and the heme group is not clearly resolved. There is a weak indication of a heme tilt and the concomitant motion of Val F5 in the difference map. However, the tilt is less clear than in the results of Anderson and Pulsinelli. Arnone et al.[13] point out that a full definition of the changes within the allosteric core "will require higher resolution studies".

The details of the constraints that prevent the overall displacement of the allosteric core in the deoxy tetramer have not been fully analyzed by energy calculations, although structural data and theoretical calculations[47] suggest that the constraining forces act on the F-helix, the FG corner, and the heme itself. It appears that the configuration with the tilted or distorted heme in the liganded subunit of the deoxy tetramer is the destablized one; that is, the strain introduced in the allosteric core cannot be dissipated in the deoxy quaternary structure due to intersubunit ($\alpha_1\beta_2$, $\alpha_2\beta_1$) constraints and so leads to the observed lower ligand affinity. The displacements found on energy minimization of the perturbed structure are an indication of the strain introduced by ligand binding. To determine whether this strain is sufficient to account for the lower affinity of the deoxy structure, a quantitative evaluation of the energy involved is required. Although the energy function and the minimization method used are not sufficiently accurate for an unequivocal conclusion,[15] the result that the minimized perturbed structure is less stable than the minimized unperturbed structure is in accord with the important role of steric strain in reducing the affinity of the deoxy tetramer. As already mentioned, this agrees with experimental studies.[41-44,48,49]

For an unliganded subunit in the oxy tetramer, the analysis described here[15] suggests that the allosteric core would have a configuration such that His F8 is symmetrically positioned relative to the heme and little or no strain is introduced on binding a ligand. The increased affinity of the oxyhemoglobin tetramer relative to the isolated chains indicates that, in fact, it is the geometry of the *unliganded* subunit that is now slightly strained.[50] This conclusion is in accord with the results of a comparison of liganded hemoglobin with unliganded bis-maleimido-methyl-ether (BME) modified (BME) hemoglobin, which is constrained to remain in the oxy quaternary structure.[11] A higher resolution BME structure determined recently confirms this result.[49]

We now turn to the question of the relation between the structural changes that occur on ligation in the deoxy tetramer and the transition to the oxy quaternary structure. Perutz and Ten Eyck,[11] Perutz,[45] and Baldwin and Chothia[46] have pointed out that the dominant quaternary structural change can be described in terms of the relative motion of the $\alpha_1\beta_1$ and $\alpha_2\beta_2$ dimers and its effect on the contacts at the $\alpha_1\beta_2$ and $\alpha_2\beta_1$ interfaces. Further, Baldwin and Chothia[46] have suggested that the contacts between the FG corner of the α-chains and the C-helix of the β-chains, $(FG\alpha_1)(C\beta_2)$ and $(FG\alpha_2)(C\beta_1)$, act as "flexible hinges" with the residues involved undergoing significant distortions. By contrast, the contacts between the FG corner of the β-chains and the C-helix of the α-chains, $(C\alpha_1)(FG\beta_2)$ and $(C\alpha_2)(FG\beta_1)$, undergo large displacements with relatively little local distortion of the residues. It has been suggested[45,46] that there are two stable positions for this "switch" region involving the dovetailing of different residues in the deoxy and oxy quaternary structure. The tertiary structural alterations already described appear to be essential for the quaternary transition to occur. Moreover, the ligation of a subunit in the deoxy tetramer, as determined by calculations,[14,15] can have a direct effect in shifting the structure along the reaction path toward the oxy form. Changes induced in the allosteric core on ligand binding in the presence of quaternary constraints lead to heme-histidine motion, which in turn results in the displacement of the FG region. The calculated distortion[14,15] of the FG corner of the α-chain, though small, is in the direction required for the alteration of the flexible-hinge region that is observed in the deoxy-to-oxy quaternary transition. Further, Val FG5 perturbs some residues in the C-helices of the α-chain that are involved in the switch region of the intersubunit contacts. Thus, the tertiary structural changes induced by ligation of a subunit in the deoxy tetramer correspond to displacements along the reaction path from the unliganded deoxy to the liganded oxy tetramer. The inverse effect is likely to be present if a ligand is removed from a subunit in the oxy tetramer.

IV. STATISTICAL THERMODYNAMIC MODEL FOR COOPERATIVITY

The crystallographic and theoretical results described in Section III serve as the basis for relating structure and function in hemoglobin by introduction of a statistical mechanical model. In so doing, one attempts to abstract the essential features of the structural changes that occur on ligation and to incorporate them into a description of the thermodynamics of ligand binding. This makes it possible to go beyond the phenomenological models for cooperativity (see Section II).

The first step in such an approach was taken by Perutz[45] in his now classic formulation of a stereochemical mechanism for hemoglobin cooperativity. Although certain aspects of the mechanism have been questioned and others modified on the basis of newer data, the essential features required for formulating a statistical mechanical model appear to be valid. The elements in the Perutz scheme are (1) deoxy and oxy quaternary conformations of different stability for the hemoglobin tetramer; (2) two tertiary structures for each individual chain, one for the unliganded and the other for the liganded monomer, and (3) the presence of interchain salt bridges (some with ionizable protons) which provide a mechanism for coupling the alterations in tertiary structure due to oxygen binding with a change in the relative stability of the two quaternary structures. Perutz[45] suggested that his scheme was able to account qualitatively for the sigmoidal oxygenation curve of hemoglobin and for the effects on it of changes in pH and the concentration of 2,3-diphosphoglycerate. However, a quantitative interpretation of the available equilibrium and kinetic data was not attempted by him.

To demonstrate how specific structural assumptions can be introduced into a statistical mechanical model for the equilibrium properties of hemoglobin, the Perutz scheme was used and tested by comparison of the results of its implementation with equilibrium data for the oxygen binding of native hemoglobin and of various mutant and modified forms.[50-52] The effect of pH, salt concentrations, 2,3-diphosphoglycerate, and dissociation were considered. Although many important details concerning the electronic and geometric changes occurring

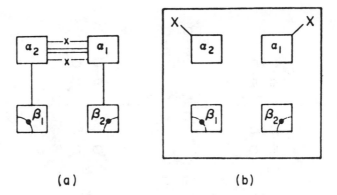

FIGURE 10. Diagrammatic representation of hemoglobin tetramer with all subunits having the unliganded tertiary structure; (a) deoxy quaternary and (b) oxy quaternary conformations. (From Szabo, A. and Karplus, M., *J. Mol. Biol.*, 72, 168, 1972. With permission.).

on oxygen binding do not appear in the model, the structural features that are included make it possible to identify certain of the model parameters with the free-energy changes of specific processes (e.g., ionization of Bohr protons, breaking of salt bridges).

To illustrate the methodology, we first describe the original statistical mechanical model. A generating function, which is a macroscopic analog of the grand canonical partition function, is used to provide a quantitative relation between the stereochemical and thermodynamic features. The generating function methodology is general and can be used to modify the model as more information becomes available. Because new data and calculations have, in fact, permitted refinements to be introduced, we then outline the results obtained with an improved model.[53]

In the simplest form of the model,[50] it provides a way of considering simultaneously oxygen binding and proton release (or, equivalently, OH^- binding). Both are encompassed in a formulation of the adsorption problem involving O_2 and OH^-, with hemoglobin acting as the adsorber. Each molecule of hemoglobin has four binding sites for O_2 (the four hemes) and four binding sites for OH^- (the NH_3^+ of the α-chain N-terminal valines and the protonated imidazoles of the β-chain C-terminal histidines). In accord with the Perutz scheme,[45] there are eight salt bridges of which four titrate in the physiological pH range; it is the latter that involve the OH^- adsorption sites. Six of the salt bridges are interchain (4 α←→ α', 2 β←→ α) and two are intrachain (2 β → β).[35,45] Supporting evidence for the identification of the Bohr groups has come from NMR,[54,55] from hydrogen exchange experiments,[56] from chemical reactivity studies,[57,58] and from studies of mutant and modified hemoglobins.[59-63] Measurements indicate that the $\alpha_1\alpha_2$ and intra-β-chain Bohr groups contribute 60 to 80% of the total alkaline Bohr effect at normal (e.g., 0.1 M Cl^-) salt concentrations.[59-66] The remainder is thought to come from other chloride-linked sites.[57,67,68]

To obtain a grand canonical partition function for the hemoglobin tetramer, a set of diagrams and diagram rules were used for evaluating the various contributing factors.[50] In Figure 10, the diagram on the left represents the deoxy quaternary structure, and the diagram on the right represents the oxy quaternary structure, both without any O_2 bound. The assumption is made that in the oxy quaternary structure, whether or not the subunits are liganded (the squares represent unliganded subunits), all of the *interchain* salt bridges are broken; the β-*intrachain* salt bridges are still present for unliganded subunits. In the deoxy quaternary structure, the four αα', the two βα, and the β-salt bridges are shown as lines. Further, we have used the convention that a line with an "x" or a "·" on it represents a salt bridge which has an ionizable proton, namely, the valines for the αα'-salt bridges and

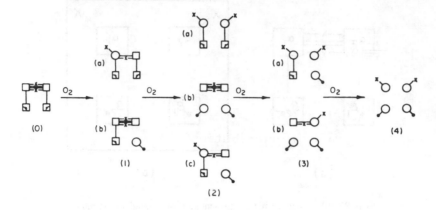

FIGURE 11.   Diagrammatic representation of the oxygenation of the hemoglobin tetramer with deoxy quaternary conformation. (From Szabo, A. and Karplus, M., *J. Mol. Biol., 72,* 170, 1972. With permission.)

the imidazoles for the β-salt bridges. The behavior of the salt bridges on the change of quaternary structure is indicated in the figure. The effect of changes in tertiary structure on the salt bridges is indicated in Figure 11 which shows all the diagrams for the deoxy quaternary structure as a function of oxygenation. Various salt bridges are seen to be broken as oxygen is bound to the chains. In particular, if oxygen is bound to an α-subunit (1a in Figure 11; a circle indicates a liganded subunit) the two salt bridges involving its C-terminal arginine are assumed to be broken; the broken salt bridge without an ionizable proton is not shown and the one with an ionizable proton is indicated as being free. The two salt bridges involving the C-terminal arginine of the *unliganded* α-chain are still present, as shown in the diagram. A corresponding description applies to ligation of a β-chain (1b in Figure 11). Here, the two salt bridges involving its C-terminal histidine are broken, the one with an ionizable proton being shown as free. It is assumed that there is no change in the remaining salt bridges originating from unliganded chains; there is no direct evidence for this assumption, but is is made to simplify the problem and to see if it can work (for a refinement of the model, see below).

As one follows Figure 11 from left to right, one sees diagrams with an increasing number of liganded chains. The right-hand-most diagram (4) corresponds to the completely liganded tetramer in the deoxy conformation. All of the salt bridges, both the interchain and intrachain, are broken. A corresponding set of diagrams (not shown) represent the possible contributions with zero to four ligands in the oxy quaternary structure.

The expression for the partition function of the system has the form

$$\Sigma(\lambda,\mu) = \sum_{i=0}^{4} \lambda^i \left(\sum_{j=0}^{4} \mu^j q_{ij}^O \right) + \sum_{i=0}^{4} \lambda^i \left(\sum_{j=0}^{4} \mu^j q_{ij}^D \right) \qquad (8)$$

where λ and μ are the concentrations of O_2 and OH^-, respectively, and q_{ij}^O and q_{ij}^D are the appropriate weighting factors for the oxy and deoxy quaternary structures, respectively, with iO_2 and jOH^- bound. It is the q_{ij}^O and q_{ij}^D which are expressed in terms of molecular parameters by use of the diagrams. Given Equation 8, the fractional saturation by oxygen or OH^- is obtained in the usual way; e.g., for oxygen,

$$\langle y_{O_2} \rangle = \frac{\lambda}{4} \frac{\partial}{\partial \lambda} [\ln \Sigma(\lambda,\mu)]_\mu \qquad (9)$$

Contribution to Σ_D

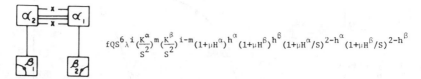

$$fQS^6\lambda^i(\frac{K^\alpha}{S^2})^m(\frac{K^\beta}{S^2})^{i-m}(1+_\mu H^\alpha)^{h^\alpha}(1+_\mu H^\beta)^{h^\beta}(1+_\mu H^\alpha/S)^{2-h^\alpha}(1+_\mu H^\beta/S)^{2-h^\beta}$$

Contribution to Σ_O

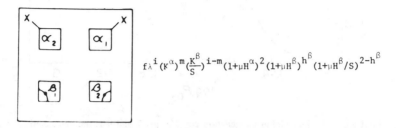

$$f\lambda^i(K^\alpha)^m(\frac{K^\beta}{S})^{i-m}(1+_\mu H^\alpha)^2(1+_\mu H^\beta)^{h^\beta}(1+_\mu H^\beta/S)^{2-h^\beta}$$

FIGURE 12. Diagrammatic rules (see text for details). (From Szabo, A. and Karplus, M., *J. Mol. Biol.*, 72, 174, 1972. With permission.)

To obtain the terms contributing to Σ, we use the diagrammatic rules shown in Figure 12. Each diagram, such as those shown in Figure 11, corresponds to a term

$$\lambda^i \sum_{j=0}^{4} \mu^j q_{ij}^O \qquad \text{or} \qquad \lambda^i \sum_{j=0}^{4} \mu^j q_{ij}^D$$

in Equation 8; that is, each diagram is a sum over structures with 0, 1, ... 4 OH^- adsorbed. In Figure 12 are given the rules for the deoxy- and oxyhemoglobin diagrams. They include the various molecular parameters that appear in the model. There is a single salt bridge parameter S, since all the salt bridges are assumed to have the same strength, for simplicity; in one form of the generalized model, salt bridges of unequal strengths are introduced (see below). There are the oxygen binding constants K^α and K^β for the individual α- and β-chains, respectively, in the absence of salt bridges or other interactions. The constants H^α and H^β are related to the pK_a of the two types of ionizable protons. Finally, there is the parameter Q which gives the intrinsic relative stability of the deoxy and oxy quaternary structures for unliganded hemoglobin in the absence of salt bridges.

The rules in Figure 12 for obtaining the partition function terms from the diagrams have the following interpretation: i is the number of O_2 bound, f is a statistical factor, h^α is the number of free crosses, h^β is the number of free dots, and m is the number of α-chains which are circles. It is assumed that if an OH^- had been adsorbed by a group with an ionizable proton, the salt bridge involving that proton cannot be formed.

For most of the parameters of the problem (S, K^α, K^β, H^α, H^β) some estimate of the proper order of magnitude can be made. Thus, K^α and K^β should be close to the free chain values, which have been measured. The H^α and H^β constants should correspond to the range of pK_a values observed for valine and histidine in proteins. For the salt bridges, there are a variety of experimental measurements for model compounds and indirect determinations for some proteins which suggest free energy values of 1 to 4 kcal. As to the parameter Q, it is more difficult to give quantitative limits. From the structural differences at the interchain contacts, it appears to be determined primarily by a difference in Van der Waals interactions of a large number of groups, though there are also electrostatic contributions.[47] It seems unlikely that the free energy associated with Q is larger than 10 kcal.

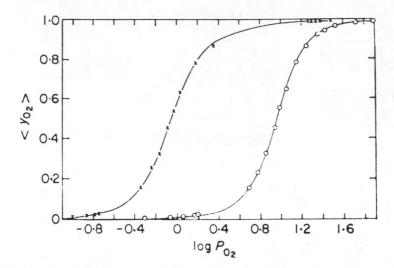

FIGURE 13.   Hemoglobin oxygenation curve at pH 7 and 9.1; the calculated values are shown by solid curves and the experimental values by (o) for pH 7 and (x) for pH 9.1. (From Szabo, A. and Karplus, M., *J. Mol. Biol.*, 72, 168, 1972. With permission.)

If one chooses values for the parameters in the physical ranges just described, one finds that the model yields qualitatively reasonable results. However, in terms of quantitative comparisons with experiment, what one has to do is to fit the parameters to a set of data. The best measurements for hemoglobin oxygenation available at the time the model was formulated (1970) were those of Roughton and Lyster.[5] They had measured O_2 binding at pH 7 and pH 9.1 for low and high enough partial pressures to yield meaningful values for the asymptotic portions of the curves. Figure 13 shows the experimental results and the model fits. The significant point about the excellent agreement in Figure 13 is that it is obtained with all of the parameters within the ranges outlined above; e.g., the salt bridges have a free energy of 2.2 kcal/mol and the K^α, K^β values are within a factor of two of the free chain values. Moreover, variation of the parameters indicated that they must be in the suggested ranges if one is to obtain satisfactory results by the fitting procedure. Further, the model has been shown to be applicable to 2,3-diphosphoglycerate binding and to mutant and modified hemoglobins.[50-52]

The structural, spectroscopic, and thermodynamic data on hemoglobin that have become available since the statistical thermodynamic model was first proposed indicated some years ago (1981) that it would be useful to reexamine the basic assumptions and to extend the applications. Of particular importance are the structural results concerning the nature of tertiary-quaternary coupling (see Section III). Comparisons of crystal structures of unliganded deoxy and liganded oxy tetramers with liganded deoxy tetramers and unliganded oxy tetramers demonstrated that the quaternary structure has an important influence on the tertiary conformation of a subunit. Further, empirical energy calculations suggested that the low affinity of the deoxy quaternary structure arises not only from the perturbation of salt bridges, but also from steric constraints imposed by contacts at the $\alpha_1\beta_2$ and $\alpha_2\beta_1$ interfaces that prevent the dissipation of ligation-induced strain in the allosteric core. A corollary of this result is the observation that formation and breaking of the salt bridges is determined in a complex way by both the ligation state of the subunits and the quaternary structure of the tetramer.

A generalization of the statistical mechanical model was formulated[53] that includes two different tertiary structures (liganded and unliganded) for each of the two quaternary structures, in accord with the data cited above. This leads directly to the possibility of constraints on ligand binding independent of the salt bridges and also permits a general formulation of the salt bridge coupling to tertiary and quaternary structural change. Although tetramer dissociation (i.e., the monomer-dimer-tetramer equilibrium) has been included in the complete model,[69] we do not consider it here because it is not essential for the analysis of cooperativity in hemoglobin. Since the oxygen binding by the monomer and dimer are noncooperative, it is necessary only to correct the measured binding curves for dissociation so as to be able to extract results for the cooperative tetramer. In the generalized model, the generating functions for the deoxy (Σ_D) and oxy (Σ_O) quaternary states can be written

$$\Sigma_D(\lambda,\mu) = \frac{QS^6}{(r')^2} [1 + \mu H^\alpha/S]^2 [1 + \mu H^\beta/S]^2$$

$$\times \left\{ 1 + \lambda K_D^\alpha \, r^2 \frac{[1 + \mu H^\alpha/(rS)]}{[1 + \mu H^\alpha/S]} \right\}^2 \times \left\{ 1 + \lambda K_D^\beta \, r^2 \frac{[1 + \mu H^\beta/(rS)]}{[1 + \mu H^\beta/S]} \right\}^2$$

$$\Sigma_O(\lambda,\mu) = [1 + \mu H^\alpha]^2 [1 + \mu H^\beta/(r'S)]^2$$

$$\times \{1 + \lambda K_O^2\}^2 \left\{ 1 + \lambda \frac{K_O^\beta}{(r'S)} \frac{[1 + \mu H^\beta]}{[1 + \mu H^\beta/(r'S)]} \right\}^2 \tag{10}$$

The new parameters are (rS), the effective strength of salt bridges in liganded chains of the deoxy tetramer; (r'S), the effective strength of the internal salt bridge of unliganded β-chains in the oxy tetramers; and K_D^α, K_D^β, K_O^α, and K_O^β are the intrinsic ligand affinities for the α- and β-chains in the deoxy (D) and oxy (O) quaternary structures. The original model is obtained for $r = 1/S$, $r' = 1$, and $K_D^\alpha = K_O^\alpha = K^\alpha$, $K_D^\beta = K_O^\beta = K^\beta$; $r = 1$, $r' = 1/S$ corresponds to pure quaternary coupling, as in the Monod, Wyman, Changeux model, and requires $K_D^\alpha < K_O^\alpha$, $K_D^\beta < K_O^\beta$. An alternative extension of the original model allows different strengths for the salt bridges that are pH dependent (S) and pH independent (S_1) in the alkaline physiological range. It is obtained from Equation 10 with QS^6 replaced by $QS_1^4S^2$, $r = 1/S_1$, $r' = 1$, $K_D^\alpha = K_O^\alpha = K^\beta$ and $K_D^\beta = K_O^\beta = K^\beta$.

In formulating the present model (Equation 10), the literature has been reviewed to determine if other assumptions of the original model require revision. That more than two quaternary structures exist has been suggested.[70,71] There is no structural evidence for this, and the data on which the suggestion was based are consistent with the original model, which involves only two quaternary structures, but an equilibrium of many species at each state of oxygenation (e.g., see Figure 11). Also, it has been suggested that cooperativity can occur in the absence of a change in quaternary structure.[72,73] Experimental results indicate that such a contribution to the cooperative oxygen binding, if it exists, is not important. It has been proposed[74-77] that the Bohr effect may involve a number of titrating groups, in addition to the salt bridges included in the Perutz scheme. The evidence for this view is not strong; i.e., the NMR assignments on which the proposal is based have proved to be equivocal[78] and the electrostatic model that indicated many histidines were involved in the Bohr effect has been revised.[79] However, it is likely that assuming only two groups are involved in the Bohr effect is an oversimplification. Further, since variations in the ionic strength and the types of anions present in solution do influence the properties of hemoglobin, such effects have to be included in a more complete model.[50-52]

The generalized model[53] was first applied to the very accurate oxygen binding measurements at a single pH made by Mills et al.[80] However, since at a single pH, the original model, the generalized model, and the Monod-Wyman-Changeux model for inequivalent chains are isomorphic, the analysis was restricted to the original model. The solution conditions of the measurement[80] were pH 7.4, 21.5°C, 0.1 M Cl$^-$ and a hemoglobin con-

Table 1
ORIGINAL MODEL RESULTS FOR MILLS ET AL.
MEASUREMENTS[a]

		Parameter values[b]		
pK_α, pK_β	$-RT \ln Q$	$-RT \ln S$	$-RT \ln K^\alpha,$ $-RT \ln K^\beta$	Variance
7.5, 6.2	6.16	-2.61	$-9.68, -9.44$	8.3×10^{-12}
7.7, 5.93	6.24	-2.60	$-9.70, -9.15$	8.4×10^{-12}
7.7, 5.93[c]	5.52	-3.04	$-11.44, -9.47$	2.1×10^{-9}

[a]   The Adair constants (corrected for statistical factors) are[80] $\Delta G_{41} = -5.46$ kcal/ mol, $\Delta G_{42} = -5.432$ kcal/mol, $\Delta G_{43} = -7.658$ kcal/mol, and $\Delta G_{44} = -8.71$ kcal/mol.

[b]   All values, except pKs and variances, are in kcal/mol.

[c]   Fit obtained with the model parameters reported by Johnson and Ackers[81] as initial values.

centration 5.35 to 382 μM in heme. A significant fraction of the hemoglobin molecules is expected to have dissociated into dimers at the lower concentrations used by Mills et al.; e.g., at 5.35 μM, 35% of the heme exist in dimeric form in oxyhemoglobin. To avoid the need to consider dimers, the model was fitted to data generated from tetramer Adair constants reported by Mills et al.[80]

The model results for the tetramer are shown in Table 1.[53] All of the fits are in good agreement with experiment and yield parameter values in the allowed ranges (see above). It is evident from the table that the information from such equilibrium measurements at a single pH is not sufficient to determine a unique parameter set. Included is a fit obtained starting with the parameter set of Johnson and Ackers,[81] which has the special property noted by them that K^α has an unreasonable value ($-RT \ln K^\alpha = -11.3$ kcal/mol). It is clear from a comparison with the other results in Table 1 that such a K^α is not required to fit the Mills et al. data. Application of the various tetramer parameters from Table 1 and the experimental dimer-tetramer equilibrium constants to the Mills et al. data at five heme concentrations yields excellent fits, confirming the validity of the analysis.[69]

Another test of the model is provided by the calculated Bohr effect at pH 7.4. With the parameter set for $pK_\alpha = 7.5$ and $pK_\beta = 6.2$ given in Table 1, there is a release of 2.4 protons per tetramer, in agreement with independent measurements;[65,66] results for the al-kaline Bohr curve over a range of pH values are considered below. Finally, the predictions for the relative fractional saturation of the α- and β-hemes ($<y>^\alpha$ vs. $<y>^\beta$) as a function of the total saturation $<y>$ were examined. The calculated results are shown in Figure 14; the maximal difference between the two curves (with $<y>^\beta$ larger than $<y>^\alpha$) is less than 5% and occurs between 10 and 20% oxygenation. This agrees with the NMR experiments of Viggiano and Ho[72,82] for stripped hemoglobin, which found $<y>^\alpha$ and $<y>^\beta$ to be equal within experimental error, and with the optical absorbance measurements of Nasuda-Kouy-ama et al.,[83] who found $<y>^\beta$ to be slightly larger than $<y>^\alpha$ throughout the entire oxygenation curve. Thus, the original model *can* explain the single pH binding data of Mills et al.,[80] the Bohr proton release at pH 7.4, and the NMR, as well as optical measurements, on chain inequivalence. This disagrees with the conclusion of Johnson and Ackers,[81] who claimed that these data disproved the original model.

The parameters that fit the Roughton and Lyster data[5] at pH 7 are in the same range but differ somewhat from those for the Mills et al. data,[80] demonstrating that differences in solution conditions can perturb the properties of hemoglobin and, therefore, the model results. In particular, inorganic phosphates and 2,3-diphosphoglycerate may have been present in the Roughton-Lyster study and were absent in that of Mills et al.

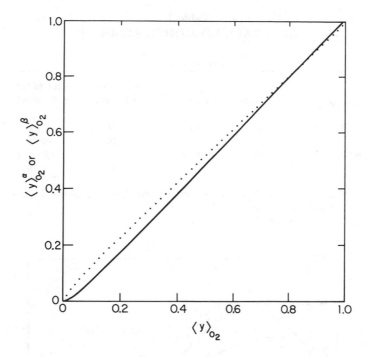

FIGURE 14. Computed relative fractional saturation of α- and β-chain hemes as a function of oxygenation. —, $\langle y \rangle^\alpha$; parameters for the original model ($pK_\alpha = 7.5$, $pK_\beta = 6.2$ set of Table 1). (From Lee, A. and Karplus, M., *Proc. Natl. Acad. Sci. U.S.A.*, 80, 7057, 1983. With permission.)

Because oxygenation data at one or even two pH values can be adequately described by the original model, it is necessary to introduce additional information to apply its generalization in a meaningful way. Use was made of the experimentally determined differential titration curves for the total proton release on oxygenation as a function of pH in the range 7.0 to 9.2 at 20°C and 0.25 M Cl$^-$ obtained by Antonini et al.[84] Although the Bohr proton release at pH 7.4 was found to be in agreement with experiment for the original model (see above), calculation of the proton release curve showed that, relative to the Antonini et al. results, it was shifted too far into the alkaline range; i.e., when the first parameter set for the Mills et al. data[80] in Table 1 is used, ΔH_{max}^+ is found at pH 7.8 instead of 7.5 and ΔH_{max}^+ is overestimated (2.6 instead of 2.1 protons released). To obtain the experimental ΔH_{max}^+, the salt bridge energy S must be reduced. With such a reduced value of S, the original model cannot fit the oxygenation data. Because the reduction in the ligand affinity of the deoxy structure is determined by S (see Σ_D in Equation 10), a small value of S means that the ligand affinity is too large. Consequently, use of a generalized model is required with $K_O^\alpha < K_D^\alpha$ and $K_O^\beta < K_D^\beta$. Fits to the Mills et al. data with two different choices for the pK values are given in Table 2; because the results are not sensitive to r and r′, we use the original values (r = 1/S, r′ = 1). Calculated and experimental proton release curves are shown in Figure 15; they agree within experimental error over most of the range, with the largest deviations between pH 7 and 7.5 as expected, due to the neglect of acid Bohr groups. As to the α-, β-chain inequivalence, the maximal difference in fractional saturation ($\langle y \rangle^\beta > \langle y \rangle^\alpha$) is 6% at an overall saturation of 25%.

For the original model with unequal salt bridge strengths, the parameters pK_α, pK_β, and S can be determined from the proton release data as above (row 1 of Table 2). A fit of the remaining parameters to the Mills et al.[80] data yields the values $-RT \ln Q = 6.393$ kcal/mol, $-RT \ln S_1 = 3.031$ kcal/mol, $-RT \ln K^\alpha = -8.675$ kcal/mol, and $-RT \ln K^\beta = -9.643$ kcal/mol, all within the allowed ranges.

Table 2
GENERALIZED MODEL RESULTS[a]

Parameter values[b]

pK_α, pK_β	$-RT \ln Q$	$-RT \ln S$	$-RT \ln K_D^\alpha$, $-RT \ln K_D^\beta$	$-RT \ln K_O^\alpha$, $-RT \ln K_O^\beta$
7.12, 6.05	2.51	-1.89	-7.79, -8.10	-8.85, -9.03
6.96, 6.75	0.99	-1.67	-7.28, -8.30	-8.94, -9.38

[a]   Fitted to Mills et al. data[80] and Antonini et al. data[84]; the original model values for r and r' were used (r = 1/S, r' = 1).

[b]   All values, except pKs, are in kcal/mol.

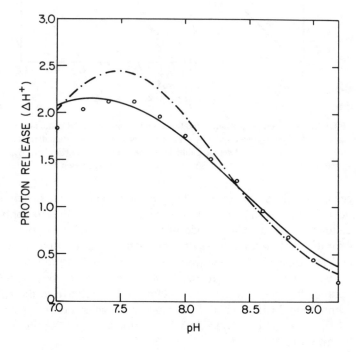

FIGURE 15.   Proton release (Δh^+) on oxygenation. (—) generalized model with $pK_\alpha = 7.12$ $pK_\beta = 6.05$; (-·-·-) generalized model with $pK_\alpha = 6.96$, $pK_\beta = 6.75$; (o-o-o) data of Antonini et al.[84] (From Lee, A. and Karplus, M., *Proc. Natl. Acad. Sci. U.S.A.*, 80, 7058, 1983. With permission.)

An important feature of the generalized model is its ability to describe different tertiary-quaternary coupling schemes within one framework by changing the values of r and r' (see Equation 10). As r controls the coupling of proton release to ligation in the deoxy state, it is expected that the pH dependence of the first Adair constant, K_1, will be most sensitive to this parameter; correspondingly, the pH dependence of K_4 is most sensitive to r'. Recently, precise measurements of the pH dependence of the oxygenation curve have been made in the limiting region of low oxygen concentration from which r can be estimated.[85] Under the simplifying assumption that singly liganded tetramers exist primarily in the deoxy quaternary structure (valid for pH $\leq$ 8.5), it can be shown that the value of r in the generalized model is coupled to the parameters S, pK_α, and pK_β and the ratio K_D^α/K_D^β. Because the overall proton release determines S, pK_α, and pK_β[53] only r and K_D^α/K_D^β need to be fitted. From measurements of the pH dependence of the first Adair constant in the range pH 6.5 to 9 at 20°C and 0.1 M Cl$^-$ and the fits from Table 2, the allowed r values are required to satisfy the condition $0 \leq \ln rS/\ln S \leq 0.3$. Such a relationship implies that the salt bridges are weakened but not necessarily broken (r = 1/S) on ligation in the deoxy quaternary structure. Further, the data and their analysis exclude the Monod-Wyman-Changeux limiting model (r = 1, r' = 1/S), in which the salt bridges are coupled purely to quaternary structure.

The variation of the Hill parameter n (see Equation 2) with pH is also related to the coupling of the salt bridges to tertiary and quaternary structure. Although it was believed from early experiments that n was independent of pH,[86] more recent work shows that n varies with pH. With the generalized model (Table 2), n decreases by 0.2 between pH 7 and 9; for the original model fits (Table 1), n decreases by 0.5 between pH 7 and pH 9. The reduction in the variation of n with pH in the generalized model is due to the presence of pH-independent constraints in the deoxy state.

V. CONCLUDING DISCUSSION

Hemoglobin cooperativity has been analyzed in both structural and statistical mechanical terms and the relation between the two has been demonstrated. The main features of the statistical mechanical model follow the mechanism proposed by Perutz; i.e., there are two quaternary structures (deoxy and oxy) for the tetramer, two tertiary structures (liganded and unliganded) for the subunits in each form of the tetramer, and structural elements that couple the stabilities of the tertiary and quaternary structures. In the original statistical mechanical model, it was assumed for simplicity that the tertiary-quaternary coupling was due entirely to a set of salt bridges of equal strength. The present analysis, in conjunction with structural and empirical energy function results, indicates that the salt bridges may be of unequal strengths and that steric constraints, not dependent on pH, are involved in the coupling and make a contribution to reducing the ligand affinity in the deoxy tetramer. To explain the approximately linear dependence of proton release on oxygenation, the original model assumed that the salt bridges were broken when a subunit bound oxygen. The present analysis shows that the salt bridge coupling may be more complicated; i.e., the salt bridge strengths can be a function of both the tertiary structure of a subunit and the quaternary structure of the tetramer.

The success of the generalized model in describing the currently available thermodynamic data with assumptions that are based on structural and spectroscopic results provides a basis for more detailed studies of structure-function relations in hemoglobin. Of particular importance is the evaluation of the structural changes induced by ligation and their energetic correlates. Calculations combined with structural data have indicated how ligation of the deoxy tetramer shifts the tertiary structure in the direction of that found in the transition to the oxy tetramer. This provides a reaction path from the ligated heme group to the surface of the subunit where the quaternary interactions occur. Details of intersubunit coupling and the effect of the quaternary transition on the tertiary structure are needed to complement these results.

Once the static structural changes involved in the cooperative mechanism have been elucidated, it will be possible to consider the dynamics of hemoglobin. During a given transition of an individual protein molecule, the reaction path, per se, is not expected to be followed exactly. As is well known from gas-phase calculations for reactions of simple molecules, the variety of initial conditions for the reactants and the availability of kinetic energy in the reacting systems leads to trajectories for the reaction that involve a range of coordinates for the particles. The trajectories are in the general neighborhood of, but not identical with, the reaction path. In protein molecules, such as the individual subunits and the entire hemoglobin tetramer, it is now known that at room temperature sizeable fluctuations (average root-mean-square displacements of 0.5 to 0.75 Å) in the atomic positions are taking place at all times.[87] Thus, the determination of a reaction path represents only the first step in understanding the detailed dynamics of a reaction. In hemoglobin there are two basic reactions to be considered, the tertiary structural change on ligation or deligation in the absence of a quaternary transition and the quaternary transition itself. For both of these it is likely that fluctuations, biased along the reaction path, are the mechanism by which the structural change occurs. In the quaternary transition, the frictional interactions with the solvent must play a role because of the large, approximately rigid-body type of displacements of the individual subunits. A description involving diffusive motion in the presence of a potential of mean force appears to be most appropriate. A formulation of the motion analogous to that used for the opening and closing of the lysozyme active-site cleft should be applicable.[88]

Although much has been learned about the cooperative mechanism of hemoglobin, there is even more that remains to be done to complete our understanding of this fascinating molecule. With new and improved experimental methods, including site-specific mutagenesis,[89] and the developments in theoretical approaches to biomolecules,[90] significant progress can be expected in the fundamental chemical analysis of cooperativity.

ACKNOWLEDGMENTS

The work reported here was supported in parts by grants from the National Science Foundation, Washington, D.C. and the National Institutes of Health, Bethesda, MD. Much of the review is taken from previously published papers, as indicated by the references, of the authors.

REFERENCES

1. **Antonini, E. and Brunori, M.,** *Hemoglobin and Myoglobin in their Reactions with Ligands,* North-Holland, Amsterdam, 1971.
2. **Fermi, G. and Perutz, M.,** in *Haemoglobin and Myoglobin: Atlas of Molecular Structures in Biology,* Phillips, D. C. and Richards, F. M., Eds., Clarendon, Oxford, 1981.
3. **Dickerson, R. E. and Geis, I.,** *Hemoglobin: Structure, Function, Evolution, and Pathology,* Benjamin/Cummings, Menlo Park, CA, 1983.
4. **Bunn, H. F. and Forget, B. G.,** *Hemoglobin: Molecular, Genetic and Clinical Aspects,* W. B. Saunders, Philadelphia, 1986.
5. **Roughton, F. J. W. and Lyster, R. L.,** Some combinations of the Scholander-Roughton syringe capillary and van Slyke's gasometric techniques, and their use in special haemoglobin problems, *Hvalradets Skr.,* 48, 185, 1965.
6. **Pauling, L.,** The oxygen equilibrium of hemoglobin and its structural interpretation, *Proc. Natl. Acad. Sci. U.S.A.,* 21, 186, 1935.
7. **Monod, J., Wyman, J., and Changeux, J. P.,** On the nature of allosteric transitions: a plausible model, *J. Mol. Biol.,* 12, 88, 1965.
8. **Koshland, D. E., Nemethy, G., and Filmer, D.,** Comparison of experimental binding data and theoretical models in proteins containing subunits, *Biochemistry,* 5, 365, 1966.
9. **Fermi, G., Perutz, M. F., Shaanan, B., and Fourme, R.,** The crystal structure of human deoxyhaemoglobin at 1.74 Å resolution, *J. Mol. Biol.,* 175, 159, 1984.
10. **Shaanan, B.,** Structure of human oxyhaemoglobin at 2.1 Å resolution, *J. Mol. Biol.,* 171, 31, 1983.
11. **Perutz, M. and Ten Eyck, L. F.,** Stereochemistry of cooperative effect in hemoglobin, *Cold Spring Harbor Symp. of Quant. Biol.,* 36, 295, 1972.
12. **Muirhead, H., Cox, J., Mazzarella, L., and Perutz, M.,** Structure and function of haemoglobin. III. A three-dimensional Fourier synthesis of human deoxyhaemoglobin at 5.5 Å resolution, *J. Mol. Biol.,* 28, 117, 1967.
13. **Arnone, A., Rogers, P., Blough, N. V., McGourty, J. L., and Hoffman, B. M.,** X-ray diffraction studies of a partially liganded hemoglobin, [α(FeII-CO)β(MnII)]₂, *J. Mol. Biol.,* 188, 693, 1986.

14. **Gelin, B. R. and Karplus, M.,** Mechanism of tertiary structural change in hemoglobin, *Proc. Natl. Acad. Sci. U.S.A.,* 74, 801, 1977.

15. **Gelin, B. R., Lee, A. W.-M., and Karplus, M.,** Hemoglobin tertiary structural change on ligand binding: its role in the cooperative mechanism, *J. Mol. Biol.,* 171, 489, 1983.

16. **Pauling, L.,** The interpretation of some chemical properties of hemoglobin in terms of its molecular structure, *Stanford Med. Bull.,* 6, 215, 1948.

17. **Pauling, L.,** Nature of the iron-oxygen bond in oxyhaemoglobin, *Nature (London),* 203, 182, 1964.

18. **Goddard, W. A., III and Olafson, B. D.,** Ozone model for bonding of an O_2 to heme in oxyhemoglobin, *Proc. Natl. Acad. Sci. U.S.A.,* 72, 2335, 1975.

19. **Olafson, B. D. and Goddard, W. A., III,** Molecular description of dioxygen bonding in hemoglobin, *Proc. Natl. Acad. Sci. U.S.A.,* 74, 1315, 1977.

20. **Case, D. A., Huynh, B. H., and Karplus, M.,** Binding of oxygen and carbon monoxide to hemoglobin. An analysis of the ground and excited states, *J. Am. Chem. Soc.,* 101, 4433, 1979.

21. **Weiss, J. J.,** Nature of the iron-oxygen bond in oxyhaemoglobin, *Nature (London),* 202, 83, 1964.

22. **Seno, Y., Otsuka, J., Matsuoka, O., and Fuckikami, N.,** Heitler-London calculation on the bonding of oxygen to hemoglobin. III. Singlet states in the case of O-O axis inclined to heme plane, *J. Phys. Soc. Jpn.,* 41, 977, 1976.

23. **Dedieu, A., Rohmer, M.-M., Viellard, H., and Viellard, A.,** Electronic and structural aspects of the dioxygen complexes of metalloporphyrins. An *ab initio* study, *Nouv. J. Chimie,* 3, 653, 1979.

24. **Kirchner, R. F. and Loew, G. H.,** Semiempirical calculations of model oxyheme: variation of calculated electromagnetic properties with electronic configuration and oxygen geometry, *J. Am. Chem. Soc.,* 99, 4639, 1977.

25. **Phillips, S. E. V.,** Structure of oxymyoglobin, *Nature (London),* 273, 247, 1978.

26. **Case, D. A. and Karplus, M.,** Stereochemistry of carbon monoxide binding to myoglobin and hemoglobin, *J. Mol. Biol.,* 123, 697, 1978.

27. **Case, D. A. and Karplus, M.,** Dynamics of ligand binding to heme proteins, *J. Mol. Biol.,* 132, 343, 1979.

28. **Karplus, M. and Porter, R. N.,** *Atoms and Molecules,* W. A. Benjamin, New York, 1970.

29. **Kuriyan, J., Wilz, S., Karplus, M., and Petsko, G. A.,** X-ray structure and refinement of carbonmonoxy (Fe II)-myoglobin at 1.5 Å resolution, *J. Mol. Biol.,* 192, 133, 1986.

29a. **Li, X.-Y. and Spiro, T. G.,** Is bound CO linear or bent in heme proteins? Evidence from resonance Raman and infra-red spectroscopic data, *J. Am. Chem. Soc.,* 110, 6024, 1988.

30. **Perutz, M. F.,** Regulation of oxygen affinity of hemoglobin: influence of structure of the globin on the heme iron, *Annu. Rev. Biochem.,* 48, 327, 1979.

31. **Warshel, A.,** Energy-structure correlation in metalloporphyrins and the control of oxygen binding by hemoglobin, *Proc. Natl. Acad. Sci. U.S.A.,* 74, 1789, 1977.

32. **Scheidt, W. R. and Reed, C. A.,** Spin-state/stereochemical relationships in iron porphyrins: implications for the hemoproteins, *Chem. Rev.,* 81, 543, 1981.

33. **Hopfield, J. J.,** Relation between structure, co-operativity and spectra in a model of hemoglobin action, *J. Mol. Biol.,* 77, 207, 1973.

34. **Gelin, B. R. and Karplus, M.,** Sidechain torsional potentials and motion of amino acids in proteins: bovine pancreatic trypsin inhibitor, *Proc. Natl. Acad. Sci. U.S.A.,* 72, 2002, 1975.

35. **Fermi, G.,** Three-dimensional Fourier synthesis of human deoxyhaemoglobin at 2.5 Å resolution: refinement of the atomic model, *J. Mol. Biol.,* 97, 237, 1975.

36. **Hoard, J. L. and Scheidt, W. R.,** Stereochemical trigger for initiating cooperative interaction of the subunits during the oxygenation of cobaltohemoglobin, *Proc. Natl. Acad. Sci. U.S.A.,* 70, 3919, 1972.

37. **Eisenberger, P., Shulman, R. G., Brown, G. S., and Ogawa, S.,** Structure-function relations in hemoglobin as determined by X-ray absorption spectroscopy, *Proc. Natl. Acad. Sci. U.S.A.,* 73, 491, 1976.

37a. **Fermi, G., Perutz, M. F., and Shulman, R. G.,** Iron distances in hemoglobin: comparison of X-ray crystallographic and extended X-ray absorption fine structure studies, *Proc. Natl. Acad. Sci. U.S.A.,* 84, 6167, 1987.

38. **Spiro, T. G. and Burke, J. M.,** Protein control of porphyrin conformation. Comparison of resonance Raman spectra of heme proteins with mesoporphyrin IX analogous, *J. Am. Chem. Soc.,* 98, 5482, 1976.

39. **Anderson, L.,** Intermediate structure of normal human haemoglobin: methaemoglobin in the deoxy quaternary conformation, *J. Mol. Biol.,* 79, 495, 1973.

40. **Asakura, T. and Sono, M.,** Optical and oxygen binding properties of spirographis, isospirographis, and 2,4-diformyl hemoglobins, *J. Biol. Chem.,* 249, 7087, 1974.

41. **Perutz, M. F., Heidner, E. J., Ladner, J. E., Beetlestone, J. G., Ho, C., and Slade, E. F.,** Influence of globin structure on the state of the heme. III. Changes in heme spectra accompanying allosteric transitions in methemoglobin and their implications for heme-heme interaction, *Biochemistry,* 13, 2187, 1974.

42. **Ogawa, S. and Shulman, R. G.,** High resolution nuclear magnetic resonance spectra of hemoglobin. III. The half-ligated state and allosteric interactions, *J. Mol. Biol.,* 70, 315, 1972.

43. **Perutz, M. F., Kilmartin, J. V., Nagai, K., Szabo, A., and Simon, S. R.,** Influence of globin structures on the state of the heme. Ferrous low spin derivatives, *Biochemistry,* 15, 378, 1976.
44. **Maxwell, J. C. and Caughey, W. S.,** An infrared study of NO bonding to heme B and hemoglobin A. Evidence for inositol hexaphosphate induced cleavage of proximal histidine to iron bonds, *Biochemistry,* 15, 388, 1976.
45. **Perutz, M. R.,** Stereochemistry of cooperative effects in haemoglobin, *Nature,* 228, 726, 1970.
46. **Baldwin, J. and Chothia, C.,** Haemoglobin: the structural changes related to ligand binding and its allosteric mechanism, *J. Mol. Biol.,* 129, 175, 1979.
47. **Lee, A. W.-M. and Karplus, M.,** to be submitted, 1989.
48. **Messana, C., Cerdonio, M., Shenkin, D., Noble, R. W., Fermi, G., Perutz, R. N., and Perutz, M. F.,** Influence of quaternary structure of the globin on thermal spin equilibria in different methamoglobin derivatives, *Biochemistry,* 17, 3652, 1978.
49. **Perutz, M. F., Fermi, G., Luisi, B., Shaanan, B., and Liddington, R. C.,** *Acc. Chem. Res.,* 20, 309, 1987.
50. **Szabo, A. and Karplus, M.,** A mathematical model for structure-function relations in hemoglobin, *J. Mol. Biol.,* 72, 163, 1972.
51. **Szabo, A. and Karplus, M.,** Analysis of cooperativity in hemoglobin. Valency hybrids, oxidations, and methemoglobin replacement reactions, *Biochemistry,* 14, 931, 1975.
52. **Szabo, A. and Karplus, M.,** Analysis of the interaction of organic phosphates with hemoglobin, *Biochemistry,* 15, 2869, 1976.
53. **Lee, A. W.-M. and Karplus, M.,** Structure-specific model of hemoglobin cooperativity, *Proc. Natl. Acad. Sci. U.S.A.,* 80, 7055, 1983.
54. **Kilmartin, J., Breen, J., Roberts, G., and Ho, C.,** Direct measurement of the pK values of an alkaline Bohr group in human hemoglobin, *Proc. Natl. Acad. Sci. U.S.A.,* 70, 1246, 1973.
55. **Brown, F. and Campbell, I.,** Cross correlation of titrating histidines in oxy- and deoxyhaemoglobin: an NMR study, *FEBS Lett.,* 65, 322, 1976.
56. **Ohe, M. and Kajita, A.,** Changes in pK_a values of individual histidine residues of human hemoglobin upon reaction with carbon monoxide, *Biochemistry,* 19, 4443, 1980.
57. **Van Beek, G.,** Oxygen-Linked Binding of Anions to Human Hemoglobin, Dissertation University of Nijmegen, The Netherlands, 1979.
58. **Garner, W. H., Bogardt, R. A., and Gurd, F. R. N.,** Determination of the pK values for the α-amino groups of human hemoglobin, *J. Biol. Chem.,* 250, 4389, 1975.
59. **Perutz, M. R., Pulsinelli, P., Ten Eyck, L., Kilmartin, J. V., Shibata, S., Iuchi, I., Miyaji, T., and Hamilton, H. B.,** Haemoglobin Hiroshima and the mechanism of the alkaline Bohr effect, *Nature (London) New Biol.,* 232, 147, 1971.
60. **Wajcman, H., Kilmartin, J., Najman, A., and Labie, D.,** Hemoglobin Cochin-Port-Royal: consequences of the replacement of the β chain C-terminal by an arginine, *Biochim. Biophys. Acta,* 400, 354, 1975.
61. **Wajcman, H., Aguilar, J., Labie, D., Poyart, C., and Bohn, B.,** Structural and functional studies of hemoglobin Barcelona ($\alpha_2\beta_2$94 Asp (FG$_1$) $\rightarrow$ His). Consequences of altering an important intrachain salt bridge involved in the alkaline Bohr effect, *J. Mol. Biol.,* 156, 185, 1982.
62. **Phillips, S. E. V., Perutz, M., Poyart, C., and Wajcman, H.,** Structure and function of haemoglobin Barcelona Asp FG1(94)β $\rightarrow$ His, *J. Mol. Biol.,* 164, 477, 1983.
63. **Perutz, M. F., Fermi, G., and Shih, T.-B.,** Structure of deoxyhemoglobin Cowtown [His Hc3(146)β $\rightarrow$ Leu]: origin of the alkaline Bohr effect and electrostatic interactions in hemoglobin, *Proc. Natl. Acad. Sci. U.S.A.,* 81, 4781, 1984.
64. **Kilmartin, J., Hewitt, J., and Wootton, J. F.,** Alterations of functional properties associated with the change in quaternary structure in unliganded haemoglobin, *J. Mol. Biol.,* 93, 203, 1975.
65. **Kilmartin, J., Fogg, J., and Perutz, M.,** Role of C-terminal histidine in the alkaline Bohr effect of human hemoglobin, *Biochemistry,* 19, 3189, 1980.
66. **Poyart, C., Bursaux, E., and Bohn, B.,** An estimation of the first binding constant of O_2 to human hemoglobin A, *Eur. J. Biochem.,* 87, 75, 1978.
67. **Rollema, H., deBruin, S. H., Jansen, L. H. M., and Van Os, G. A.,** The effect of potassium chloride on the Bohr effect of human hemoglobin, *J. Biol. Chem.,* 250, 1333, 1975.
68. **Perutz, M. F., Kilmartin, J. V., Nishikura, K., Fogg, J. H., Butler, P. J. G., and Rollema, H. S.,** Identification of residues contributing to the Bohr effect of human haemoglobin, *J. Mol. Biol.,* 138, 649, 1980.
69. **Lee, A. W.-M.,** Theoretical Modelling of Hemoglobin Cooperativity: A Statistical-Thermodynamic and Empirical Energy Function Study, Ph.D. thesis, Harvard University, Cambridge, MA, 1984.
70. **Minton, A. and Imai, K.,** The three-state model: A minimal allosteric description of homotropic and heterotropic effects in the binding of ligands to hemoglobin, *Proc. Natl. Acad. Sci. U.S.A.,* 71, 1418, 1974.

71. **Fung, L. and Ho, C.,** A proton nuclear magnetic resonance study of the quaternary structure of human hemoglobins in water, *Biochemistry*, 14, 2526, 1975.

72. **Viggiano, G. and Ho, C.,** Proton nuclear magnetic resonance investigation of structural changes associated with cooperative oxygenation of human adult hemoglobin, *Proc. Natl. Acad. Sci. U.S.A.*, 76, 3673, 1979.

73. **Asakura, T. and Lau, P.-W.,** Sequence of oxygen binding by hemoglobin, *Proc. Natl. Acad. Sci. U.S.A.*, 75, 5462, 1978.

74. **Russu, I., Ho, N., and Ho, C.,** A proton nuclear magnetic resonance investigation of histidyl residues in human normal adult hemoglobin, *Biochemistry*, 21, 5031, 1982.

75. **Flanagan, M., Ackers, G., Matthew, J., Hanania, G., and Gurd, F.,** Electrostatic contributions to the energetics of dimer-tetramer assembly in human hemoglobin: pH dependence and effect of specifically bound chloride ions, *Biochemistry*, 20, 7439, 1981.

76. **Matthew, J., Hanania, G., and Gurd, F.,** Electrostatic effects in hemoglobin: hydrogen ion equilibria in human deoxy- and oxyhemoglobin A, *Biochemistry*, 18, 1919, 1979.

77. **Matthew, J., Hanania, G., and Gurd, F.,** Electrostatic effects in hemoglobin: Bohr effect and ionic strength dependence of individual groups, *Biochemistry*, 18, 1928, 1979.

78. **Shih, D. T.-B., Perutz, M. F., Gronenborn, A. M., and Clore, G. M.,** Histidine proton resonances of carbonymonoxyhaemoglobins A and cowtown in chloride-free buffer, *J. Mol. Biol.*, 195, 453, 1987.

79. **Matthew, J. M., Gurd, F. R. N., Bertrand Garcia-Mareno, E., Flanagan, M. A., March, K. L., and Shire, S. J.,** pH-Dependent processes in proteins, *CRC Crit. Rev. Biochem.*, 18(2), 91, 1985.

80. **Mills, F., Johnson, M., and Ackers, G.,** Oxygenation-linked subunit interactions in human hemoglobin: experimental studies on the concentration dependence of oxygenation curves, *Biochemistry*, 15, 5350, 1976.

81. **Johnson, J. and Ackers, G.,** Thermodynamic analysis of human hemoglobins in terms of the Perutz mechanism: extensions of the Szabo-Karplus model to include subunit assembly, *Biochemistry*, 21, 201, 1982.

82. **Viggiano, G., Ho, N., and Ho, C.,** Proton nuclear magnetic resonance and biochemical studies of oxygenation of human adult hemoglobin in deuterium oxide, *Biochemistry*, 18, 5238, 1979.

83. **Nasuda-Kouyama, A., Tachibana, H., and Wada, A.,** Preference of oxygenation between α and β subunits of haemoglobin results of multidimensional spectroscopic observation, *J. Mol. Biol.*, 164, 451, 1983.

84. **Antonini, E., Wyman, J., Brunori, M., Fonticelli, C., Bucci, E., and Rossi-Fanelli, A.,** Studies on the relations between molecular and functional properties of hemoglobin. V. The influence of temperature on the Bohr effect in human and in horse hemoglobin, *J. Biol. Chem.*, 240, 1096, 1965.

85. **Lee, A. W.-M., Karplus, M., Poyart, C., and Bursaux, E.,** Analysis of proton release in oxygen binding by hemoglobin: implications for the cooperative mechanism, *Biochemistry*, 27, 1285, 1988.

86. **Antonini, E., Wyman, J., Rossi-Fanelli, A., and Caputo, A.,** Studies on the relations between molecular and functional properties of hemoglobin. III. The influence of salts on the Bohr effect in human hemoglobin, *J. Biol. Chem.*, 237, 2773, 1962.

87. **Karplus, M. and McCammon, J. A.,** Dynamics of proteins: elements and function, *Annu. Rev. Biochem.*, 53, 263, 1983.

88. **McCammon, J. A., Gelin, B. R., Karplus, M., and Wolynes, P. G.,** The hinge-bending mode in lysozyme, *Nature*, 262, 325, 1976.

89. **Ulmer, K.,** Protein engineering, *Science*, 219, 666, 1983.

90. **Brooks, C. B., III, Karplus, M., and Pettitt, B. M.,** Proteins: a theoretical perspective of dynamics, structure and thermodynamics, *Adv. Chem. Phys.*, LXXI, 1988.

91. **Pulsinelli, P. D.,** personal communication.

Chapter 3

ASPARTATE TRANSCARBAMYLASE FROM *ESCHERICHIA COLI*

Guy Hervé

TABLE OF CONTENTS

"Si deux théories expliquent également bien un phénomène il convient d'adopter la plus simple."

Guillaume d'OCCAM (1295-1349)

I. INTRODUCTION

The amount of information which has been obtained on *E. coli* aspartate transcarbamylase (ATCase) since the beginning of its study, 20 years ago, is impressive. However, many aspects of its regulatory properties are still obscure. This enzyme catalyzes the first reaction of the pyrimidine pathway, that is, the carbamylation of the amino group of aspartate by carbamylphosphate (Figure 1). The reaction is feedback inhibited by the end product CTP, and it is stimulated by a purine, ATP. This antagonism is one of the phenomena which, in the cell, will tend to balance the production of purines and pyrimidines. In addition, ATCase shows cooperativity in the utilization of its substrate aspartate. The experimentation on ATCase is facilitated by its stability and by the existence of suitable *E. coli* mutations, which allow the production of large quantities of this enzyme, especially when the ATCase operon is located in plasmids.[1,2] This possibility of producing gram quantities of enzyme extends to the mutant forms which are now currently obtained by *in vitro* mutagenesis.

II. STRUCTURE OF THE UNLIGANDED ENZYME

Some early work showed that ATCase contains two kinds of polypeptide chains which play different roles[3] and can be isolated as *catalytic subunits,* which catalyze the carbamylation of the amino group of aspartate, but do not exhibit regulatory properties, and *regulatory subunits* which do not possess any catalytic activity, but bear the regulatory sites for the effector binding.

The three-dimentional structure of ATCase is now known with a resolution of 2.5 Å thanks to the crystallographic work made in Lipscomb's laboratory.[4,5] This enzyme is made of two catalytic trimers (catalytic subunits) which are in contact and held together by three regulatory dimers (regulatory subunits), so the native form of the enzyme contains six chains of each type. The spatial arrangement of these subunits is schematically represented in Figure 2. From its center, the enzyme extends 50 Å along the molecular three-fold axis and up to 75 Å in the direction of the regulatory chains along the molecular two-fold axis.

The amino acid sequence of the two types of chains is known.[6-8] Their remarkable features are the following: the catalytic chain contains only two tryptophan residues and a single cysteine which is near the catalytic site, but does not participate directly in the catalytic process.[9] The regulatory chain does not contain tryptophan, but it possesses four cysteine residues which are clustered in its C-terminal region and are linked to a zinc atom (one per regulatory chain) which is not involved in the catalytic property, but whose presence is an absolute requirement for the association of the catalytic and regulatory subunits.[10]

The secondary structure of one catalytic chain associated to one regulatory chain (C1-R1 interaction in Figure 2), is schematically represented in Figure 3. As described by Honzatko et al.,[4] the catalytic chain shows an entanglement of α-helices and presents two domains (equatorial and polar). The catalytic site is located in the region called "phosphate crevice". Actually, this site lies at the interface between two neighbor catalytic chains belonging to the same trimer (see Section II.A). The regulatory chain is also made of two domains (Figure 3): the allosteric domain where the regulatory sites are located, which is also the binding area for the formation of the regulatory dimer, the zinc domain which corresponds to the C-terminal region of the regulatory chain where the four cysteine residues which bind the zinc atom are clustered. This sulfhydryl-zinc tetrahedral feature ensures an adequate structuration of this C-terminal region of the regulatory chain for its interaction with the catalytic subunit (C1-R1 interaction). The distance between one regulatory site and the nearest catalytic site is about 60 Å.

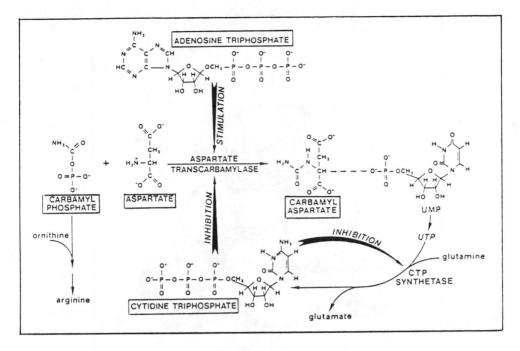

FIGURE 1.   Regulation of ATCase activity in *E. coli* pyrimidine pathway.

The three-dimensional arrangement of the catalytic and regulatory subunits, which is schematically represented in Figure 2, involves several contacts of both ionic and apolar nature.[4,5,11] Each catalytic chain appears to be in contact with three other catalytic chains and two regulatory chains. Each regulatory monomer is in contact with one other regulatory chain and two catalytic chains. In the unliganded enzyme, the two catalytic trimers are also in contact.

- The R1-R6 interface between the allosteric domains within a regulatory dimer involves the C-terminal strands of these regulatory chains, the helix H1′ and the adjacent strand S2′.

- There is a complex interaction between R1, C1, and C4 which involves residues 231 and 239 of the equatorial domain of C1 (between strand 9 and 10), residues 163 to 166 of the polar domain of C4 (between strand S6 and helix H6), and residues 111 to 114 of the zinc domain of the regulatory chain R1 (between strands S6′ and S7′).

- The R1-C1 interface involves an interaction between Arg113 (helix H4), Asn132 (between strand S5 and helix H5), and Ser11 (strand S1) of the polar domain of the catalytic chain with Glu141 (strand S9′) on the regulatory chain. This interface involves also an interaction between Glu117 of the catalytic chain (helix H4) and Lys138, Tyr139 of the regulatory chain (between strands S8′ and S9′).

- Within a catalytic trimer, chains are in contact through interactions between their polar and equatorial domains. These interactions involve especially strand S3 and helix H3 of a given polar domain with helix H9 and the loop connecting strand S10 and helix H8 of the equatorial domain belonging to the neighbor chain.

In addition to this, there are important interactions between the covalently linked domains within the two types of subunits. The interactions between residues across the allosteric and zinc domains of a regulatory chain are purely hydrophobic and involve helices H1′ and H2′ on one side and strand S6′ and helix H3′ on the other side. In the catalytic chain, the

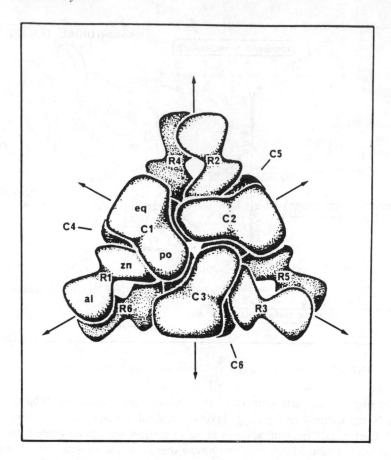

FIGURE 2.   Quaternary structure of ATCase as viewed down the threefold axis.
C = catalytic chain; R = regulatory chain; eq: equatorial domain of C; po: polar
domain of C; al: allosteric domain of R; Zn: zinc domain of R. (From Krause,
K. L., Volz, K. W., and Lipscomb, W. N., *Proc. Natl. Acad. Sci. U.S.A.*, 82,
1643, 1985. With permission.)

association of the polar and equatorial domains involves a mixture of polar and nonpolar
interactions.

Currently performed experiments are aimed at detailing the implication of these different
interactions to the various regulatory mechanisms exhibited by the enzyme.

A. The Catalytic Site

The catalytic site is located at the interface between two neighboring catalytic chains in
the same trimer and involves amino acid side chains belonging to these two partners. This
catalytic site could be delineated through the crystallographic study at 2.5 Å resolution of
ATCase fully liganded with a bisubstrate analog.[12-14] The analog used in this study, like in
many other experimental approaches concerning ATCase, was *N*-(phosphonacetyl)-*L*-as-
partate (PALA), a potent inhibitor of ATCase which was devised by Collins and Stark.[15]
The structure of this bisubstrate analog is probably very close to that of the transition state
of the substrates.[16] Figure 4 shows the amino acid side chains that appear to be in contact
with this inhibitor in the catalytic site of the native enzyme.[14] The involvement of two
neighboring subunits for the constitution of the catalytic site is also supported by the results
of complementation experiments.[17,18] Although the results of some biochemical experiments
were already pointing to the implication of His134[16,19-22] and Lys84,[19,23,24] current investigations using site-directed mutagenesis and pH-dependence studies are aimed at elucidating

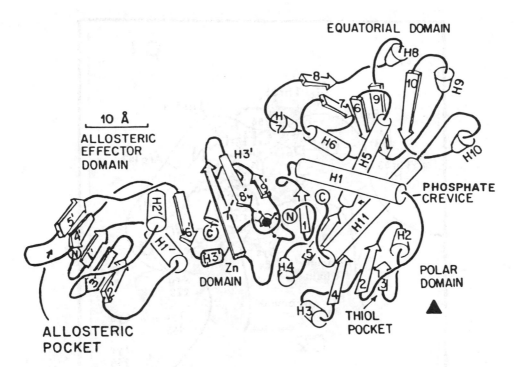

FIGURE 3. Schematic trace of one catalytic-regulatory unit of ATCase. The view is down the threefold axis. The black dot represents the zinc atom. (From Krause, K. L., Volz, K. W., and Lipscomb, W. N., *J. Mol. Biol.*, 193, 527, 1987. With permission.)

the exact role of these amino acid side chains. Such specific replacements have already demonstrated the involvement of these two residues.[25] This conclusion also results from the analysis of the pH-dependence of the isolated catalytic subunits.[26]

B. The Regulatory Site

The regulatory site is located close to the area of association of the two regulatory chains constituting the dimeric regulatory subunit. ATP and CTP bind competitively to this site in a way which involves the N-terminal strand of the regulatory chain. Information obtained by enzyme kinetics,[27] X-ray crystallography,[22] and NMR (nuclear magnetic resonance) studies[28] indicate that both nucleotides bind the regulatory sites in the anticonformation.

The binding of ATP and CTP to the regulatory sites was shown to be anticooperative.[29-34] This phenomenon might be related to the recently discovered synergistic inhibition of ATCase activity by CTP and UTP.[100]

III. PROPERTIES OF THE ISOLATED CATALYTIC AND REGULATORY SUBUNITS

Several methods based on the reaction of mercurial compounds with the cysteine residues of the regulatory chains allow the reversible dissociation of the native enzyme into isolated catalytic and regulatory subunits.[35,36]

A. The Catalytic Subunit

The isolated trimer of catalytic chains (catalytic subunit) is stable and catalytically active, but does not show any regulatory property. Its saturation curve for aspartate is hyperbolic, and its activity is insensitive to the effectors ATP and CTP.[3] It must be noted, however,

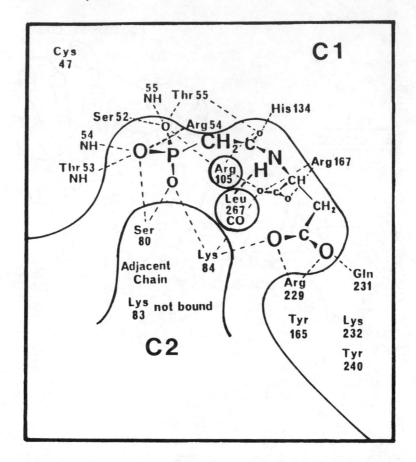

FIGURE 4.   The catalytic site of ATCase. This figure represents the positioning of the transition state analog PALA in the catalytic site of ATCase. C1 and C2 are two neighboring catalytic chains within a catalytic trimer. Arg105 and Leu267 belong to C1. (From Volz, K. W. and Lipscomb, W. N., *Biochem. Biophys. Res. Commun.*, 136, 822, 1986. With permission.)

that very high concentrations of these nucleotides provoke a nonspecific inhibition through binding at the catalytic site in competition with carbamylphosphate.[37,38] In fact, many small molecules which contain a phosphate group behave in the same way.[39]

Although there has been some controversy about the catalytic mechanism of ATCase studied in the isolated catalytic subunits,[37,40,41] there is a large amount of evidence that, under the conditions generally used, these catalytic subunits act through an ordered Bi-Bi mechanism in which carbamylphosphate binds first, followed by aspartate, with release of the products in the order of carbamylaspartate then phosphate.[37,38,42-44] A direct demonstration of this mechanism was recently provided through equilibrium isotope exchange kinetic experiments.[45] Early experiments dealing with enzyme kinetics[37] and NMR studies[46] brought Jacobson and Stark[47] to predict that the reaction involves the interaction of a group belonging to the enzyme with the carbonyl group of carbamylphosphate, enhancing the polarization of the latter, thus favoring its nucleophilic attack by the amino group of aspartate (Figure 5). Such a mechanism, which was further supported by ^{13}C NMR studies,[20] is consistent with the present knowledge of the catalytic site,[14] and there is an increasing number of indications that this group is probably His134 (Figure 4). In addition to the fact that this residue has the right location to play such a role, this conclusion is supported by the properties of a series of substrate analogs,[39] by the behavior of modified enzymes in which this His134

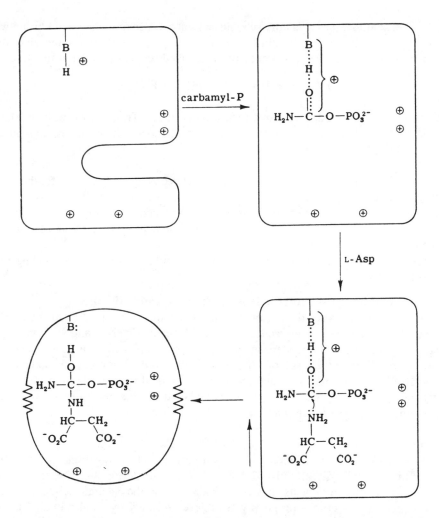

FIGURE 5. Proposed mechanism for ATCase reaction. Carbamylphosphate binds and interacts with a BH⁺ group of the protein through its carbonyl group, an interaction which results in an increased polarization of this group. Aspartate binds and its α-amino group is in a position to react with the bound and activated carbamylphosphate. (From Jacobson, G. R. and Stark, G. R., *The Enzymes*, Vol. 9, 3rd ed., Boyer, P. D., Ed., Academic Press, New York, 1973, 255. With permission.)

residue was replaced by site-directed mutagenesis,[18,25,100] and by the analysis of the pH-dependence of the catalytic subunits.[26] Arg105 and Thr55 are also probably involved in this polarization process.[16]

Through molecular modeling of the tetrahedral intermediate of the reaction, it has been recently proposed that its breakdown is facilitated by a proton transfer between the amino group of aspartate and a terminal oxygen of carbamylphosphate.[16]

The conformational change which is promoted by the binding of the substrates extends to the vicinity of the two tryptophan residues which are located in the equatorial domain, providing an experimentally useful signal to follow the occupancy of the catalytic sites by substrates or analogs.[42,48,49]

B. The Regulatory Subunit

The isolated regulatory subunits do not exhibit any catalytic properties, but they still bear the regulatory sites for the competitive binding of the effectors ATP and CTP or their analogs.[38,50] In solution, the regulatory subunits present a dimer-monomer equilibrium which is considerably shifted toward the dimeric form in the presence of saturating zinc ions.[51] This pattern is complicated by a tendency of the isolated regulatory subunit to aggregate.

IV. REGULATORY PROPERTIES

It is essentially as a model of allosteric regulation that ATCase is extensively studied, since this enzyme shows the different types of interactions which can be exhibited by regulatory enzymes.

- Positive homotropic cooperative interactions between the catalytic sites for the binding of the substrate aspartate.
- Negative and positive heterotropic interactions between regulatory and catalytic sites, allowing for the feedback inhibition of the enzyme activity by CTP and for the stimulation of this activity by ATP
- Negative homotropic interactions between the regulatory sites in the binding of the two effectors

A. Homotropic Cooperative Interactions between the Catalytic Sites

Most of the experimental work done on ATCase is aimed at understanding the mechanisms of the homotropic cooperative interactions between the catalytic sites, and substantial progress was made recently in the understanding of this process. There is a general consensus that these interactions are the result of a transition from a conformation of the enzyme which has a low affinity for aspartate (T-state) to a conformation which has a high affinity for this substrate (R-state), and a great deal of information could be obtained on these two extreme forms, both in terms of structure and behavior.

1. Structural Transition

On the basis of accurate sedimentation experiments[52] and of low-angle X-ray solution scattering,[53] it was primarily established that ATCase swells during the allosteric transition and the sedimentation experiments showed, in addition, that the catalytic subunits condense during this process.[52] A crucial progress was achieved recently by Krause et al.[13] in determining the three-dimensional structure of the R form of the enzyme saturated with PALA, with a resolution of 2.5 Å, enabling its comparison with the structure of the T form established a few years before in the same laboratory with a similar accuracy.[4] This comparison shows that during the allosteric transition of ATCase, the two catalytic trimers move apart along the threefold axis by 12 Å, that is, about $1/_{10}$ of the enzyme diameter. At the same time, they rotate around this axis by 5° each. As a consequence, the regulatory subunits reorientate around their twofold axis by 15°. Plate 1* shows the comparison of these two extreme structures of the enzyme. This quaternary structure change is associated with alterations of the inter-subunits and interdomain interactions. Interestingly, there is no significant change in the R1-R6 and C1-R1 interfaces. It is the complex interaction between C1-C4 and R1 which is the most extensively modified during this transition. This is mainly associated with an important movement of the Tyr240 loop (between strands S9 and S10). As a consequence of this, the loops belonging to the C1 and C4 superposed catalytic chains now stack on top of each other instead of side by side, preventing the reversion to the T-state by a simple collapse along the threefold axis.[13] The overall consequence of these structural changes is that the two domains of each catalytic chain come closer together, ensuring a better contact with the substrates or their analogs. The change in the backbone of one catalytic chain is shown in Plate 2*, according to Krause et al.[13]

---

* Plates 1 and 2 appear following page 120.

During this structural transition some side chain-side chain interactions are lost and some others are established. In this regard, signficant results were obtained by site-directed mutagenesis affecting the Tyr240 loop,[54] suggesting that interactions between Asp271 (N-end of helix H8) and Tyr240 on one hand, and between Glu239 of C1 and Tyr165 of C4 (between strand S6 and helix H6) on the other hand would stabilize the T-state, while interaction between Glu239 of C1 (belonging to a catalytic subunit) and both Lys164 and Tyr165 of C4 (belonging to the other catalytic subunit) would stabilize the R-state.[54]

More recently, a very elegant investigation involving site-directed mutagenesis[55,56] allowed the demonstration that upon substrate binding, the stabilization of the R structure involves the establishment of ionic bonding between the equatorial and polar domains of the catalytic chain, namely between Glu50 and Arg167, Arg234, thus tightening together the carbamylphosphate and aspartate binding domains. This tertiary structure change would induce the quaternary structure change through disruption of intercatalytic chain interaction (C1-C4), involving Glu239 on one side, Lys164 and Tyr165 on the other side, and conversely (Figure 6). However, the very low activity of these mutants indicates that additional interactions and structural features are involved in the allosteric transition.

2. Dynamics of the Allosteric Transition

Thus a considerable knowledge has been attained concerning the structure and the biochemical properties of the two extreme forms of the enzyme having low and high affinity for aspartate. Similarly to what was achieved in the case of hemoglobin[57] and glycogen phosphorylase,[58] these two extreme forms could be stabilized through alteration of the protein-solvent interactions.[59] The generally accepted interpretation of these experiments is that the stabilization of the T-state results from the strengthening of the intersubunit ionic bonds upon decrease of the dielectric constant of the solvent, and that further decrease of the solvent polarity favors the R-state through allowing the exposure of a larger surface of the protein to the solvent. This last interpretation is in accordance with the fact that the transition toward the R-state increases by 300 square Å the surface of ATCase which is exposed to the solvent.[101] It has been reported above that substrate binding alters the environment of the tryptophan residues present in the catalytic chains.[42,48,49] No additional change of this environment is provoked by the quaternary structure change.[48]

Through chemical quench studies, Kihara et al.[60] have estimated that the allosteric transition in ATCase takes 10 ms at 4°C. This transition is in some way concerted since the maximal effect is obtained before full saturation of the catalytic sites. Early reports indicated that the binding of four to five molecules of PALA or carbamylphosphate and succinate was necessary to complete the conformational change.[52,61-63] However, it must be kept in mind that in a population of enzyme molecules, succinate or PALA will tend to bind to those molecules which are already converted to R by previous binding, and even if the binding of a single molecule of substrate or substrate analog would be sufficient to entirely convert an enzyme molecule into the R-state, the values reported above would correspond to the number of substrate analogs molecules which must be bound on the average in order to ensure that each enzyme molecule has bound one molecule of ligand. This possibility is suggested by more recent experiments using small concentrations of PALA under conditions where this compound is entirely bound.[64]

Regardless of the fact that ATCase is made of two catalytic trimers, the six catalytic sites seem to be all equivalent in the phenomenon of cooperativity. This is suggested by an elegant experiment using hybrid enzymes containing catalytic chains which were inactivated by pyridoxylation of the catalytic sites.[65] Two hybrids, one containing two active sites in one catalytic subunit and none in the other, and the second containing one active site in each catalytic subunit showed the same degree of cooperativity.

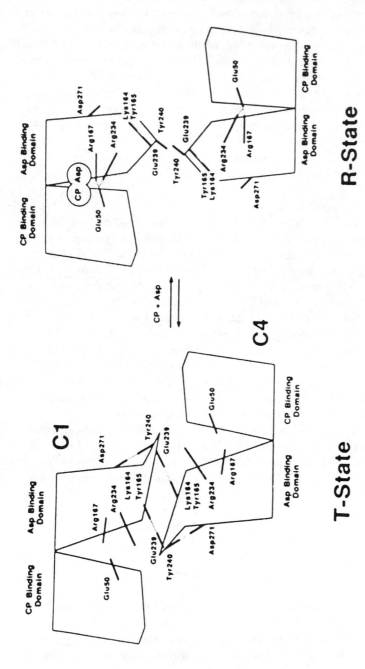

FIGURE 6. Changes in ionic bonding associated to the allosteric transition in ATCase. This identification of the ionic bonds which are important in the stabilization of the T and R forms of the enzyme results from site-directed mutagenesis experiments. (From Ladjimi, M. M. and Kantrowitz, E. R., *Biochemistry*, 27, 276, 1988. With permission.)

It has been known for a long time that the pH-dependence of the reaction catalyzed by ATCase changes when the concentration of the substrate aspartate is raised. At low aspartate concentration, the rate of reaction is maximal at pH = 6.8, whereas a maximum activity is observed at pH = 8.2 in the presence of high concentrations of this substrate.[66,67] Through the use of a series of aspartate analogs able or unable to promote the allosteric transition, it was concluded that this phenomenon is linked to the homotropic cooperative interactions between the catalytic sites,[68] suggesting that the two extreme T- and R-states of the enzyme differ by the pK(s) of one or several groups involved in aspartate binding and/or catalysis. However, regarding this phenomenon, it is interesting to note that a mutant was recently obtained which exhibits uncoupling between pH-dependence subunit interactions and cooperativity.[69] The rate of reaction catalyzed by the noncooperative isolated catalytic subunits is maximal at pH = 8.2 regardless of the aspartate concentration.

A detailed study of the pH-dependence for activity of the isolated catalytic subunits allowed the characterization of three pKs of groups involved in substrates binding and catalysis, pointing to the implication of His134.[26]

Although the allosteric transition in ATCase appears to be in someway concerted, it cannot be explained on the basis of a simple two-state equilibrium which would be shifted through exclusive binding of aspartate to the R-state. The change in pH-dependence for activity which is described above already suggested that at low aspartate concentration not only is the T-state predominant, but it participates predominantly in the reaction. More direct evidence is provided by the fact that two pseudosubstrates of ATCase, cysteine sulfinate[70] and L-alanosine[71] whose carbamylation is catalyzed by ATCase, are unable to promote the allosteric transition, despite the fact that these compounds have a higher affinity for the R-state. A similar situation is found when studying the utilization of carbamylaspartate in the reverse reaction.[72]

If the K_m of ATCase for aspartate is a good estimation of the dissociation constant of this substrate,[37] the two extreme states of ATCase would differ by a factor of 20 in their affinity for aspartate,[52,73] a value which corresponds to an apparent intersubunit interaction energy of 1.8 kcal/mol.[74]

It should be noted that it is still possible that the low affinity form of ATCase, whose properties are discussed above, might not exactly correspond to the T-state which is prevailing in the total absence of ligand. Thus, the question of the true activity of this T-state is not entirely answered yet. In particular, the respective contributions of protein "breathing" and allosteric transition to the properties of the catalytic site is one of the important aspects which is presently under investigation.

B. Heterotropic Interactions between Catalytic and Regulatory Sites

Through their binding to the regulatory sites the effectors modulate the activity of ATCase. It has been considered for many years that ATP and CTP were the only efficient effectors of the enzyme. However, early studies showed that deoxyATP is as good an activator as ATP.[75] Amazingly, it is only recently that it was discovered that although UTP does not have a significant influence on the ATCase activity by itself, it acts strongly in synergy with CTP.[100] Early information suggested that ATP and CTP bind competitively to the same regulatory sites,[50,76] an interpretation which was refined by crystallographic studies which show that ATP and CTP reside on overlapping sites.[4] Both CTP and ATP bind in the *anti*conformation.[4,27,28] Both nucleotides bind to the regulatory subunits with two binding constants,[29-32,34,50] which differ by about one order of magnitude, a phenomenon which is interpreted as negative cooperativity between the two regulatory sites of the regulatory dimer.[38,76-78] These negative interactions involve a local conformational change independent of the allosteric transition, since it operates in the isolated regulatory dimers[38,50] and in modified forms of ATCase which are "frozen" in R-state and do not exhibit the allosteric transition.[34]

Like many small molecules bearing a phosphate group, at high concentrations the nucleotides can also bind nonspecifically to the catalytic sites in competition with carbamylphosphate.[22,33,37-38,79]

Interestingly, it is the structure of the regulatory subunits which imposes the polarity of the effector influence. This property was demonstrated by elegant experiments using hybrids of catalytic and regulatory subunits coming from *Escherichia coli* and *Serratia marcensens*.[80] Curiously, in the latter organism, CTP is an activator. It could be shown that the regulatory subunits of *S. marcensens* impose a positive influence by CTP when associated to the catalytic subunits of both organisms, while the *E. coli* regulatory subunits impose a negative influence of CTP when associated to both types of catalytic subunits.

The first models which were proposed to account for the regulatory properties of ATCase postulated that the nucleotides were acting on the same transition as the one involved in the homotropic cooperative interactions between the catalytic sites for aspartate binding.[81-82] However, over the years, a lot of evidence has accumulated, indicating that such is not the case. This evidence includes

1. A series of modified forms of ATCase were obtained in which the homotropic cooperative interactions are specifically abolished. These "frozen" R forms are still sensitive to the effectors ATP and CTP.[48,67,83-87]

2. A similar result could be obtained by heat treatment.[88]

3. Although L-alanosine and L-cysteine sulfinate are used as pseudosubstrates by ATCase without any detectable cooperativity, these reactions are both sensitive to the two effectors.[70-71]

4. The disconnection of homotropic and heterotropic interactions was also obtained in a series of elegant experiments in which the T- and R-states of ATCase were "frozen" through the use of chemical cross-linking reagents.[73,89-91] The two extreme forms obtained are Michaelian, show the expected low and high affinities for aspartate, but are still sensitive to the effectors ATP and CTP to the same extent.

5. The use of a covalently bound spin label allows the detection of only one type of regulatory interaction. The detection of the other type of interaction requires the use of a different probe.[29]

6. When both bromo-CTP and succinate are added to ATCase in the presence of saturating carbamylphosphate, two relaxation processes are observed which are associated with two distinct conformational transitions of the enzyme induced by bromo-CTP and succinate, respectively.[92] Such could not be the case if these two ligands acted by shifting the same conformational equilibrium in opposite directions. In such a case, only the quickest transition would be observed. A similar results was obtained when studying the binding of 6-mercapto-9-β-D-ribofuranosylpurine 5′-triphosphate and succinate.[93]

7. Upon binding of carbamylphosphate and succinate, a spectral perturbation of the zinc ion binding site is observed by circular dichroism spectroscopy. No perturbation of this site is observed upon CTP binding.[61]

8. Saturating amounts of ATP do not provoke the shift of the optimum pH of activity that accompanies the homotropic cooperative interactions between the catalytic sites.[68]

9. The two extreme states of ATCase cannot be distinguished on the basis of their affinity for ATP, indicating that ATP cannot shift an equilibrium between these two states.[30,34]

10. In addition, it was shown that the percentage effects of ATP and CTP vary linearly with the number of regulatory sites occupied by these nucleotides.[31,78]

1. The "Primary-Secondary Effects" Model

On the basis of these observations, the action of the effectors ATP and CTP (heterotropic interactions) has been interpreted in terms of a "primary-secondary effects" model in which

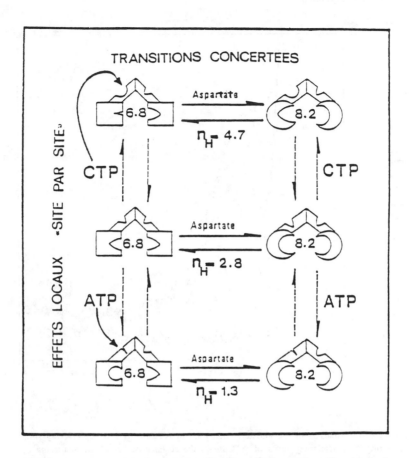

FIGURE 7. The "primary-secondary effects" model for ATCase. The enzyme is schematically represented by two catalytic chains linked together through a regulatory dimer. Squares and circles correspond respectively to the T- and R-states of the catalytic chains which are involved in the homotropic cooperative interactions between the catalytic sites. Arrows emanating from ATP and CTP indicate the binding of these nucleotides to the regulatory sites; n_H refers to the experimentally determined Hill coefficients; pH 6.8 and 8.2 are the experimentally determined values of the optimum pH of the T- and R-states. Horizontal arrows: concerted T $\rightleftharpoons$ R transition associated to the homotropic cooperative interactions between the catalytic sites for aspartate binding. Vertical arrows: "site by site" local conformational changes involved in the heterotropic interactions for feedback inhibition by CTP and activation by ATP.[34,68] (From Tauc, P. et al., *J. Mol. Biol.*, 155, p. 155, 1982. With permission.)

the nucleotides act by altering the affinity of the catalytic sites for aspartate through local conformational changes. The model also describes the coupling between the homotropic and the heterotropic interactions.[34,68] It also accounts for the fact that ATP entirely reverses the effect of CTP and conversely,[68] a phenomenon which is also consistent with the competitive binding of these two nucleotides to the regulatory sites. This model is presented in Figure 7. When ATP binds to a regulatory site, it promotes a local conformational change that increases the affinity for aspartate of the catalytic site that is under its influence. This is the "primary effect". In the presence of a given concentration of aspartate, this effect will increase the level of saturation of the catalytic sites by this substrate and, consequently, will shift the T$\rightleftharpoons$R equilibrium toward the R form. This is the "secondary effect". The feedback inhibitor CTP acts in a similar way. When it binds to a regulatory site, it promotes a local conformational change that decreases the affinity for aspartate of the catalytic site that is under its influence (primary effect). In the presence of a given concentration of aspartate,

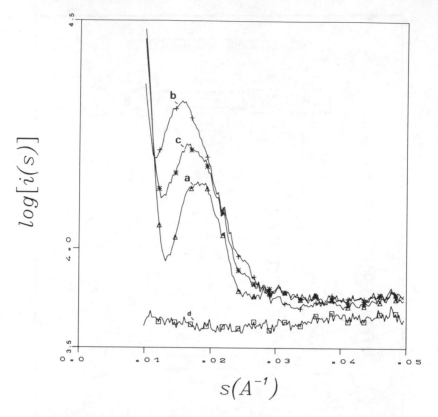

FIGURE 8. Analysis of the allosteric transition in ATCase by low-angle X-ray solution scattering.[94] a: Scattering curve of ATCase in the absence of ligand. b: Scattering curve obtained in the presence of saturating transition state analog PALA. c: midtransition obtained in the presence of nonsaturating amounts of either PALA (tightly bound) or carbamylphosphate and succinate (weakly bound). These conditions are used to show that ATP has no direct influence on the T ⇌ R equilibrium, and CTP has only a small effect.[94] d: solvent. (From Hervé, G. et al., *J. Mol. Biol.*, 185, 189, 1985. With permission.)

this effect will decrease the degree of saturation of the catalytic sites by aspartate and, consequently, will shift the T ⇌ R equilibrium toward the T form (secondary effect). Several parameters of the model were already quantitatively estimated.

While this model accounts satisfactorily for the effect of ATP, the possibility of a small additional direct influence of CTP on the T ⇌ R equilibrium could not be excluded and was actually observed in a further investigation.

2. X-Ray Solution Scattering Experiments

Since the quaternary structure change associated to the allosteric transition in ATCase is important, it can be monitored by X-ray solution scattering experiments.[53] This method allowed the direct verification of the primary-secondary effects model for the heterotropic interactions. Subsaturating concentrations of substrate analogs were used in order to bring the T/R ratio to 1 (Figure 8). When this situation was reached under the influence of entirely bound PALA, as predicted by the model, ATP did not have any detectable direct influence on this ratio, and CTP had only a very small effect. On the contrary, an important effect of the two nucleotides was observed when the T ⇌ R equilibrium was adjusted by the addition of succinate at equilibrium in the presence of saturating carbamylphosphate.[94] Recent equilibrium isotope exchange kinetics experiments lead Hsuanyu and Wedler to confirm these conclusions, and to show that the effectors alter the rate constant for aspartate binding.[95]

3. The "Primary Effects" of CTP and ATP

The "primary effect" of the two effectors is defined as a local conformational change since each binding concerns only $1/6$ of the entire enzyme molecule. However, it must be kept in mind that the effect of binding these nucleotides to one regulatory site is propagated at60 Å away, since this is the distance to the nearest catalytic site. This conformational change appears to be very discrete, since it cannot be detected in the crystals analyzed with a resolution of 2.5 Å.[4] It is likely that these primary effects involve the R1-C1 interface. Regarding this phenomenon, it is interesting to remark that this interface is not affected during the allosteric transition,[13] an observation which is consistent with the conclusion that the T- and R-states are equally sensitive to the primary effects of ATP and CTP.[34,68]

In accordance with the primary-secondary effects model, there are numerous indications that ATP and CTP do not act in a reverse way on the same transition and that their primary effects correspond to different mechanisms:

1. ATP appears to act by increasing the rate of aspartate association and the maximum velocity, whereas CTP seems to alter the reaction enthalpy.[44]
2. ATP and CTP alter in different ways the fluorescence and energy transfer for both intrinsic and extrinsic fluorescent probes.[49,96]
3. The affinity of the enzyme for CTP, but not its affinity for ATP, is altered upon PALA binding.[30,34]
4. ATP and CTP influence differently some phenylalanine residues.[97]
5. A hybrid molecule made up from partially proteolyzed catalytic subunits and normal regulatory subunits is still activated by ATP, but insensitive to CTP.[98]
6. Upon treatment by tartryl diazide in the absence of ligands, ATCase retains some sensitivity toward CTP while showing little or no activation by ATP.[91]
7. Calorimetric determinations of the thermodynamic parameters associated to the binding of the effectors showed major differences in the interaction of ATP and CTP with ionizable groups in the protein.[32,79]
8. The mutant pAR5-ATCase in which amino acid replacements are localized in the C-terminal region of the regulatory chain[99] is normally sensitive to the activator ATP, but is insensitive to the feedback inhibitor CTP, in spite of the fact that this last nucleotide binds to the regulatory sites with an unchanged affinity.[87] Since the modification is localized close to the C1/R1 interaction domain the behavior of this mutant suggests that the primary effects of these two effectors involve, at least partially, different features of this interface.

V. CONCLUSION

Thus, ATCase appears to manifest a greater variety of conformational changes than at first supposed. Some of these conformational changes involve the entire molecule, while some of them can simultaneously alter different parts of the enzyme in different ways. Among the local conformational changes, those involved in the primary effects of the nucleotides are especially intriguing and might involve different subunit-subunit interactions. The present combination of methods going from physics (crystallography, X-ray solution scattering,...) to genetics (site-directed mutagenesis,...) should provide an interesting view of this complexity. Some aspects of ATCase behavior are strikingly similar to reported properties of other allosteric enzymes. These similarities are discussed in Chapter 12.

REFERENCES

1. **Gerhart, J. C. and Holoubek, H.,** The purification of aspartate transcarbamylase of *Escherichia coli* and separation of its protein subunits, *J. Biol. Chem.,* 242, 2886, 1967.
2. **Nowlan, S. F. and Kantrowitz, E. R.,** Superproduction and rapid purification of *Escherichia coli* aspartate transcarbamylase and its catalytic subunit under extreme derepression of the pyrimidine pathway, *J. Biol. Chem.,* 260, 14712, 1985.
3. **Gerhart, J. C. and Schachman, H. K.,** Distinct subunits for the regulation and catalytic activity of aspartate transcarbamylase, *Biochemistry,* 4, 1054, 1965.
4. **Honzatko, R. B., Crawford, J. L., Monaco, H. L., Ladner, J. E., Edwards, B. F. P., Evans, D. R., Warren, S. G., Wiley, D. C., Ladner, R. C., and Lipscomb, W. N.,** Crystal and molecular structures of native and CTP-liganded aspartate carbamoyl transferase from *Escherichia coli, J. Mol. Biol.,* 160, 219, 1982.
5. **Ke, H. M., Honzatko, R. B., and Lipscomb, W. N.,** Structure of unligated aspartate carbamoyltransferase of *Escherichia coli* at 2.6 Å resolution, *Proc. Natl. Acad. Sci. U.S.A.,* 81, 4037, 1984.
6. **Weber, K.,** New structural model of *E. coli* aspartate transcarbamylase and the amino-acid sequence of the regulatory polypeptide chain, *Nature (London),* 218, 1116, 1968.
7. **Hoover, T. A., Roof, W. D., Foltermann, K. F., O'Donovan, G. A., Bencini, D. A., and Wild, J. R.,** Nucleotide sequence of the structural gene (PyrB) that encodes the catalytic polypeptide of aspartate transcarbamylase of *Escherichia coli, Proc. Natl. Acad. Sci. U.S.A.,* 80, 2462, 1983.
8. **Konigsberg, W. H. and Henderson, L.,** Amino acid sequence of the catalytic subunit of aspartate transcarbamylase from *Escherichi coli, Proc. Natl. Acad. Sci. U.S.A.,* 80, 2467, 1983.
9. **Vanaman, T. C. and Stark, G. R.,** A study of the sulfhydrile groups of the catalytic subunit of *Escherichia coli* aspartate transcarbamylase. The use of enzyme-5-thio-2-nitrobenzoate mixed disulfides as intermediates in modifying enzyme sulfhydryl groups, *J. Biol. Chem.,* 245, 3565, 1970.
10. **Nelbach, M. E., Pigiet, V. P., Gerhart, J. C., and Schachman, H. L.,** A role for zinc in the quaternary structure of aspartate transcarbamylase from *Escherichia coli, Biochemistry,* 11, 315, 1972.
11. **McCarthy, M. P. and Allewell, N. M.,** Thermodynamics of assembly of *Escherichia coli* aspartate transcarbamylase, *Proc. Natl. Acad. Sci. U.S.A.,* 80, 6824, 1983.
12. **Krause, K. L., Volz, K. W., and Lipscomb, W. N.,** Structure at 2.9 Å resolution of aspartate carbamoyltransferase complexed with the bisubstrate analogue N-(phosphonacetyl)-L-aspartate, *Proc. Natl. Acad. Sci. U.S.A.,* 82, 1643, 1985.
13. **Krause, K. L., Volz, K. W., and Lipscomb, W. N.,** The 2.5 Å structure of aspartate carbamoyltransferase complexed with the bisubstrate analogue N-(phosphonacetyl)-L-aspartate, *J. Mol. Biol.,* 193, 527, 1987.
14. **Volz, K. W., Krause, K. L., and Lipscomb, W. N.,** The binding of N-(phosphonacetyl)-L-aspartate to aspartate transcarbamoyltransferase of *Escherichia coli, Biochem. Biophys. Res. Commun.,* 136, 822, 1986.
15. **Collins, K. D. and Stark, G. R.,** Aspartate transcarbamylase. Interaction with the transition state analogue N-(phosphonacetyl)-L-aspartate, *J. Biol. Chem.,* 246, 6599, 1971.
16. **Gouaux, J. E., Krause, K. L., and Lipscomb, W. N.,** The catalytic mechanism of *Escherichia coli* aspartate carbamoyltransferase: a molecular modelling study, *Biochem. Biophys. Res. Commun.,* 142, 893, 1987.
17. **Robey, E. A. and Schachman, H. K.,** Regeneration of active enzyme by formation of hybrids from inactive derivatives: implications for active sites shared between polypeptide chains of aspartate transcarbamylase, *Proc. Natl. Acad. Sci. U.S.A.,* 82, 361, 1985.
18. **Wente, S. R. and Schachman, H. K.,** Shared active sites in oligomeric enzymes: model studies with defective mutants of aspartate transcarbamylase produced by site-directed mutagenesis, *Proc. Natl. Acad. Sci. U.S.A.,* 84, 31, 1987.
19. **Greenwell, P., Jewett, S. L., and Stark, G. R.,** Aspartate transcarbamylase from *Escherichia coli.* The use of pyridoxal 5'-phosphate as a probe in the active site, *J. Biol. Chem.,* 248, 5994, 1973.
20. **Roberts, M. F., Opella, S. J., Schaffer, M. H., Philipps, H. M., and Stark, G. R.,** Evidence from ^{13}C-NMR for protonation of carbamyl-P and N-(phosphonacetyl)-L-aspartate in the active site of aspartate transcarbamylase, *J. Biol. Chem.,* 251, 5976, 1976.
21. **Gueguen, P., Padron, M., Perbal, B., and Hervé, G.,** Incorporation of amino acid analogs during the biosynthesis of *Escherichia coli* aspartate transcarbamylase, *Biochim. Biophys. Acta,* 615, 59, 1980.
22. **Honzatko, R. B. and Lipscomb, W. N.,** Interactions of phosphate ligands with *Escherichia coli* aspartate carbamoyltransferase in the crystalline state, *J. Mol. Biol.,* 160, 265, 1982.
23. **Lauritzen, A. M. and Lipscomb, W. N.,** Modification of three active site lysine residues in the catalytic subunit of aspartate transcarbamylase by D- and L-bromosuccinate, *J. Biol. Chem.,* 257, 1312, 1982.
24. **McMurray, C. H., Evans, D. R., and Sykes, B. D.,** A nuclear magnetic resonnance study of the binding of substrate analogues to a modified aspartate transcarbamylase, *Biochem. Biophys. Res. Commun.,* 48, 572, 1972.

25. **Robey, E. A., Wente, S. R., Markby, D. W., Flint, A., Yang, Y. R., and Schachman, H. K.,** Effect of amino acid substitutions on the catalytic and regulatory properties of aspartate transcarbamylase, *Proc. Natl. Acad. Sci. U.S.A.,* 83, 5934, 1986.

26. **Léger, D. and Hervé, G.,** Allostery and pKa changes in aspartate transcarbamylase from *Escherichia coli,* I. Analysis of the pH dependence in the isolated catalytic subunits, *Biochemistry,* 27, 4293, 1988.

27. **London, R. and Schmidt, P. G.,** A model for nucleotide regulation of aspartate transcarbamylase, *Biochemistry,* 11, 3136, 1972.

28. **Banerjee, A., Levy, H. R., Levy, G. C., and Chan, W. C.,** Conformations of bound nucleoside triphosphate effectors in aspartate transcarbamylase. Evidence for the London-Schmidt model by transferred nuclear overhauser effects, *Biochemistry,* 24, 1593, 1985.

29. **Buckman, T.,** Spin-labeling studies of aspartate transcarbamylase. I. Effect of nucleotide binding and subunit separation, *Biochemistry,* 9, 3255, 1970.

30. **Winlund-Gray, C., Chamberlin, M. J., and Gray, D. M.,** Interaction of aspartate transcarbamylase with regulatory nucleotides, *J. Biol. Chem.,* 248, 6071, 1973.

31. **Matsumoto, S. and Hammes, G. E.,** An equilibrium binding study of the interaction of aspartate transcarbamylase with cytidine 5′-triphosphate and adenosine 5′-triphosphate, *Biochemistry,* 12, 1388, 1973.

32. **Allewell, N., Friedland, J., and Niekamp, K.,** Calorimetric analysis of aspartate transcarbamylase from *Escherichia coli:* binding of cytosine-5′-triphosphate and adenosine-5′-triphosphate, *Biochemistry,* 14, 224, 1975.

33. **Suter, P. and Rosenbusch, J. P.,** Asymmetry of binding and physical assignments of CTP and ATP sites in aspartate transcarbamoylase, *J. Biol. Chem.,* 252, 8136, 1977.

34. **Tauc, P., Leconte, C., Kerbiriou, D., Thiry, L., and Hervé, G.,** Coupling of homotropic and heterotropic interactions in *Escherichia coli* aspartate transcarbamylase, *J. Mol. Biol.,* 155, 155, 1982.

35. **Gerhart, J. C. and Holoubeck, H.,** The purification of aspartate transcarbamylase of *Escherichia coli,* and separation of its subunits, *J. Biol. Chem.,* 242, 2886, 1967.

36. **Yang, Y. R., Kirschner, M. W., and Schachman, H. K.,** Aspartate transcarbamoylase (*Escherichia coli*): preparation of subunits, *Methods Enzymol.,* 51, 35, 1978.

37. **Porter, R. W., Modebe, M. O., and Stark, G. R.,** Aspartate transcarbamylase: kinetic studies of the catalytic subunit, *J. Biol. Chem.,* 244, 1846, 1969.

38. **Issaly, I., Poiret, M., Tauc, P., Thiry, L., and Hervé, G.,** Interactions of Cibacron blue F3GA and nucleotides with *Escherichia coli* aspartate carbamoyltransferase and its subunits, *Biochemistry,* 21, 1612, 1982.

39. **Dennis, P. R., Krishna, M. V., Di Gregorio, M., and Chan, W. C.,** Ligand interactions at the active site of aspartate transcarbamylase from *Escherichia coli, Biochemistry,* 25, 1605, 1986.

40. **Schaffer, M. H. and Stark, G. R.,** Aspartate transcarbamylase is not a ping-pong enzyme, *Biochem. Biophys. Res. Commun.,* 46, 2082, 1972.

41. **Heyde, E., Nagabhushanam, A., and Morrison, J. F.,** Mechanism of the reaction catalyzed by the catalytic subunit of aspartate transcarbamylase. Kinetic studies with carbamylphosphate as substrate, *Biochemistry,* 12, 4718, 1973.

42. **Collins, K. D. and Stark, G. R.,** Aspartate transcarbamylase; studies of the catalytic subunits by UV difference spectroscopy, *J. Biol. Chem.,* 244, 1869, 1969.

43. **Wedler, F. C. and Gasser, F. J.,** Ordered substrate binding and evidence for a thermally induced change in mechanism for *E. coli* ATCase, *Arch. Biochem. Biophys.,* 163, 57, 1974.

44. **Wedler, F. C. and Gasser, F. J.,** Modes of modifier action in *E. coli* ATCase, *Arch. Biochem. Biophys.,* 163, 69, 1974.

45. **Hsuanyu, Y. and Wedler, F.,** Kinetic mechanism of native *E. coli* aspartate transcarbamylase, *Arch. Biochem. Biophys.,* 259, 316, 1987.

46. **Schmidt, P. G., Stark, G. R., and Baldeschwieler, J. D.,** Aspartate transcarbamylase; A nuclear magnetic resonnance study of the binding of inhibitors and substrates to the catalytic site, *J. Biol. Chem.,* 244, 1860, 1969.

47. **Jacobson, G. R. and Stark, G. R.,** Aspartate transcarbamylase, in *The Enzymes,* Vol. 9, 3rd ed., Boyer, P. D., Ed. Academic Press, New York, 1973, 255.

48. **Kerbiriou, D., Hervé, G., and Griffin, J.,** An aspartate transcarbamylase lacking catalytic subunits interactions. Study of conformational changes by ultraviolet and circular dichroism spectroscopy, *J. Biol. Chem.,* 252, 2881, 1977.

49. **Royer, C. A., Tauc, P., Hervé, G., and Brochon, J. C.,** Ligand binding and protein dynamics: a fluorescence depolarization study of aspartate transcarbamylase from *Escherichia coli, Biochemistry,* 26, 6472, 1987.

50. **Changeux, J. P., Gerhart, J. C., and Schachman, H. L.,** Allosteric interactions in aspartate transcarbamylase; binding of specific ligands to the native enzyme and its isolated subunits, *Biochemistry,* 7, 531, 1968.

51. **Cohlberg, J. A., Pigiet, V. P., and Schachman, H. K.,** Structure and arrangement of the regulatory subunits in aspartate transcarbamylase, *Biochemistry,* 11, 3396, 1972.

52. **Howlett, G. J. and Schachman, H. K.,** Allosteric regulation of aspartate transcarbamylase. Changes in the sedimentation coefficient promoted by the bisubstrate analogue N-(phosphonacetyl)-L-aspartate, *Biochemistry,* 16, 5077, 1977.

53. **Moody, M. F., Vachette, P., and Foote, A. M.,** Changes in the X-ray solution scattering of aspartate transcarbamylase following the allosteric transition, *J. Mol. Biol.,* 133, 517, 1979.

54. **Middleton, S. A. and Kantrowitz, E. R.,** Importance of the loop at residues 230-245 in the allosteric interactions of *Escherichia coli* aspartate carbamoyltransferase, *Proc. Natl. Acad. Sci. U.S.A.,* 83, 5866, 1986.

55. **Ladjimi, M. M., Middleton, S. A., Kelleher, K. S., and Kantrowitz, E. R.,** The relationship between domain closure and binding, catalysis and regulation in *E. coli* aspartate transcarbamylase, *Biochemistry,* 27, 268, 1988.

56. **Ladjimi, M. M. and Kantrowitz, E. R.,** A possible model for the concerted allosteric transition as deduced from site-directed mutagenesis studies, *Biochemistry,* 27, 276, 1988.

57. **Vitrano, E., Cupane, A., and Cordone, L.,** Oxygen affinity of bovine haemoglobin. Relevance of electrostatic and hydrophobic interactions, *J. Mol. Biol.,* 180, 1157, 1984.

58. **Dreyfus, M., Vandenbunder, B., and Buc, H.,** Stabilization of a phosphorylase b active conformation by hydrophobic solvents, *FEBS, Lett.,* 95, 185, 1978.

59. **Dreyfus, M., Fries, J., Tauc, P., and Hervé, G.,** Solvent effects on allosteric equilibria: stabilization of T and R conformations of *Escherichia coli* aspartate transcarbamylase by organic solvents, *Biochemistry,* 23, 4852, 1984.

60. **Kihara, H., Barman, T. E., Jones, P. T., and Moody, M. F.,** Kinetics of the allosteric transition of aspartate transcarbamylase. Chemical quench studies, *J. Mol. Biol.,* 176, 523, 1984.

61. **Griffin, J. H., Rosenbusch, J. P., Blout, E. R., and Weber, K. K.,** Conformational changes in aspartate transcarbamylase. II. Circular dichroism evidence for the involvement of metal ions in allosteric interactions, *J. Biol. Chem.,* 248, 5057, 1973.

62. **Blackburn, M. N. and Schachman, H. K.,** Allosteric regulation of aspartate transcarbamylase. Effect of active site ligands on the reactivity of sulfhydryl groups of the regulatory subunits, *Biochemistry,* 16, 5084, 1977.

63. **Johnson, R. S. and Schachman, H. K.,** Propagation of conformational changes in Ni(II)-substituted aspartate transcarbamylase: effect of active-site ligands on the regulatory chains, *Proc. Natl. Acad. Sci. U.S.A.,* 77, 1995, 1980.

64. **Foote, J. and Schachman, H. K.,** Homotropic effects in aspartate transcarbamylase. What happens when the enzyme binds a single molecule of the bisubstrate analog N-phosphonacetyl-L-aspartate, *J. Mol. Biol.,* 186, 175, 1985.

65. **Gibbons, I., Ritchey, J. M., and Schachman, H. K.,** Concerted allosteric transition in hybrids of aspartate transcarbamylase containing different arrangements of active and inactive sites, *Biochemistry,* 15, 1324, 1976.

66. **Gerhart, J. C. and Pardee, A. B.,** Aspartate transcarbamylase, an enzyme designed for feedback inhibition, *Fed. Proc. Fed. Am. Soc. Exp. Biol.,* 23, 727, 1964.

67. **Kerbiriou, D. and Hervé, G.,** Biosynthesis of an aspartate transcarbamylase lacking co-operative interactions. I. Disconnection of homotropic and heterotropic interactions under the influence of 2-thiouracil, *J. Mol. Biol.,* 64, 379, 1972.

68. **Thiry, L. and Hervé, G.,** The stimulation of *Escherichia coli* aspartate transcarbamylase activity by adenosine triphosphate. Relations with the other regulatory conformational changes; a model, *J. Mol. Biol.,* 125, 515, 1978.

69. **Ladjimi, M. M. and Kantrowitz, E. R.,** Catalytic-regulatory subunit interactions and allosteric effects in aspartate transcarbamylase, *J. Biol. Chem.,* 262, 312, 1987.

70. **Foote, J., Lauritzen, A. M., and Lipscomb, W. N.,** Substrate specificity of aspartate transcarbamylase. Interaction of the enzyme with analogs of aspartate and succinate, *J. Biol. Chem.,* 260, 9624, 1985.

71. **Baillon, J., Tauc, P., and Hervé, G.,** L-alanosine: a non cooperative substrate for *Escherichia coli* aspartate transcarbamylase, *Biochemistry,* 24, 7182, 1985.

72. **Foote, J. and Lipscomb, W. N.,** Kinetics of aspartate transcarbamylase from *Escherichia coli* for the reverse direction of reaction, *J. Biol. Chem.,* 256, 11428, 1981.

73. **Enns, C. A. and Chan, W. C.,** Conformational states of aspartate transcarbamylase stabilized with a cross-linking reagent, *J. Biol. Chem.,* 254, 6180, 1979.

74. **Wyman, J.,** Linked functions and reciprocal effects in haemoglobin: a second look, *Adv. Protein Chem.,* 19, 223, 1964.

75. **Gerhart, J. C. and Pardee, A. B.,** The enzymology of control by feedback inhibition, *J. Biol. Chem.,* 237, 891, 1962.

76. **London, R. E. and Schmidt, P. G.**, A nuclear magnetic resonnance study of the interaction of inhibitory nucleosides with *Escherichia coli* aspartate transcarbamylase and its regulatory subunits, *Biochemistry*, 13, 1170, 1974.

77. **Cook, R. A.**, Subunit interactions in aspartate transcarbamylase, *Biochemistry*, 11, 3792, 1972.

78. **Tondre, C. and Hammes, G. G.**, Interaction of aspartate transcarbamylase with 5-bromocytidine 5'-tri,di and monophosphate, *Biochemistry*, 13, 3131, 1974.

79. **Burz, D. S. and Allewell, N. M.**, Interaction of ionizable groups in *Escherichia coli* aspartate transcarbamylase with adenosine and cytidine-5'-triphosphate, *Biochemistry*, 21, 6647, 1982.

80. **Folterman, K. F., Beck, D. A., and Wild, J. R.**, *In vivo* formation of hybrid aspartate transcarbamylase from native subunits of divergent members of the family enterobacteriaceae, *J. Bacteriol.*, 167, 285, 1986.

81. **Changeux, J. P. and Rubin, M. M.**, Allosteric interactions in aspartate transcarbamylase. III. Interpretation of experimental data in terms of the model of Monod, Wyman and Changeux, *Biochemistry*, 7, 553, 1968.

82. **Howlett, G. J., Blackburn, M. N., Compton, J. G., and Schachman, H. K.**, Allosteric regulation of aspartate transcarbamylase. Analysis of the structural and functional behavior in terms of a two state model, *Biochemistry*, 16, 5091, 1977.

83. **Kerbiriou, D. and Hervé, G.**, An aspartate transcarbamylase lacking catalytic subunit interactions. II. Regulatory subunits are responsible for the lack of co-operative interactions between catalytic sites. Drastic feed-back inhibition does not restore these interactions, *J. Mol. Biol.*, 78, 687, 1973.

84. **Kantrowitz, E. R., Jacobsberg, L. B., Landsfear, S. M., and Lipscomb, W. N.**, Interaction of tetraiodofluorescein with a modified form of aspartate transcarbamylase, *Proc. Natl. Acad. Sci. U.S.A.*, 74, 111, 1977.

85. **Kantrowitz, E. R. and Lipscomb, W. N.**, Functionally important arginine residues of aspartate transcarbamylase, *J. Biol. Chem.*, 252, 2873, 1977.

86. **Landfear, S. M., Lipscomb, W. N., and Evans, D. R.**, Functional modifications of aspartate transcarbamylase induced by nitration with tetranitromethane, *J. Biol. Chem.*, 253, 3988, 1978.

87. **Ladjimi, M. M., Ghelis, C., Feller, A., Cunin, R., Glansdorff, N., Piérard, A. and Hervé, G.**, Structure-function relationship in allosteric aspartate transcarbamylase from *E. coli*. II. Involvement of the C-terminal region of the regulatory chain in homotropic and heterotropic interactions, *J. Mol. Biol.*, 186, 715, 1985.

88. **Weitzman, P. D. and Wilson, I. B.**, Studies on aspartate transcarbamylase and its allosteric interaction, *J. Biol. Chem.*, 241, 5481, 1966.

89. **Enns, C. A. and Chan, W. C.**, Stabilization of the relaxed state of aspartate transcarbamylase by modification with a bifunctional reagent, *J. Biol. Chem.*, 253, 2511, 1978.

90. **Chan, W. C. and Enns, C. A.**, Aspartate transcarbamylase: loss of homotropic but not heterotropic interactions upon modification of the catalytic subunit with a bifunctional reagent, *Can. J. Biochem.*, 57, 798, 1979.

91. **Chan, W. C. and Enns, C. A.**, Hybrid transcarbamoylase containing cross-linked subunits, *Can. J. Biochem.*, 59, 461, 1981.

92. **Hammes, G. G. and Wu, C. W.**, Relaxation spectra of aspartate transcarbamylase. Interaction of the native enzyme with aspartate analogs, *Biochemistry*, 10, 1051, 1971.

93. **Wu, C. W. and Hammes, G. G.**, Relaxation spectra of aspartate transcarbamylase interactions of the native enzyme with an adenosine 5'-triphosphate analog, *Biochemistry*, 12, 1400, 1973.

94. **Hervé, G., Moody, M. F., Tauc, P., Vachette, P., and Jones, P. T.**, Quaternary structure changes in aspartate transcarbamylase studied by X-ray solution scattering; signal transmission following effector binding, *J. Mol. Biol.*, 185, 189, 1985.

95. **Hsuanyu, Y. and Wedler, F.**, Effectors of *E. coli* aspartate transcarbamylase differentially perturb aspartate binding rather than the T-R transition, *J. Biol. Chem.*, 263, 4172, 1988.

96. **Wong, G. C.**, Ph.D. thesis, University of California, Davis, CA. University Microfilms, Ann Arbor, Michigan,

97. **Moore, A. C. and Brown, T. B.**, Binding of regulatory nucleotides to aspartate transcarbamylase: nuclear magnetic resonnance studies of selectively enriched carbone-13 regulatory subunits, *Biochemistry*, 19, 5768, 1980.

98. **Heyde, E., Nagabhushanam, A., and Venkataraman, S.**, Enzyme forms produced from aspartate transcarbamylase by digestion with trypsin, *Biochem. J.*, 135, 125, 1973.

99. **Cunin, R., Jacobs, A., Charlier, D., Crabeel, M., Hervé, G., Glansdorff, N., and Piérard, A.**, Structure-function relationship in allosteric aspartate transcarbamylase from *Escherichia coli*. I. Primary structure of a pyrI gene encoding a modified regulatory subunit, *J. Mol. Biol.*, 186, 707, 1985.

100. **Wild, J.**, personal communication.

101. **Cherfils, J. and Janin, J.**, personal communication.

Chapter 4

GLYCOGEN PHOSPHORYLASE *b*

Louise N. Johnson, Janos Hajdu, K. Ravindra Acharya, David I. Stuart, Paul J. McLaughlin, Nikos G. Oikonomakos, and David Barford

TABLE OF CONTENTS

"It is characteristic of Science that solved problems appear trivial regardless of the magnitude of the effort spent in their solution."

Gregorio Weber

I. INTRODUCTION

A. History and Scope of the Review

The story of the allosteric control of glycogen phosphorylase began with the observations of Carl and Gerty Cori in 1936,[2] that in a dialyzed muscle extract, the presence of adenylic acid was necessary for the phosphorylytic breakdown of glycogen.[1] Although it was not known at that time, this was the first demonstration of allosteric activation of an enzyme. Four years later, the enzyme, glycogen phosphorylase, was purified from rabbit skeletal muscle, and its properties were analyzed in some detail.[2]

Originally it was assumed that the nucleotide was a coenzyme in the reaction. One year later, however, a new form of the enzyme was isolated which did not require the presence of the nucleotide for activity. This active form (termed phosphorylase *a*), could be converted into the other form which depended on adenylic acid for activity (termed phosphorylase *b*).[3] The enzyme catalyzing the interconversion was named "prosthetic group removing enzyme" (PR for short). However, in spite of numerous attempts, it was impossible to detect AMP in phosphorylase *a*. The breakthrough came in 1959 with the work of Fischer and co-workers[4] who discovered that PR actually catalyzed the hydrolysis of a specific phosphoserine residue in phosphorylase *a*, thereby converting it into phosphorylase *b*.

Madsen and Cori[5] determined the subunit structure of both forms showing that, *in vitro*, phosphorylase *b* was a dimer of presumably two identical subunits and phosphorylase *a* was a tetramer. This was the first description of an oligomeric enzyme. The role of AMP remained a puzzle for nearly 30 years. The work of Cori and his collegues[1-3,5,6,10] had shown that phosphorylase *b* had a specific binding site for AMP, but that the nucleotide did not participate in the reaction (evidence reviewed in Reference 6). These observations were important for the development of the theory of allosteric proteins and cellular control systems.[7] In a kinetic study published by Helmreich and Cori[6] in 1964, it was shown that binding of AMP led to a mutual increase in affinity for substrate and vice versa. This work provided a solution to the riddle of the role of AMP in activation of phosphorylase and was one of the key papers cited by Monod, Wyman, and Changeux[8] in their general model for allostery. At this time, in an early demonstration of sensitivity of antibodies to different protein conformations, it was also shown that the conformation of phosphorylase *b* in the presence of both glycogen and AMP was apparently different from the conformation of the enzyme in the presence of either ligand separately.[46]

Shortly, before the discovery of the seryl phosphate, the enzyme was found to contain one molecule of pyridoxal-5-phosphate (PLP) per subunit.[9] Removal of this group leads to inactivation; activity could be restored by mixing the apoenzyme with PLP.[10] As we now know, PLP is the real prosthetic group in both phosphorylase *a* and *b*. However, its precise role in the catalytic reaction has been a long-standing mystery (Section III.E).

Under physiological conditions, a substantial proportion of the phosphorylase in the cell is bound to the glycogen particles along with most of the other enzymes associated with glycogen metabolism.[11-19] *In vitro*, in the absence of glycogen and effectors, phosphorylase *a* is a tetramer and phosphorylase *b* is a dimer (reviewed in Reference 20). The phospho-dephospho hybrid (*ab*) enzyme, which is an intermediate during the phosphorylation and dephosphorylation of the protein,[21] exhibits intermediate properties. Activation of the *b* form either by AMP binding or by covalent modification (phosphorylation) shifts the dimer-tetramer equilibrium toward the tetramer side. However, the tetrameric form exhibits little activity and a very low affinity for glycogen. Binding to glycogen dissociates all tetrameric forms of the enzyme into dimers. It is most probable that only the dimeric form of the enzyme is capable of binding to the large glycogen molecule.

Perhaps the main driving force in working on phosphorylase has been the fact that it comprises an excellent model for the study of almost any aspect of enzyme action and regulation. After half a century of extensive work, it still can produce many "firsts" and surprises. This was the first enzyme to be shown (1) to exhibit a subunit structure; (2) to undergo allosteric activation and inhibition; (3) to be covalently interconvertible by a phosphorylation-dephosphorylation process; (4) to utilize the 5'-phosphate of the cofactor pyridoxal-5-phosphate in the catalytic mechanism; (5) to exhibit a glycogen storage site separate from the catalytic site; and (6) recently, to be using fast crystallographic techniques, phosphorylase *b* was the first enzyme where an active enzyme substrate complex could be analyzed in three dimensions as it was converted into products.

In this review we give an account of our current understanding of the catalytic and control mechanisms of muscle glycogen phosphorylase *b* with special emphasis on the results from the crystal structure determination of the rabbit muscle enzyme. There have been several excellent reviews of phosphorylase,[18,20,22,23] of phosphorylase *b*,[24] of phosphorylase *a*,[25-27] of the role of the cofactor pyridoxal phosphate,[28,29] of the interconverting enzymes[30] (see also Reference 27 and references in the same volume), and of noncovalent control.[31,32]

B. Physiological Role

Glycogen phosphorylase (α-1,4-glucan: orthophosphate glycosyl transferase E.C.2.4.1.) catalyzes the first step in the intracellular degradation of glycogen

$$(\alpha\text{-1,4-glucoside})_n + P_i \rightarrow (\alpha\text{-1,4-glucoside})_{n-1} + \text{Glc-1-P}$$

The equilibrium ratio [Glc-1-P/P$_i$] for this reaction at pH 6.8 is 0.28, but *in vivo* the enzyme functions in the direction of glycogen degradation because levels of P$_i$ are relatively high (2 to 3 mmol kg^{-1} wet tissue) and those of Glc-1-P are low (.05 mmol kg^{-1} wet tissue in resting mouse hamstring muscle[32,33]). A separate enzyme, glycogen synthase, utilizing uridine diphosphate glucose (UDPG) is available for glycogen synthesis. The role of phosphorylase is to supply Glc-1-P from the storage polysaccharide; a function connected in muscle with the energy demands of contraction, in heart and brain with the provision of fuel during brief periods of anoxia and in the liver with the maintenance of blood sugar levels. The importance of skeletal muscle glycogen phosphorylase is demonstrated by patients with McArdles syndrome, an inborn error of metabolism caused by lack of glycogen phosphorylase in muscle.[34] Patients have cramps and weakness in exercise, which become progressively severe in adulthood. The disease is rare, but following the introduction of noninvasive ^{31}P NMR techniques for diagnosis,[35] the number of reported cases is likely to rise.

In normal resting muscle, the predominant form of the enzyme is phosphorylase *b*, although there may be significant ($\sim$10%) levels of phosphorylase *a*.[32] Phosphorylase *b* is activated by AMP and IMP (the deamination product of AMP through the action of AMP aminohydrolase) and by high levels of inorganic phosphate or similar dianions. It is inhibited by ATP, ADP, Glc-6-P, and UDPG. In addition, *in vitro*, a number of other compounds have been shown to affect activity. For example, spermine and polyamines[36-38] augment AMP or IMP activation as do certain organic solvents,[39,40] while NADH and NADPH are inhibitors.[41,42] Other effectors have been summarized by Dombradi.[20] As was originally shown by the Coris,[2] glucose is an inhibitor. In muscle, there is no free glucose, but this inhibition may be important for the liver where phosphorylase is involved in regulation of blood glucose levels.[43,44] Nucleotides and nucleosides at high concentration ($\sim$2 m*M*) and a number of aromatic compounds derived from purines (e.g., caffeine) are now known to bind at a second effector site in phosphorylase (Section II.J) and exhibit additional regulation by inhibition of the enzyme, an inhibition which is synergistic with glucose. Finally, glycogen itself can be considered as a further effector promoting the dimeric state and contributing to activation (Section II.G). A summary of selected values for kinetic and equilibrium binding constants is given in Table 1. These demonstrate only some of the simpler aspects of the complex heterotropic interactions that are likely to be important for the *in vivo* regulation.

AMP activation of phosphorylase *b* results in homotropic effects for nucleotide binding and heterotropic interactions with substrates (P_i or Glc-1-P). For example, the K_a for P_i is 23.0 mM at 0.015 mM AMP and 1.5 mM at 0.05 mM AMP. Likewise, the K_m for AMP is 0.18 mM at 2 mM P_i and 0.043 mM at 10 mM P_i.[6] Insight into a possible route whereby activation is achieved has come from studies with IMP.[47] IMP activation of phosphorylase *b* results in an increase in V_{max} without affecting the affinity for Glc-1-P. (The K_m for GlP is 33 mM in the presence of 1 mM IMP and 4.7 mM in the presence of 1 mM AMP.[47]) The homotropic cooperativity of Glc-1-P is strong (n = 2) and independent of IMP concentration. These differences in AMP and IMP activation have led to the suggestion that the AMP-induced allosteric transition of phosphorylase *b* consists of two stages: an enhancement of affinity for substrate (K system) and an enhancement of activity (V system). Thus, IMP only affects the latter stage of the allosteric transition. In a combined calorimetric and equilibrium dialysis study, Mateo et al.[53] have shown that the homotropic cooperativity of AMP (n = 1.4) is more pronounced than that of IMP (n = 1.2).

1. Role of Noncovalent and Covalent Activation

In response to nervous or hormonal stimulation, phosphorylase *b* is converted to phosphorylase *a* through the action of phosphorylase kinase in the presence of Mg/ATP. Phosphorylase *a* is active in the absence of AMP, although addition of AMP can produce a 10 to 20% increase in activity. Phosphorylase *a* is no longer inhibited by ATP, ADP, or glucose-6-P (although weak inhibition has been reported by Glc-6-P),[54] but its activity is still modulated by glycogen and by glucose and purine nucleotides or nucleosides. Assuming a two state system, the equilibrium constant L($= T_o/R_o$) for phosphorylase *b* is at least 3000 and for phosphorylase *a* is between 3 and 10.[27]

The relative contributions to glycogenolysis produced by phosphorylase *a* activation and phosphorylase *b* noncovalent nucleotide activation are difficult to assess.[31] It is known from studies on intact muscle that activation of phosphorylase *a* is very fast. With the onset of tetanic contraction in mouse muscle, phosphorylase *a* activity reaches 50% of the total phosphorylase activity in 2 s.[33,55] The activity continues to rise for about 10 s and thereafter declines markedly to a level below the resting level. Some of these properties can be simulated with glycogen particles which can be flash activated by addition of ATP/Mg and Ca^{2+}.[14,15,18,31] The general consensus is that phosphorylase *a* activity is most important in the initial stages in response to a stimulus, but that phosphorylase *b* noncovalent activation may become significant during prolonged exercise. In support of this, Radda and his colleagues[56] have shown that the administration of 2-deoxy glucose impairs the rate of glycogen breakdown in ischemic heart. They demonstrated that this was due to accumulation of 2-deoxy glucose-6-P which specifically inhibits phosphorylase *b* but not phosphorylase *a*.

The most clear-cut evidence for the role of noncovalent activation of phosphorylase *b* comes from studies with phosphorylase kinase deficient I-strain mice which are able to degrade muscle glycogen in response to adrenalin administration or electrical stimulation despite their inability to form phosphorylase *a*.[55,57] These mice are able to degrade greater amounts of glycogen than normal mice and perform better in some endurance tests. Their glycogenolysis may be mediated by IMP, since the muscles show higher IMP contents than normal mice and form sufficient IMP during exercise to activate phosphorylase *b*.[33,58,59] Although the affinity of IMP-activated phosphorylase *b* for substrate ($K_m(P_i)$ = 40 mM) is poor,[27] this may be augmented by increased levels of spermine.[38] Spermine (the main polyamine in mouse muscle) increase V_{max} for 5'-IMP activated phosphorylase *b* three- to fivefold and also increases K_i for Glc-6-P. Significant glycogenolysis could also be achieved if, in working muscle, the enzyme had access to a sequestered store of AMP.[31,59] In the

Table 1
DISSOCIATION AND INHIBITION CONSTANTS FOR PHOSPHORYLASE b

Metabolite	K_D, Dissociation constants (mM) for metabolite				
	Free enzyme	Enzyme—AMP complex	Enzyme—AMP—Pi complex	Enzyme—AMP—glycogen complex	Enzyme—AMP—Glc-1-P complex
A substrates					
Glycogen[a]		4.6[b]	0.2[b]	—	0.9[b]
Pi	93[c]	15[b]	—	2.2[b]	—
Glc-1-P	9.5[c]	2.7[c], 7.4[b]	—	3.0[b]	—

K_I, inhibition constants (mM)

	Free enzyme	
B activators[d]		
AMP	0.35[c], 0.085[e]	0.9[f] in competition with AMP
IMP	2.0[e]	
C inhibitors		
ATP	3.0[e]	0.9[g] in competition with AMP in presence of 10 mM Pi
ADP	0.13[e]	
Glc-6-P	0.019[h]	0.2[i], 0.9[j] in competition with AMP
UDPG		1.4[k] in competition with Glc-1-P

a   Glycogen concentrations expressed as mM glucose units. See Sections II.G and III.A for a discussion of the glycogen storage site.

b K_D from kinetic studies.[45]

c K_D from PLP fluorescence.[24]

d   See Section II.I for a discussion of nucleotide binding to the inhibitor site.

e K_D from ESR measuremen:s.[48]

f   See Reference 47.

g   See Reference 6.

h   See Reference 51.

i   See Reference 49.

j   See Reference 50.

k   See Reference 52.

muscles of I-strain mice, glycogen degradation lags progessively further behind that of normal mice muscles for a period of about 8 s.[55] While such a lag may not be deleterious in the shielded environment of the laboratory, it could be unfortunate for the animal in the wild.

In the past, there has been discussion as to why glycogenolysis is negligible in resting muscle since the total AMP concentation (estimated to be between 0.05 and 0.2 mmol kg^{-1}) is sufficient to activate phosphorylase *b*. A partial answer was provided by the inhibition with Glc-6-P (present at a concentration of about 0.2 mmol kg^{-1})[33,49] and ATP (present at a concentration of about 3 to 7 mmol kg^{-1}),[33,49] a classical feedback mechanism, and the observation that when complexed in the glycogen particle, phosphorylase *b* exhibits a lower affinity for nucleotides.[17] Recent ^{31}P NMR studies[60-62] on frog gastrocnemius muscle have shown that the concentration of free AMP is very much less than the total AMP; sufficiently low (0.1 μmol kg^{-1}), in fact, as to result in negligible activation. These estimates are based on the assumption that the adenylate kinase and creatine kinase reaction are at equilibrium in muscle. They show, also, that free ADP levels are so low (0.02 mmol kg^{-1}) that ADP is unlikely to be an effective inhibitor in resting muscle. UDPG is also too low (0.03 mmol kg^{-1}) to be effective.[63] In fatigued muscle, there is a pronounced rise in free AMP (measured in amphibian muscle), in IMP (measured in mouse muscle), and also in the inhibitors Glc-6-P and ADP. Levels of ATP do not change dramatically. It is here that the heterotropic interactions of phosphorylase *b* are likely to be important for activation in the face of inhibition. As the levels of phosphate rise in working muscle, so the affinity for AMP is increased which produces greater activity and greater affinity for substrate. It is the purpose of this review to seek a stereochemical explanation for these phenomena.

II. STRUCTURAL STUDIES

A. Crystallization

As with all protein crystallographic studies, the growth of large crystals with suitable unit cell parameters was the initial hurdle for structure determination of glycogen phosphorylase. In 1972, a favorable crystal form of rabbit muscle phosphorylase *b* was discovered in Oxford.[64] The crystals are tetragonal space group P4$_3$2$_1$2, unit cell dimensions a = b = 128.5 Å; c = 116.3 Å with one phosphorylase subunit, molecular weight (97,400) per asymmetric unit. The two subunits of the physiologically active dimer are related by a crystallographic twofold axis of symmetry and thus are chemically and stereochemically identical. At first, it was thought that the crystals had been obtained in the presence of AMP, since crystallization had followed the usual procedure for microcrystallization used in the purification of the enzyme.[65] However, later work[66] showed that the AMP had been converted to IMP by trace amounts of AMP aminohydrolase adenylate deaminase in the crystallization mixtures. This was a fortunate accident, for the tetragonal crystal form can only be obtained in the presence of 2m*M* IMP, or, with more difficulty, in the absence of nucleotide. The crystal structure has been solved at 6 Å, 3 Å, then 2.25 Å using conventional methods.[66-68] Most recently, the resolution has been extended to 1.9 Å with new data measured at the Synchrotron Radiation Source at the Science and Engineering Research Council's laboratory at Daresbury, U.K. The structure has been refined by restrained least squares crystallographic refinement, and the current reliability index R factor agreement is 0.188. The refinement is not yet complete, but most of the structure for some 6705 protein atoms and nearly 692 water molecules is well established.[69]

In 1976, Fletterick et al.[70] found that crystals of phosphorylase *a*, in the presence of the inhibitor 50 m*M* glucose, could also be grown in the same tetragonal space group as phosphorylase *b*. The structure was solved at 6 Å, and the resolution was extended in the same year to 3 Å[71] and later to 2.5 Å.[72] The structure has subsequently been refined by least squares refinement and energy minimization at 2.1 Å.[73] The current precision of the structure determinations of phosphorylase *a* and *b* are roughly comparable.

For both crystals, the enzyme is in, or very close to, the T-state. IMP is a weak activator of phosphorylase b (Section I.B), and there is evidence that the conformation of the enzyme in the presence of IMP is close to the nucleotide-free conformation.[18,74] Glucose is a competitive inhibitor of phosphorylase a,[75] that promotes the T-state of the enzyme[76,77] and which prevents the aggregation of dimers to tetramers.[78]

A study of the activity of the T-state crystals used in the X-ray diffraction studies was carried out by Kasvinsky and Madsen in 1976.[79] The studies were performed with microcrystals mildly cross-linked with glutaraldehyde, since noncross-linked crystals crack and dissolve in the presence of substrates. It was found that crystallization reduced the maximal velocity by 11- to 50-fold for phosphorylase b and 50- to 100-fold for phosphorylase a depending on whether maltoheptaose or Glc-1-P was the variable substrate. Essentially no differences were found between the K_m values for either substrate in the soluble or crystalline states. Thus, as is often the case with enzyme crystals, the thermodynamic equilibrium properties of the system are unchanged by crystallization, but the dynamic properties are reduced, probably because conformational changes are restricted in the crystal lattice. Comparative measurements with glycogen and maltoheptaose showed that the smaller substrate enters the interior of the microcrystal and activity is not confined to the surface activity alone. Interestingly, the Hill coefficients for the homotropic cooperativity of maltoheptaose observed for both solutions and crystals of phosphorylase a were similar, indicating that some part of allosteric response for phosphorylase a might be possible in the crystal. No heterotropic cooperativity was observed in the IMP activated phosphorylase b either in the crystal or in solubion; this could be a result of glutaraldehyde cross-linking.[80]

Crystals of the R-state enzyme have not been easy to obtain, partly because of the change in solubility and aggregation properties that accompany activation. With phosphorylase b there is the additional complication of obtaining a homogeneous R-state preparation in the presence of AMP. Attempts to crystallize rabbit muscle phosphorylase b in the presence of AMP and substrates have resulted in only small crystals or crystals with very large unit cells. This has prompted a return to the monoclinic form obtained in the presence of ammonium sulfate.[81-84] These crystals have the enzyme in the tetrameric state. Crystallographic analysis is going well.[238] Crystals of apophosphorylase b reconstituted with pyridoxal pyrophosphate, a possible transition state analog which stabilizes the R-state,[85] have been obtained by Fletterick and his colleagues and a detailed analysis is in progress. Another possibility has been to change species. Muscle phosphorylases from shark,[86] pig,[87] and liver phosphorylase a[88] have been prepared in the R-state, but no detailed analyses are available. Crystals of the nonregulatory active E. coli maltodextrin phosphorylase appeared promising, but were found to suffer from anisotropic disorder which made high resolution studies difficult.[89] A detailed structural analysis of the nonregulatory potato phosphorylase is in progress.[239]

B. Primary, Secondary, and Tertiary Structure

The amino acid sequence of rabbit muscle glycogen phosphorylase was determined by chemical methods in 1977.[90-93] This definitive work has played a key role in the interpretation of the X-ray crystal structure. The precision of the crystallographic work has only recently been good enough to assess agreement between the chemical and structural data. To date, there is excellent agreement, and there is only one small correction, first observed in the structure of phosphorylase a,[240] and then in phosphorylase b. There is an extra isoleucine residue inserted after residue 307 making the sequence in this region DIIRR. The recent determination of the complete cDNA sequence of rabbit muscle phosphorylase has revealed seven further corrections, all in the N-terminal portion of the molecule, at positions 30, 32, 42, 55, 57, 88, and 112.[241] In the sequence shown in Figure 1, these small corrections have been made, and the residue numbers increased accordingly. The enzyme contains 842 amino

```
            10        20        30        40        50        60        70
r   SRPLSDQEKRKQISVRGLAGVENVTELKKNFNRHLHFTLVKDRNVATPRDYYFALAHTVRDHLVGRWIRTQQHYY
p   TLSEKIHHPITEQGGESDLSSFAPDAASITSSIKYHAEFTPVFSPERFELPKAFFATAQSVRDSLLINWNATYDIYE
e                SQPIFNDKQFQEALSRQWQRYGLNSAAEMTPRQWWLA----VSEALAEMLRAQPFAKP
       80        90        100       110       120       130       140       150
r   EKDPKRIYYLSLEFYMGRTLQNTMVNLALENACDEATYQLGLDMEELEEIEEDAGLGNGGLGRLAACFLDSMATL
p   KLNMKQAYYLSMEFLQGRALLNAIGNLELTGAFAEALKNLGHNLENVASQEPDAALGNGGLGRLASCFLDSLATL
e   VANQRHVNYISMEFLIGRLTGNNLLNLGWYQDVQDSLKAYDINLTDLLEEIDPALGNGGLGRLAACFLDSMATV
        160       170       180       190       200       210       220
r   GLAAYGYGIRYEFGIFNQKICGGWQMEEADDWLRYGNPWEKARPEFTLPVHFYGRVEHTSQGAK-WVDTQVVLAMP
p   NYPAWGYGLRYKYGLFKQRITKDGQEEVAEDWLEIGSPWEVVRNDVSYPIKFYGKVSTGSDGKRYWIGGEDIKAVA
e   GQSATGYGLNYQYGLFRQSFVDGKQVEAPDDWHRSNYPWFRHNEALDVQVGIGGKVTK--DGR--WEPEFTITGQA
        230       240       250       260       270       280       290       300
r   YDTPVPGYRNNVVNTMRLWSAKAPN-DFNLKDFNVGGYIQAVLDRNLAENISRVLYPNDNFFEGKELRLKQEYFVV
p   YDVPIPGYKTRTTISLRLWSTQVPSADFDLSAFNAGEHTKACEAQANAEKICYILYPGDESEEGKILRLKQQYTLC
e   WDLPVVGYRNGVAQPLRLWQATHAH-PFDLTKFNDGDFLRAEQQGINAEKLTKVLYPNDNHTAGKKLRLMQQYFQC
        310       320       330       340       350       360       370
r   AATLQDIIRRFKSSKFGCRDPVRTNFDAFPDKVAIQLNDTHPSLAIPELMRVLVDLERLDWDKAWEVTVKTCAYT
p   SASLQDIISRFE-RRSGDR----IKWEEFPEKVAVQMNDTHPTLCIPELMRILIDLKGLNWNEAWNITQRTVAYT
e   ACSVADILRRH---HLAGR-----ELHELADYEVIQLNDTHPTIAIPELLRVLIDEHQMSWDDAWAITSKTFAYT
     380       390       400       410
r   NHTVLPEALERWPVHLLETLLPRHLQIIYEINQRFLNRV------------------------------------
p   NHTVLPEALEKWSYELMQKLLPRHVEIIEAIDEELVHEIVLKYGSMDLNKLEEKLTTMRILENFDLPSSVAELFIKP
e   NHTLMPEALERWDVKLVKGLLPRHMQIINEINTRFKTLV------------------------------------
                                        420       430       440       450
r   -----------------------------------AAAFPGDVDRLRRMSLVEEGAVKRINMAHLCIAGSH
p   EISVDDDTETVEVHDKVEASDKVVTNDEDDTGKKTSVKIEAAAEKDIDKKTPVSPEPAVIPPKKVRMANLCVVGGH
e   -----------------------------------EKTWPGDEKVWAKLAVVHD---KQVHMANLCVVGGF
        460       470       480       490       500       510       520
r   AVNGVARIHSEILKKTIFKDFYELEPHKFQNKTNGITPRRWLVLCNPGLAEIIAERIG-EEYISDLDQLRKLLSYV
p   AVNGVAEIHSEIVKEEVFNDFYELWPEKFQNKTNGVTPRRWIRFCNPPLSAIITKWTGTEDWVLKTEKLAELQKFA
e   AVNGVAALHSDLVVKDLFPEYHQLWPNKFHNVTNGITPRRWIKQCNPALAALLDKSLQ-KEWANDLDQLINLVKLA
        530       540       550       560       570       580       590
r   DDEAFIRDVAKVKQENKLKFAAYLEREYKVHINPNSLFDVQVKRIHEYKRQLLNCLHVITLYNRIKK----EPNKFVV
p   DNEDLQNEWREAKRSNKIKVVSFLKEKTGYSVVPDAMFDIQVKRIHEYKRQLLNIFGIVYRYKKMKEMTAAERKTNFV
e   DDAKFRDLYRVIKQANKVRLAEFVKVRTGIDINPQAIFDIQIKRLHEYKRQHLNLLRILALYKEIRE----NPQADRV
     600       610       620       630       640       650       660       670
r   PRTVMIGGKAAPGYHMAKMIIKLITAIGDVVNHDPVVGDRLRVIFLENYRVSLAEKVIPAADLSEQISTAGTEASG
p   PRVCIFGGKAFATYVQAKRIVKFITDVGATINHDPEIGDLLKVVFVPDYNVSVAELLIPASDLSEHISTAGMEASG
e   PRVFLFGAKAAPGYYLAKNIIFAINKVADVINNDPLVGDKLKVVFLPDYCVSAAEKLIPAADISEQISTAGKEASG
     680       690       700       710       720       730       740       750
r   TGNMKFMLNGALTIGTMDGANVEMAEEAGEENFFIFGMRVEDVDRLDQRGYNAQEYYDRIPELRQIIEQLSSGFF
p   TSNMKFAMNGCIQIGTLDGANVEIREEVGEENFFLFGAQAHEIAGLRKERADGKFVPDERFEEVKEFVR-SGAFG
e   TGNMKLALNGALTVGTLDGANVEIAEKVGEENIFIFGHTVKQVKAILAKGYDPVKWRKKDKVLDAVLKELESGKY
        760       770       780       790       800       810       820
r   SPKQPDLFKDIVNMLMHH--DRFKVFADYEEYVKCQERVSALYKNPREWTRMVIRNIATSGKFSSDRTIAQYAREIW
p   SYNYDDLIGSLEGNEGFGRADYFLVGKDFPSYIECQEKVDEAYRDQKRWTTMSILNTAGSYKFSSDRTIHEYAKDIW
e   SDGDKHAFDQMLHSIGKQGGDPYLVMADFAAYVEAQKQVDVLYRDQEAWTRAAILNTARCGMFSSDRSIRDYQARIW
        830       840
r   GVEPSRQRLPAPDEKIP
p   NIEAVEIA
e   QAKR
```

FIGURE 1.   The amino acid sequence for glycogen phosphorylase *b* from rabbit muscle (r),[90-93] potato p,[172] and *E. coli* (e).[173,174] An extra Ile has been inserted after residue 307 in the rabbit muscle phosphorylase sequence and the numbers increased accordingly (see text). Seven further corrections from the c-DNA sequence for rabbit muscle phosphorylase have also been included.

acids and has a subunit molecular weight of 97,434 including the pyridoxal phosphate, the *N*-acetyl group, and the extra isoleucine.

Mammalian phosphorylases show a high degree of conservation. Sequence comparisons of partial cDNAs from rat, rabbit, and human muscle phosphorylases in regions that encode parts of the enzyme located near and encompassing the C-terminus show that there is

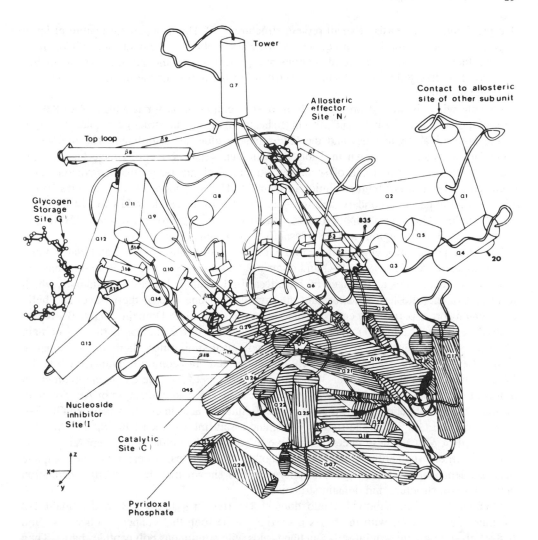

FIGURE 2. A schematic diagram of the phosphorylase *b* monomer viewed down the crystallographic y axis. α-Helices and β-strands are represented by arrows and cylinders, respectively, and are labeled according to the assignments of Table 2. Glucose-1-phosphate and pyridoxal phosphate (partially obscured by the structural elements of the protein) are shown bound at the catalytic site C, AMP at the allosteric effector site N, maltopentaose at the glycogen storage site G, and adenosine at the nucleoside inhibitor site I. The C-terminal domain (domain 2) is shown shaded.

approximately 96% conservation in amino acid sequence.[94] Genomic sequences hybridizing to the human muscle phosphorylase cDNA have been mapped to the human chromosome number 11.[95]

Sequences surrounding the pyridoxal phosphate site for both dogfish[96] and yeast phosphorylase[97,244] show a high degree of homology with the rabbit muscle enzyme. The yeast enzyme is phosphorylated to form phosphorylase *a* on a threonine residue, and the sequence around this residue shows no similarity with the muscle phosphorylase.[244] Conservation of muscle phosphorylase sequences among different species is high. Moreover, the recent cDNA sequence of human liver phosphorylase shows 80% homology of the deduced amino acid sequence with rabbit muscle phosphorylase.[245] The sequences of the nonregulatory phosphorylases are discussed in Section II.J.

A schematic diagram of the subunit structure is shown in Figure 2. The structure is relatively compact with a radius of 30 Å. The surface area is about 1.4 times greater than that predicted on the basis of small protein structures.[68,98] This is a general feature of large oligomeric proteins and is a consequence of their assembly from domains and subdomains, and the fact that there are more identations and projections on the surface than in smaller proteins. It is likely that these features contribute to the ability of the molecule to respond to external stimuli.

Phosphorylase is an α/β protein with an overall composition for residues 19 to 839 of 52% of amino acids in α-helix and 20% in β-sheets. In the crystal structure of phosphorylase *b*, the N-terminal 18 residues and the C-terminal residue are disordered. The regions of secondary structure are shown in Table 2. Although small corrections have been made, especially at the start and end of secondary structural elements, no major revisions have occurred as the result of high resolution refinement. The long loop between residues 506 to 525 with no apparent secondary structure has a region of 10 residues 514 to 524 arranged in a 3_{10} helix. Although phosphorylase is classified as an α/β protein, residues 150 to 260 contain only β-sheet regions and residues 713 to 836 contain only α-helices.

The large polypeptide chain may be divided into two domains (Figure 2 and Plates 3 and 4*). Domain 1 (residues 19 to 484) and domain 2 (residues 485 to 839). Both domains are based on β-sheet cores and flanked by α-helices. In domain 1, helices α-2 and α-8, which form part of the allosteric effector site, are arranged on one side of the β-sheet and helix α-6, which forms part of the catalytic site, is on the other side. Domain 2 has a βα repeat motive (residues 562 to 711) whose topology is identical to the nucleotide binding domain of lactate dehydrogenase.[71,99,100] Despite the topological similarities, there are no sequence similarities. Domains 1 and 2 are closely interconnected. In addition to numerous Van der Waals interactions, there are over 30 polar interactions between side chain and main chain atoms of each domain. The more important of these includes an interaction of Arg 569 with various residues (Section III.F), a glutamic acid-glutamic acid interaction between Glu 296 (α-8 helix) and Glu 385 at the start of β-14, and a salt link between Glu 382 and Arg 770 (at the end of α-26). These interactions serve to connect the domains and may relay information from one metabolite binding site to another. A recent paper describing the complete gene sequence of human muscle phosphorylase has shown that there is little correlation between exon structure and domain structure.[246]

Domain 1 may be subdivided into domain 1-1 (residues 19 to 283) and domain 1-2 (residues 284 to 484), with the break occurring in the loop that connects helices α-7 and α-8. Both subdomains exhibit self-contained folds with continuous polypeptide chains. They are connected by the parallel β-strands β-1 and β-12, and helix α-8 fits neatly into the concave cavity formed by the twisted sheet which extends across both domains.

The ligand binding sites are displayed in the subunit structure and in the dimer in Figure 2 and Plates 3 and 4, respectively. The essential cofactor pyridoxal phosphate is located at the center of the molecule. From binding studies in the crystal, four ligand binding sites per subunit have been identified. The catalytic site (site C) is situated at the center of the molecule where the structural domains (domains 1-1, 1-2, and 2) come together and is close to the pyridoxal phosphate. The allosteric effector AMP site (site N) is some 32 Å from site C and, thus, well fulfills the prediction of Monod et al.[8] for "indirect interactions between distinct specific binding sites" as the basis for regulatory function. It utilizes residues solely from domain 1 and is located at the subunit-subunit interface, a feature common to all protein allosteric sites that have been examined crystallographically (e.g., hemoglobin, phosphofructokinase, aspartate transcarbamylase). A second nucleoside or nucloside-inhibitor binding site (site I) has been identified at the entrance to the active site channel and is some 12 Å from site C. The glycogen storage site (site G) is situated on the surface of the enzyme and is about 30 Å and 39 Å from the catalytic site and allosteric effector site, respectively. This is an additional control site and serves to attach the enzyme to glycogen. The stereo-

* Plates 3 and 4 appear following page 120.

Table 2
THE SECONDARY STRUCTURAL ELEMENTS OF GLYCOGEN PHOSPHORYLASE *b*

Structural element	Residue numbers	Number of amino acids	Comment
α-1	20—38	19	N-terminal helix
α-2	47—78	32	Previously called A—A'; breaks at 57, 58, and 62
β-1	81—86	6	Strand 1, domain 1
β-2	89—92	4	Antiparallel to β-3
α-3	94—102	9	
α-4	104—115	12	
α-5	118—124	7	
β-3	129—131	4	Antiparallel to β-2
α-6	134—150	17	"Glycine" helix to active site
β-4	153—160	8	Strand 2, domain 1
β-4b	162—163	2	Antiparallel to β-11b
β-5	167—171	5	Antiparallel to β-6
β-6	174—178	5	Antiparallel to β-5
β-7	191—193	3	Strand 5, domain 1
β-8	198—209	12	Top loop; antiparallel to β-9
β-9	212—223	12	Top loop; antiparallel to β-8
β-10	222—232	11	Strand 4, domain 1
β-11	237—247	11	Strand 3, domain 1
α-7	261—274	14	Tower helix
β-11b	276—279	4	Antiparallel to β-4b
α-8	289—314	26	Previously called B—B'; breaks at 296, 298, 309
α-8b	328—333	6	
β-12	332—339	8	Strand 6, domain 1
α-9	344—355	12	
α-10	360—372	13	
β-13	371—376	6	Strand 7, domain 1
β-14	385—389	5	Antiparallel to β-16
α-11	388—396	9	
α-12	396—418	23	Glycogen storage helix
α-13	420—429	9	
β-15	430—432	3	Antiparallel to β-16
β-16	437—441	5	Antiparallel to β-14 and β-15
α-14	440—448	9	
β-17	451—454	4	Strand 8, domain 1
α-15a	456—466	11	Irregular
α-16b	468—475	8	Helix broken at F468
β-18	478—484	7	Strand 9, domain 1
π	488—495	8	
α-16	496—507	12	P497 breaks helix 16
3_{10}	514—525	12	3 turns 3_{10} helix
α-17	527—553	27	Break at K544
β-19	562—570	9	β-A of nucleotide binding domain
α-18	575—593	19	α-B of nucleotide binding domain
β-20	600—608	9	β-B of nucleotide binding domain
α-19	613—631	18	α-C of nucleotide binding domain
β-21	639—645	7	β-C of nucleotide binding domain
α-20	649—657	9	α-D of nucleotide binding domain
β-22	661—665	5	β-D of nucleotide binding domain
α-21	676—684	9	α-E of nucleotide binding domain Break at K680
β-23	687—691	5	β-E of nucleotide binding domain
α-22	695—704	10	α-F of nucleotide binding domain
β-24	709—711	3	β-F of nucleotide binding domain
α-23	714—725	12	

Table 2 (continued)
THE SECONDARY STRUCTURAL ELEMENTS OF GLYCOGEN PHOSPHORYLASE *b*

Structural element	Residue numbers	Number of amino acids	Comment
α-24	728—735	8	
α-25	735—747	13	
α-26	758—768	11	
α-27	776—792	17	
α-28	793—806	14	
α-29	812—825	14	

Note: Helices are defined according to Kabsch and Sander,[254] i.e., as overlaps of minimal helices where a minimal helix is defined by two consecutive n-turns. For example, an α-helix of minimal length 4 from residues i to i + 3 requires two hydrogen bonds NH(i − 1)...0 = C(i + 3) and NH(i)...0 = C(i + 4). However, we also include as the first residue (i − 1) the residue whose CO is regularly bonded to NH along the helix and the last residue whose NH is regularly hydrogen bonded to CO along the helix according to IUPAC-IUB recommendations. β-strands are defined according to Kabsch and Sander[254] except we also include a terminal residue in the assignment if it participates in the β-sheet hydrogen bonding pattern.

chemistry of these sites and their biochemical and physiological significance are discussed in more detail in later sections.

The distribution of these metabolite binding sites on the dimer (Plates 3 and 4) is such that the catalytic, inhibitory, and glycogen storage sites are on one side (nearest the viewer in Plate 3) and the allosteric site is on the far side of the molecule. Thus, attachment of phosphorylase to the glycogen particle will help localize the substrate (a different glycogen chain) for the catalytic site while the allosteric effector site is free to interact with nucleotides or Glc-6-P. These features have been noted also for phosphorylase *a*.[101]

C. Subunit-Subunit Contacts

Interactions, and changes in interactions, at the subunit-subunit interface are an essential feature of the allosteric response. When we examine the dimer interface in phosphorylase *b*, we find the interactions between subunits are relatively few. There are two regions, the cap (residues 36 to 45) and the tower (residues 260 to 276), that make significant excursions away from the main body of the subunit. For the cap interactions (Figure 3), there is one good ionic interaction between Glu 195 and the cap Lys 41′ of the other subunit. (Superscript prime denotes residues from the other subunit.) In addition, there are three important subunit-subunit hydrogen bonds (Gln 72 (α-helix) to Asn 44′ (cap) through a water molecule; Arg 193 to main chain CO atoms of the cap residues (Leu 39′ and to Val 40′) and several Van der Waals interactions between residues of the α-2-helix (Val 64, Gly 65, Ile 68) and residues of the symmetry related cap (Val 40′, His 36′). Interactions in this region form an essential component of the AMP and Glc-6-P allosteric sites (Section II.H). The second set of interactions made by the contacts of the tower are rather more extensive (Figure 4). Most of these result from the antiparallel packing of the two symmetry related tower helices. At the N-terminus of the tower helix (α-7), Tyr 262 is in Van der Waals contact with Pro 281′ and Phe 166′ (β-5 strand); Ile 263 and Val 266 form Van der Waals interactions with Tyr 280′ and Val 278′ (C terminal α-7-helix), respectively. In the central region of α-7, Asn 270 hydrogen bonds to Asn 274′. Each of these asparagines is involved in a network of hydrogen bonds in their own subunits. All of these interactions are then doubled by operation of the twofold axis of symmetry. Thus, the interface is comprised of a polar core flanked by nonpolar ends, a fairly unique constellation of residues than would impose certain restrictions on any relative movement of one helix with respect to the other. The Tyr 262 to

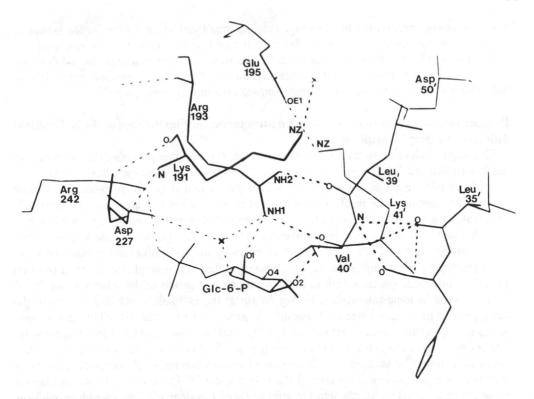

FIGURE 3. Subunit-subunit interactions of the cap region of phosphorylase *b*. The major polar interactions are between the side chain of Arg 193 and the main chain oxygens of residues 39′ and 40′ and between the side chains of Glu 195 and Lys 41′. Some of the interactions of Glc-6-P in this region are also shown. Water molecules are indicated by a cross. Hydrogen bonds are dotted.

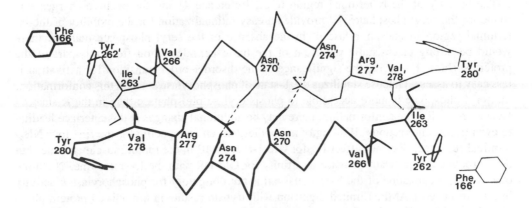

FIGURE 4. Subunit-subunit interactions for the tower regions of phosphorylase *b*. The diagram illustrates the polar core which involves residues Asn 270, Asn 274, and Arg 277 and their symmetry related equivalents and the nonpolar ends of the tower helix (see text).

Pro 281′ and Phe 166′ interaction allows a linkage between the tower and the catalytic site as discussed in Section II.F.

There are twice as many apolar groups as polar groups buried at the subunit interface. When the excluded area is plotted by means of a "footprint" routine,[102] it is found that the areas excluded from solvent, which represent approximately 7% of the total accessible surface area of the monomer, form a band around a central area from which solvent is not excluded (Plate 5*). In phosphorylase *a*, it has been computed that the solvent cavity can hold the equivalent of 150 to 180 water molecules.[86] The mode of association of the subunits in phosphorylase is different from that of other oligomeric proteins (e.g., insulin or hemoglobin) that exhibit more continuous and intimate surface area interactions.

D. Comparison of Phosphorylase *b* and *a* Structures; the Significance of the N-Terminal Tail and the Seryl Phosphate Site

The major difference in the structures of phosphorylase *b* and phosphorylase *a* occurs in the N-terminal tail.[70] In phosphorylase *b*, the N-terminal 18 residues are not located in the electron density, even in the highly refined structure, and it is fair to assume that these residues are mobile. This result was anticipated by the studies of Carty, Tu, and Graves[103] who had shown that a phosphorylated tetradecapeptide corresponding to sequence residues 5 to 18 could confer phosphorylase-*a*-like properties on intact phosphorylase *b*, presumably by binding at a specific site which is not occupied by the N-terminal tail in phosphorylase *b*. In phosphorylase *a*, amino acids 4 to 17 are localized. The seryl phosphate at position 14 makes important contacts both to its own subunit and across to the other subunit.[72,73,86] It is involved in ionic interactions to Arg 69 (from the α-2-helix) and Arg 43′ (from the cap region of the symmetry-related subunit). In addition, Lys 9 and Arg 10 make hydrogen bonds to main chain carbonyl oxygens of Tyr 113′ and Gln 114′, and Gly 116′, respectively. The importance of the extra interactions for the stabilization of the phosphorylase *a* dimers has been reviewed by Madsen.[27] NMR evidence indicates that the seryl phosphate is titratable and therefore presumably accessible to the bulk solvent.[104] The effects of various salts on phosphorylase activity indicate that the interaction of covalently bound phosphate with the protein is sensitive to the ionic environment.[105] The seryl phosphate binding site is distinct from the AMP binding site (Section II.H). The separation is about 15 Å, essentially from the outside of the α-2-helix to the inside. Both effectors make contacts to the α-2-helix and to the cap region, but to different residues.

The basicity of the N-terminal region (e.g., Reference 4) and the positive charges surrounding the seryl phosphate[27,101] provide an easy rationalization for the mobility of the N-terminal region in phosphorylase *b*. In the absence of the seryl phosphate dianion, there would be strong electrostatic repulsion of the basic peptide region from this part of the protein surface. However, the significance of the disorder-order transition for activation is less easy to assess. Glucose stabilizes the T-state of phosphorylase *a* and in this conformation, the seryl phosphate is more accessible to phosphorylase phosphatase than in the R-state.[107] Hence, in the R-state conformation, there may be additional changes at this interface leading to even tighter interactions. Binding of antibodies, which are specific for the first four NH_2-terminal residues, does not affect K_m for Glc-1-P or AMP in the rabbit muscle enzyme, but does produce a significant increase in affinity for Glc-1-P with the liver enzyme.[106] On the other hand, localization of the N-terminal tail is not obligatory for phosphorylase *b* activity in the presence of AMP. Limited digestion with trypsin results in a modified protein phosphorylase *b*′ in which the N-terminal 16 residues have been removed.[23,103] Phosphorylase *b*′ may still be activated by AMP (but not, of course, by phosphorylase kinase), but there is little inhibition by glucose-6-phosphate.[108] Phosphorylase *b*′ can also be activated by addition of a phosphorylated tetradecapeptide (corresponding to sequence residues 5 to 18) to give a phosphorylase *a*-like enzyme.[109] In many of its properties, phosphorylase *b*′ resembles the R-state of the enzyme. It has high affinity for AMP and for inorganic phosphate (approximately .02 and 3 mM, respectively), and these affinities are independent of the

---

* Plate 5 appears following page 120.

concentration of the other.[108] It is as if removal of the basic N-terminus has relieved one of the important constraints that stabilize T-state phosphorylase b.

Overall, the structures of T-state phosphorylase a and b are remarkably similar. However, there are other small but significant changes elsewhere in the structure in addition to the N-terminal tail. A detailed description of the structural changes that accompany phosphorylation was completed after this manuscript was submitted.[246]

E. Pyridoxal Phosphate

Muscle contains approximately 30 mg/100 ml of intercellular fluid of phosphorylase, and since muscle comprises 40% of the body mass, there is more pyridoxal phosphate associated with phosphorylase than all the other vitamin B_6 dependent enzymes put together.[28] Under fairly extreme conditions, the Schiff base that links the cofactor to Lys 680 of the enzyme can be reduced with $NaBH_4$ without significantly affecting activity.[110] Thus, the function of the cofactor is quite different to that found in conventional pyridoxal phosphate-containing enzymes. It is now known to involve the 5′-phosphate group (Section III.E). In phosphorylase, the pyridoxal phosphate absorbs at 333 nm, is excited for fluorescence at 330 nm, and emits at 530 nm. These rather unusual spectral properties and their significance for the Schiff base and its environment have been reviewed.[23,28] They are thought to arise chiefly from the buried hydrophobic environment of the cofactor in the enzyme. Recent ^{19}F NMR and UV(ultraviolet) studies[111] on phosphorylase reconstituted with 6-fluoropyridoxal phosphate have shown that the Schiff base is in the neutral enolamine form with the pyridine nitrogen unprotonated, a result consistent with earlier studies. A detailed description of the cofactor site in phosphorylase b has recently been published.[248]

In the crystal structure of phosphorylase b, the pyridoxal phosphate is buried in the center of the molecule at a site which is some 15Å from the surface of the enzyme. The buried nonpolar nature of the pyridoxal component is in good agreement with deductions made from spectroscopic measurements. The site for the phosphate component is polar. Calculations of solvent contact areas show that the cofactor is approximately 90% buried[112] with the C2′ methyl group providing the largest single atomic contribution to the area change. Lys 680, the Schiff base lysine, is the fourth residue of α-18 (α-18 corresponds to α-E of the nucleotide binding domain in the dehydrogenases). The long lysine side chain and the span of the aromatic ring allow the 5′-phosphate of the cofactor to interact with the amino terminal positive charge of the α-helix (Figure 5). This characteristic phosphate helix dipole interaction[113] is likely to be strong in the buried environment. The plane of the pyridoxal ring lies above the β-sheet of the nucleotide binding domain and is associated with three aromatic residues; Tyr 90 on the B face, Trp 491 approximately perpendicular to the A face, and Tyr 648 approximately parallel to the A face (Figure 5). (The A face and B face are those viewed so that the numbering system reads clockwise and anticlockwise, respectively). In addition, the following residues make Van der Waals interactions Gly 134, Gly 135, Arg 649, Val 650 (packing against the B face), Ala 653, and Thr 676. The Schiff base is almost planar and trans, as is found in the single crystal structure of pyridoxal phosphate oxime,[114] and there is a good hydrogen bond between the 3′-OH group and the Schiff base nitrogen. The N1 atom of the ring interacts via a water molecule to the OH of Tyr 90 and to the main chain carbonyl of Asn 133 which in turn is linked to the NH1 of Arg 569.

The phosphate ester group has its preferred conformation, which is also that observed in the single crystal studies,[114] in which the torsion angles C4-C5-C5′-O5′ and C5-C5′-O5′-P are both trans. The phosphate ester interacts via a water molecule to the main chain carbonyl oxygen of Val 567 and OH of Tyr 648. The phosphate group itself is highly solvated, a feature which has only recently emerged from the high resolution refinement. The phosphate oxygen O33, in addition to two hydrogen bonds to the main chain NH of Thr 676 and Gly 677 (helix dipole), also interacts with two water molecules; O11 interacts via a further water

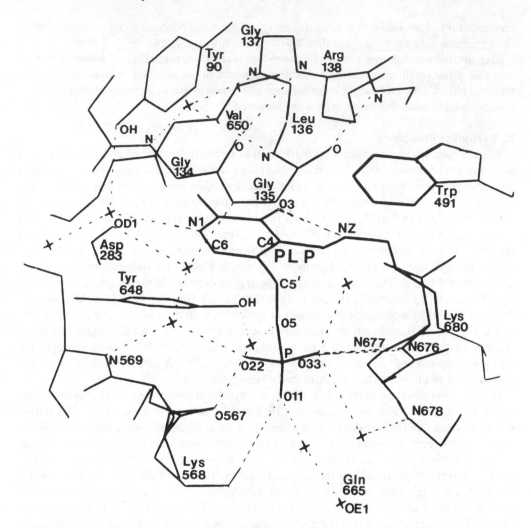

FIGURE 5.   The major contacts to the pyridoxal phosphate cofactor in glycogen phosphorylase *b*. Water molecules are indicated by crosses.

molecule to Gln 665 and indirectly to Lys 568; O22 interacts via a water molecule to the main chain NH of Arg 569. This last oxygen is the only phosphate oxygen accessible to substrate. There is a wide space adjacent to the cofactor 5′-phosphate in which substrate bind and which has allowed some elegant reconstitution studies with pyridoxal phosphate modified with bulky substituents at the 5′ position.[115,116]

The disposition and geometry of the cofactor is different from that observed in aspartate aminotransferase, the only other pyridoxal phosphate-containing enzyme whose structure is known.[117,118] In that enzyme, the cofactor is also associated with an α/β domain, but packs against the opposite side of the β-sheet to that observed in phosphorylase. The 5′-phosphate also makes a helix dipole interaction but, in addition, it is firmly held by basic groups. The torsion angles for the phosphate ester represent a strained cis trans conformation. These differences and those also associated with the environment of the N1 atom and the Schiff base reflect the different roles of the cofactor in the two enzymes. There appears to be no evolutionary relationships.

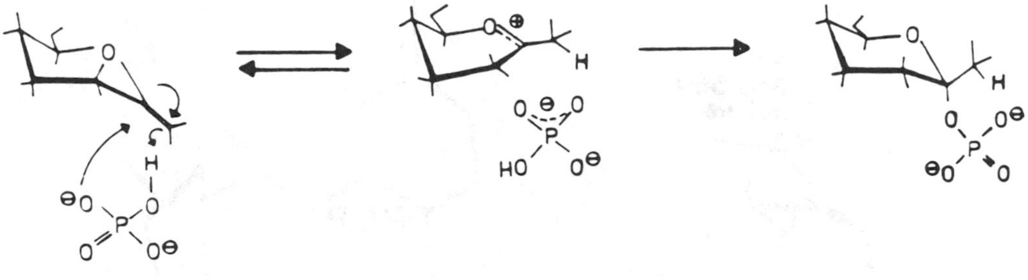

FIGURE 6.   A proposed scheme for attack of phosphate on heptenitol to form heptulose-2-phosphate.

F. The Catalytic Site (C)

The catalytic site is located in the center of the molecule where the structural domains come together. It is close to the pyridoxal phosphate site and the C-terminal end of the β-sheet of the nucleotide binding domain. Access from the bulk solvent can only be achieved through a narrow channel which is some 12 Å long and severely restricted in diameter in one region. In phosphorylase b crystals, the substrate Glc-1-P binds well at this site with small conformational changes.[119] No binding of either oligosaccharide or inorganic phosphate substrates has been detected even at very high concentrations. Formation of products was observed, however, in experiments[120] in which mixtures of substrates (either maltoheptaose and phosphate in the presence of IMP or maltotriose and glucose-1-phosphate in the presence of AMP) were diffused rapidly into crystals and data collected within 1 h using the bright Synchrotron Radiation Source in Daresbury, U.K. Since these experiments were performed under conditions in which there was very little soluble enzyme, it is assumed that the reaction has taken place in the crystal (as had already been demonstrated by Kasvinsky and Madsen[79]). This suggests that oligosaccharide and phosphate can visit the active site but in the crystal their equilibrium binding is weak and/or transient.

For phosphorylase b, the most informative studies have been those with heptenitol and heptulose-2-phosphate.[121] Heptenitol has been introduced as a substrate for phosphorylase by Klein, et al.[122-124] as part of a study with glycosylic substrates in which the potential anomeric carbon atom is linked via an electron rich bond. Phosphorylase catalyzes the utilization of 2,6-anhydro-1-deoxy-D-gluco-hept-1-enitol (heptenitol) in the presence of phosphate to form 1-deoxy-α-D-gluco-heptulose 2-phosphate (heptulose-2-phosphate) (Figure 6). Heptenitol is used exclusively as a substrate for the degradative pathway and the reaction itself does not require oligosaccharide primer.[124] Heptulose-2-phosphate is a dead end product and a potent inhibitor of the enzyme (K_i = 14 μM). It is the strongest competitive inhibitor known for rabbit muscle phosphorylase b. These characteristics suggest that the compound has certain features of a transition state analog. In the crystals of phosphorylase b, heptulose-2-phosphate was formed by incubation of heptenitol and phosphate in the presence of AMP and oligosaccharide (as activators) for 50 h.[121] Later work has shown that significant product is formed in the crystal after 8-h incubation without the presence of oligosaccharide.[120]

The "glucopyranosyl" moiety of heptulose-2-phosphate makes extensive hydrogen bonds to polar residues in the buried active site so that the hydrogen bonding potential of every polar group of atoms is satisfied. These are shown in Figure 7. Both Glu 672 and His 377 make small but obvious movements away from the ligand in order to optimize contact distances that would otherwise be too short. The C7 methyl group in the β-configuration makes few Van der Waals interactions. The torsion angle O5-C1-O1-P, which is approximately + 117° in Glc-1-P, is − 136° in heptulose-2-phosphate, a feature which is a consequence of the CH$_3$ group in the β-configuration and which places the phosphate close to the 2-OH of the sugar.[121,123] In the enzyme complex, the phosphate of heptulose-2-phosphate

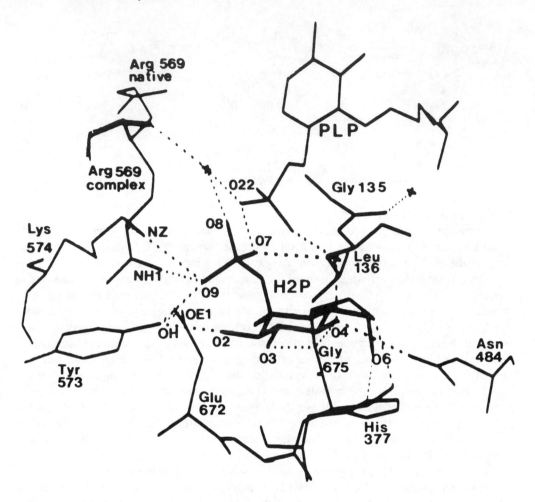

FIGURE 7.   A simplified view of the major contacts between phosphorylase *b* and heptulose-2-phosphate at the catalytic site. Asp 283 and Asn 284 are displaced and have been left out of the diagram. Arg 569 is shown in the native enzyme position. In the complex, The Arg 569 side chain moves in to make contact with the substrate phosphate.

is close to the pyridoxal phosphate (phosphorus to phosphorus distance of 5.1 Å). The interaction of the two phosphate oxygens is both direct, as originally envisaged,[121] but also involves a water molecule (Figure 7). This water, in turn, is hydrogen bonded to main chain NH of Arg 569. The phosphate of heptulose-2-phosphate is also stabilized by interactions to Gly 135 (main chain NH) at the start of α-6, to Lys 574 (NZ), and to Arg 569 (NH1). In the native structure, Asp 283 and Asn 284 also make contacts, but the difference map indicates that these residues move and are replaced by Arg 569. This is discussed later.

Comparison of the positions of heptulose-2-phosphate and Glc-1-P shows that their glucosyl components are similar but there is a marked difference in their phosphate positions. The phosphate of glucose-1-phosphate is 7.4Å from the 5′-phosphate of pyridoxal phosphate and makes rather poor contacts with the enzyme.

Glucose-1,2-cyclic phosphate is a strong inhibitor of phosphorylase[125,126,136] and stabilizes the R-state.[127] It was suggested that the cyclic ester could represent a rigid analog of one of the rotational isomers of glucose-1-phosphate. The glucosyl component of glucose 1,2-cyclic phosphate binds in a similar position to the glucosyl component of heptulose-2-phosphate, but the phosphate position is found to be intermediate between that observed for heptulose-2-phosphate and Glc-1-P. Thus, those compounds which for steric reasons have their phosphate groups closest to the cofactor 5′-phosphate also bind tightest,[124] suggesting that the phosphate site seen in heptulose-2-phosphate is close to the phosphate recognition site for the catalytic pathway. This has been supported by recent experiments on catalysis in the crystal in which by utilizing the control properties of the enzyme and making X-ray diffraction measurements very rapidly (45 min), it was possible to observe a heptenitol-phosphate enzyme-substrate complex.[120,128]

Binding studies with heptenitol, glucose, Glc-1-P, glucosamine-1-P, UDPG, and deoxy-norjirimicin (1-deoxy-5-amino-D-glucose) show the glycosyl recognition site to exhibit small variability with relative affinities easily rationalized in terms of the contacts to the enzyme. For example, deoxynorjirimycin binds weakly as shown by Ariki and Fukui.[129] In Glc-1-P, the ring oxygen is within H bonding distance of His 377 and from main chain NH Leu 136. These interactions would be much less favorable with the ring N atom of deoxy-norjirimycin. Glucosamine-1-P, which is a component of liver glycogen[130] and a substrate for phosphorylase,[131] binds tightly[132] because of the proximity of the 2-amino group to Glu 672.

The details of binding of glucose, a number of glucose analogs, and glucose-1,2-cyclic phosphate have been described for phosphorylase a.[73,133,134] A shift of about 1 Å was noted between the positions of the T-state inhibitor glucose and the R-state inhibitor glucose-1,2-cyclic phosphate with the glucose further into the catalytic site. Recently, a comparison of affinities of a number of deoxy and fluorodeoxy sugars has shown a good correlation between observed and predicted hydrogen bond strengths for the phosphorylase-glucose complex.[249]

The phosphorylase a studies have given an indication of a conformational response in the presence of substrates.[135] The dimer extends in length by 3 to 4 Å and the loop 283 to 285 which partially closes the active site moves. In the T-state, Asp 283 is held in the inactive conformation by hydrogen bonds between the adjacent Asn 284 and the glucose ligand. In addition to this major change, shifts in segments 1 to 75, 102 to 123, 242 to 297, and 386 to 401 were also observed. In phosphorylase b in the heptulose-2-phosphate complex, the whole of the loop from Pro 281 to Gly 288 moves and Arg 569 swings in close to the position previously occupied by Asp 283.

The crucial role of Arg 569 was anticipated by Vandenbunder et al.[136,137] They showed that in the absence of nucleotide, Arg 569 is inactive towards arginine-directed reagents (butanedione and phenylglyoxal). This could either be because access to the Arg was hindered or because it was protonated. (The reagents react with neutral arginines.[277]) On activation of the enzyme, Arg 569 becomes reactive, and this reaction can be protected by Glc-1-P. In addition, Miller and his colleagues[138] have shown that certain arginine compounds inhibit phosphorylase, most likely by competition by an unprotonated guanidino group for the binding site of Arg 569. The role of Arg 569 is discussed further in Section III.F.

G. The Glycogen Storage Site (G)

Electron microscopy of striated muscle thin sections has shown that most of the glycogen occurs in particles, average diameter 400 Å. These are localized in the sarcoplasm, close to the sarcoplasmic reticulum at the level of the I band of the sarcomere and often associated in ordered arrays near the transverse tubule system.[12] The discovery that phosphorylase is tightly bound to glycogen particles but able to be activated and controlled (Section I.A) provided a puzzle that was eventually solved by the demonstration of a second glycogen binding site distinct from the catalytic site.[67,71,139,140] This also explains the observations that preincubation of phosphorylase with glycogen enhances activity[141] and promotes dissociation of inactive tetramers to active dimers.[142]

In a detailed kinetic and crystallographic study, Kasvinsky et al.[139] showed that the dissociation constant for maltoheptaose at the catalytic site is 20-fold greater than at the storage site. The latter is similar to the dissociation constant for glycogen and has been estimated to be 1 mM[139] or 2.6 mM.[143] Prior occupation of oligosaccharide at the storage site was shown to be an obligatory part of the kinetic mechanism.[139] Studies with phosphorylase covalently modified at this site suggest that occupation results in an approximate eightfold increase in activity.[143]

The most informative structural studies have come from binding experiments with maltoheptaose in the crystal both with phosphorylase a[144] and phosphorylase b.[145] Goldsmith et al.[144] have described the complex with phosphorylase a at 2.5 Å resolution and used the results to propose a model for glycogen itself. Further details of the interactions have been given by Fletterick.[73] The protein-carbohydrate interactions for phosphorylase b have been described in a recent review.[250]

The glycogen storage site is situated on the surface of the molecule well-removed from other sites and the subunit-subunit interface. In phosphorylase b, the site is composed of a major site and a minor site. At the major site, the interactions are to residues of the α-12-helix and the spur formed by the small antiparallel sheet (β-15 to β-16). There are five glucosyl residue subsites labeled S1 to S5. The minor site consists of only two glucosyl residues (B1 and B2), and it lies above the nonreducing end of the major site, making contacts to the top of α-12, the turn between α-9 and α-10, and between β8 and β-9. Figure 8 illustrates the possible hydrogen bonds between the enzyme and the five glucosyl residues at the major site as observed in the phosphorylase b-maltoheptaose complex. It is apparent that the sugar residue in subsite S3 makes the most extensive contacts. Indeed, every available functional group on this sugar makes a possible hydrogen bond to the protein. The left-handed helix curls in toward the protein and then curls away (Figure 2). McLaughlin[145] has likened this interaction to the gripping of the oligosaccharide helical coil (like a roll of Sellotape) between a thumb (β-15 to β-16-sheet) and a forefinger (α-12). In subsite S3, the plane of the sugar ring apposes the protein surface with good nonpolar contacts to Val 431 and Tyr 404. Tyr 404 fits into the groove formed by the glycosidic linkage with the lone pair electrons on the oxygen directed away from the aromatic residue.

At the minor site, the dissaccharide makes interactions with the loop between β-8 and β-9 from above, while Asp 360, Trp 361, and Val 354 contribute the floor of the site. Model building studies by McLaughlin[145] suggest that the minor site could be linked to the major site by inserting one residue at the reducing end of the minor site which would link to the nonreducing end sugar of the major site by a (1 to 6) glycosidic linkage. It is interesting that the minor site does not appear to be occupied when allosteric activators are bound at site N.[140] Likewise, ^{31}P NMR studies[127] have shown that addition of oligosaccharides weakens the binding of AMPS, an effect which is reversed by binding of glucose-1,2-cyclic phosphate at the catalytic site.

On binding maltoheptaose to phosphorylase b, there are some pronounced local changes. Glu 433 and Lys 437 move to optimize contacts with subsites S2 and S3, while Gln 408 moves to alleviate a too close contact with subsite S4. In phosphorylase a, a significant movement of Tyr 404 has been noted.[73,144] In phosphorylase b, this residue is already in the correct position to accept the oligosaccharide.

The route through which binding at the oligosaccharide site can be communicated to the allosteric effector site is likely to depend on the general conformational response of the enzyme as a whole, but a possible direct route can also be traced. The movements of Glu 433 and Lys 437 are likely to perturb the small antiparallel β-sheet β-14—β-16—β-15 (see Figure 2). β-14 carries residue Glu 385 which forms an internal interaction with Glu 296 of the α-8-helix. This helix carries two important arginine residues, Arg 309 and Arg 310, that contribute to the phosphate recognition site of AMP (next section). A route to the catalytic site can also be traced. Glu 382, which is in the loop between β-13 and β-14, interacts with the main chain N of Phe 286 and with Arg 770 (Figure 10). Thus, movement

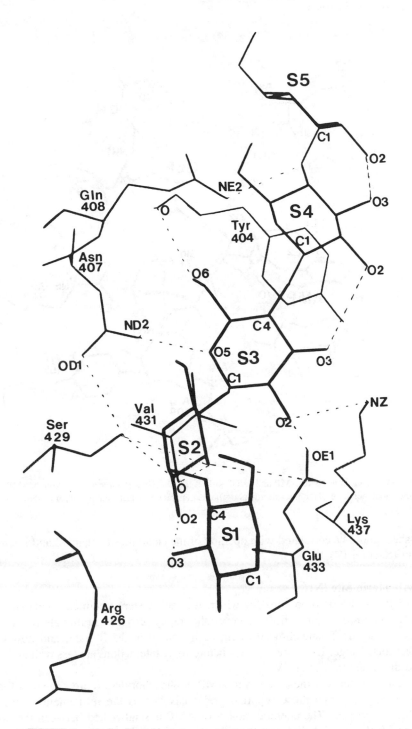

FIGURE 8.   The binding of maltoheptaose at the major glycogen storage site. Only
5 glucosyl residues are observed to be firmly bound. Residues 404, 407, 408 are from
α-12-helix.

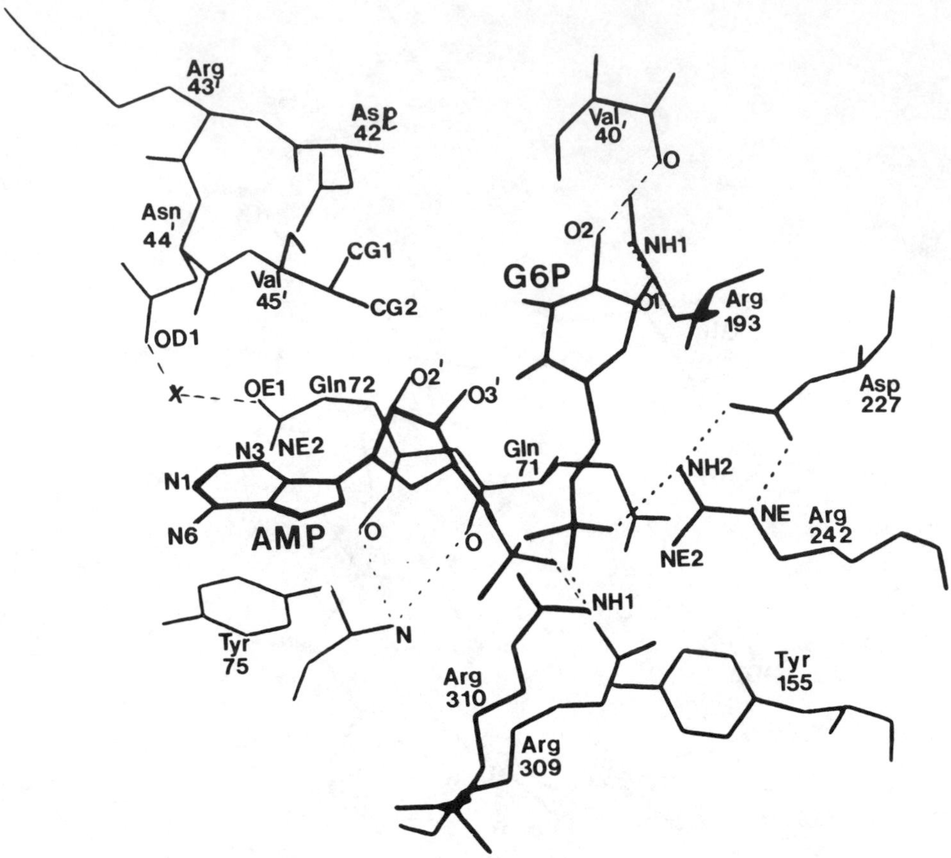

FIGURE 9.   The major contacts to AMP and to Glc-6-P at the allosteric effector site. Although AMP and Glc-6-P are shown together here, their phosphorus-phosphorus separation is just under 2 Å, and they do not bind simultaneously.

of the β-sheet could be correlated with opening of the catalytic site channel and interdomain movement (Section IV).

H. The Allosteric Site (N)

Some of the allosteric properties for which we seek a stereochemical explanation were described in the introduction. The X-ray crystallography provides a rationale for why AMP affinity is low in the T-state, how it might be increased in the R-state, and how Glc-6-P antagonizes this change. The more complex heterotropic interactions are less well understood. These are discussed in Section IV.

The essential features of the contacts to AMP in phosphorylase *b* are shown in Figure 9. These differ slightly from those reported previously,[41,74] as the refinement of the native enzyme has progressed. The adenine moiety of AMP is sandwiched between the aromatic ring of Tyr 75 (from α-2-helix) and the side chain of Val 45′ (from the cap). The closest contact to the tyrosine is about 3.5 Å and to the valine about 4.6 Å. The two side chains contribute about half of the total area changes observed for all protein atoms on binding AMP.[41] Gln 72 is almost within hydrogen bonding distance of the N3 atom of the base. The ribose makes surprisingly few contacts with the native (T-state) enzyme. There are no definite hydrogen bonds, but the O2′ hydroxyl is directed toward the side chain of Asp 42′. This residue and the whole of the loop 41 to 45′ shift on binding AMP so that contacts are likely to be made. Studies in solution with AMP analogs have shown that the O2′ hydroxyl contributes both to binding and allosteric activation.[146,147] Gln 71 and Gln 72 (α-2-helix) are also in positions in which they might contribute hydrogen bonds but the distances are too long. The phosphate of AMP makes a specific interaction with the side chain of Arg 310 (α-8-helix). Two other arginine, Arg 242 (β-11) and Arg 309 (α-8) are in the vicinity. They are over 5 Å away from the phosphate in the present structure. Both Arg 309 and Arg 310 shift toward the phosphate. These movements are likely to be concerted for Arg 309 is in ionic contact with Asp 306 which, in turn, is linked directly to Arg 242 and indirectly (through Tyr 157) to Arg 310. Tyr 155 (β-4) lines the floor of the phosphate site without making specific contacts. The importance of both Tyr 75 and Tyr 155 for the AMP binding site had been anticipated by chemical modification experiments.[148,149] In addition to these, there are a number of other aromatic residues in the vicinity, Trp 67, Tyr 74, Tyr 83, Tyr 157, Phe 196, and Trp 244.

In the phosphorylase a-AMP complex, contacts between the ribose and Asp 42′ and Gln 71 and a substantial movement of Arg 309 toward the phosphate have been noted.[73] A detailed description of both AMP and ATP binding to phosphorylase a[251] was published after this manuscript was submitted. This work and the subsequent comparison of the structures of phosphorylase a and b[247] has shown how small changes in side chain conformations at the subunit interface lead to a high affinity AMP site in phosphorylase a.

IMP is a weak activator and binds with lower affinity than AMP at the allosteric effector site (Section I.B). The crystals of phosphorylase b are grown in the presence of 2 mM IMP. The native electron density shows that IMP is bound with a low occupancy at this concentration. At 100 mM IMP a 4 Å study revealed no large differences in the mode of binding of IMP compared with AMP,[74] but a detailed high resolution study has yet to be completed. An analysis of AMP contacts reveals no features which would distinguish AMP from IMP. In AMP, the N1 atom is some 4.4 Å from the side chain Asn 44′. Asn 44′ is hydrogen bonded through a water molecule to Gln 72, which, in turn, is hydrogen bonded to Glu 76. Movement of the cap region might enable a direct interaction to be made between Asn 44′ and the nucleotide, but it is not an interaction that would discriminate between AMP and IMP. Chemical modification experiments with an AMP analog[150] suggested that Cys 318 might be near the N6 atom of the purine. In the crystal, Cys 318 is over 5 Å away. It is in a poorly ordered region involved in a lattice contact. This observation together with those described above suggest that movements of the cap region 41′ to 45′, the Gln 71 and 72 side chains from α-2, the Arg 309 and 310 side chains from α-8, and possibly the region around Cys 318, are all likely to contribute to the increase in affinity for AMP as a result of homotropic or heterotropic interactions. Thus, it is possible to see why AMP cracks the crystals since several of these interactions are at lattice contacts. However, it is less easy to understand whey IMP does not produce these changes. A possible explanation may lie in the fact that IMP binds with an almost equal affinity to site I and may inhibit its own activation (Section II.I).

The 5′ phosphate of AMP is essential for the allosteric response.[146,147,150a,151] Adenosine does not bind at the allosteric site in the crystal nor does it activate phosphorylase b. Madsen and Withers[152] have demonstrated the requirement for a nucleotide with a dianionic phosphate. Thus, adenosine 5′-phosphorothioate (AMPS), which exists solely as a dianion at pH 6.8, is a slightly better activator than AMP[77,153-155] and in the crystal binds more tightly and produces a greater conformational response of the arginine residues. At high concentrations (approximately 150 mM) phosphate and other anions high in the Hofmeister series are able to activate phosphorylase b in the absence of AMP.[156,157] In the crystal structure, phosphate (or arsenate or iodate) binds at a site some 2 Å away (P to P distance) from the 5′-phosphate of AMP. In this position, P_i still makes contacts to Arg 310, but it is now considerably nearer Arg 242. Thus, if we disregard possible movement of Arg 309, the higher affinity of this adjacent site for phosphate can be readily rationalized. The results demonstrate that, despite some degree of flexibility in torsion angles, AMP has just the wrong dimensions to span the high affinity inorganic phosphate site and the adenine-ribose recognition sites. As a molecular ruler, it fits poorly into the site generated by the T-state structure.

Glucose-6-P is an important physiological inhibitor of phosphorylase *b* (Section I.B). Kinetics studies indicate that Glc-6-P inhibits AMP activation according to the concerted model[158] but, despite their differences in structure, the two ligands are partially competitive[50] and their effects are highly cooperative.[51,151,159,160] In the absence of other ligands the K_D for Glc-6-P is 19 μM, whereas the efficacy of Glc-6-P as an inhibitor of AMP activation is represented by a $K_i \cong 0.3$ to 0.9 mM (Table 1). The crystal studies show that the inhibitor site for Glc-6-P is adjacent to and only partially overlaps the AMP binding site[161] (Figure 9). The phosphate of Glc-6-P occupies a similar position to that observed for inorganic phosphate and hence it makes strong interactions with Arg 310 and Arg 242. The position of the glucosyl moiety is quite different to the ribose-adenine of AMP. The glucose ring stacks against the side chain of Trp 67 (the two rings are nearly perpendicular) and the O2 hydroxyl forms a good hydrogen bond with the carbonyl oxygen of Val 40′. Both O1 and O5 are involved in an interaction with Arg 193. The α-anomer has been shown to bind preferentially by NMR studies.[51]

AMP binding leads to activation; Glc-6-P binding leads to inhibition. The structural results suggest that Glc-6-P, by virtue of its position and contacts, is effective in preventing the changes needed to tailor the site for AMP. It keeps the α-2- and α-8-helices apart and causes a different structural change in the cap region. The inhibitor ATP, on the other hand, causes minimal damage and binds in the crystal with a higher occupancy than AMP[74] indicating that ATP binding, with its α- and β-phosphates spanning the two phosphate recognition sites, is compatible with T-state stabilization. The allosteric site can also accommodate larger molecules such as NADH[41] and UDPG[242] in folded conformations.

I. The Nucleoside Inhibitor Site (I)

The existence of two AMP binding sites per phosphorylase subunit, one an activator site with high affinity and the other an inhibitor site with low affinity, was originally proposed on the basis of calorimetric[162,163] and equilibrium dialysis[151] measurements. The second nucleotide site was confirmed and the location established by X-ray studies with phosphorylase *a*[71,164,165] and phosphorylase *b*.[41,74] Combined crystallographic and kinetic experiments with phosphorylase *a* by Kasvinsky et al.[164] showed that a number of nucleotides, nucleosides, and methylated oxypurines were also able to inhibit the enzyme by binding at this site (K_i(AMP) = 6 mM, K_i(IMP) = 1.5 mM K_i(caffeine) = 0.1 mM) and that glucose and caffeine inhibited the binding of glucose-1-P in a synergistic manner.[76] It was suggested, and supported with experimental results on isolated hepatocytes,[166] that the second site might have a physiological regulatory role in liver phosphorylase as a negative heterotropic effector site. Studies with phosphorylase *b*[53,167] have shown that in this form of the enzyme the inhibitor site displays similar K_D values to the K_i values observed for phosphorylase *a* (e.g., K_D(AMP) = 5.2 mM K_D(IMP) = 1.7 mM) and synergistic inhibition by glucose and caffeine is also observed.[168] A physiological role for this site in the muscle has not yet been demonstrated, but the possible significance for the control of both phosphorylase *a*[166] and phosphorylase *b*[40] in the liver has been discussed.

The inhibitor site is located at the entrance to the catalytic site and involves two loops of chain, one from the N-terminal domain and the other from the C-terminal domain. The essential contacts to AMP in phosphorylase *b* (Figure 10) involve Phe 285 and Tyr 613 which stack on either side of the base. There are no strong hydrogen bonds to the protein. The ribose and phosphate are poorly localized and make little contact to the enzyme. Observations for binding at this site in the crystal have also been obtained for adenosine,

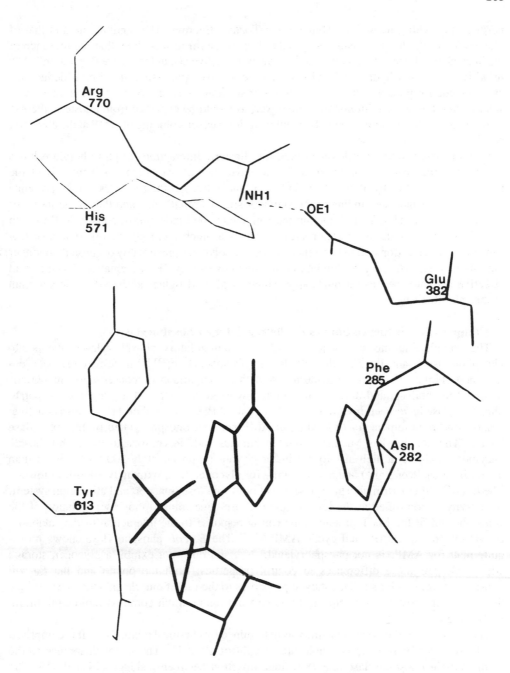

FIGURE 10. The major contacts to AMP at the nucleoside inhibitor site.

IMP, caffeine, and NADH. However, UDPG does not bind here. As Sprang and his colleagues[165] have commented, the specificity is fairly relaxed and a number of aromatic compounds bind. From binding and thermodynamic analysis, it was concluded that the predominant source of stabilization is the attractive dispersion forces between the polarizable systems of the ligand and protein. Soman and Philip[169] have shown a direct correlation between the ability of various aromatic compounds to inhibit and their hydrophobicity.

In addition to the kinetic and crystallographic studies, NMR[77] and cross-linking[170,171] experiments support the notion that purines and their analogs inhibit phosphorylase *a* synergistically with glucose by stabilizing the T-state structure. The binding site and that of glucose share the loop of chain 281 to 285. The structural results show that the presence of the intercalating ligand at the purine site also keeps the active site in the "closed" position[26,73,164] in which there is a hydrogen bond between the glucose and Asn 284. Similar deductions can be made for phosphorylase *b*. Substrates or substrate analogs which stabilize the R-state weaken binding at the inhibitor site. Moreover, as would be expected from allosteric theory, stabilization of the T-state through I-site interactions weakens binding of AMP at the allosteric site.[168]

Further insight into the role of this site for allosteric interactions in phosphorylase *b* has come from the work of Uhing et al.[40] who have shown that organic solvents affect the regulatory properties by altering the binding of nucleotides at the two sites. In the presence of organic solvents, the binding of the activator site is strengthened and that at the inhibitor site is weakened. Thus IMP, which in the native enzyme binds almost equally well to both the allosteric and inhibitor effector sites, becomes a much more effective activator at low substrate concentrations, in the presence of organic solvents, presumably because it no longer inhibits its own activation by binding at the inhibitor site. AMP in the native enzyme is an effective activator because it binds approximately 50-fold tighter at the allosteric site than at the inhibitor site.

J. Comparative Sequence Studies of Nonregulatory Phosphorylases

The complete amino acid sequences of two nonregulatory phosphorylases, the potato phosphorylase[172] and the *E. coli* maltodextrin phosphorylase,[173,174] have recently been determined (Figure 1). These provide a fascinating comparison of conservation in regions involved in catalysis and diversity in regions involved in control. The properties of potato phosphorylase have been reviewed by Fukui et al.[175] Potato phosphorylase is always active; it shows no requirement for or binding of AMP nor is it phosphorylated by phosphorylase kinase. The enzyme has Michaelis-Menten kinetics and, in contrast to the rabbit muscle enzyme, it exhibits a high affinity for linear glycans (approximately 1000-fold higher than the rabbit muscle enzyme) and a low affinity for glycogen (approximately 3000-fold lower). The *E. coli* maltodextrin phosphorylase is part of the maltose transport and utilization system. Its activity is controlled at the level of the maltose gene transcription by regulation of the concentration of the MalT protein in the cell in response to the interaction of the catabolite activation protein (CAP) and cyclic AMP.[176,177] The *E. coli* phosphorylase shows no requirement for AMP, is not phosphorylated,[178,179] and does not exhibit a glycogen storage site.[143] Despite these differences in control properties, both the potato and the *E. coli* phosphorylases exhibit similar catalytic properties to the rabbit muscle enzyme. They utilize the pyridoxal phosphate cofactor in the same way[28] and exhibit common rapid equilibrium Bi-Bi kinetics.[175,180-184]

The sequence of the potato enzyme was determined by chemical methods.[172] It is comprised of 916 residues with a subunit molecular weight of 103,916. The major difference to the rabbit muscle phosphorylase is a 78 residue insertion between residues 414 and 415. The percentage of identical residues is 43% which rises to 51% if the insertion is ignored. The sequence of the *E. coli* phosphorylase was determined from the cloned MalP gene sequence.[173,174] There are 796 amino acids which indicates that the polypeptide chain is longer than originally suggested from an estimate of 80,000 molecular weight. The percentage of residues identical with the rabbit muscle enzyme is 41%.

A comparison of the sequences with the structure of rabbit muscle phosphorylase shows a remarkable conservation of the two β-sheet cores and a rather greater variation in the residues involved in the α-helices which are mostly on the outside. There is an almost 100% conservation of amino acids in the vicinity of the pyridoxal phosphate and the catalytic site which supports the notion that these enzymes share a common mechanism. The N-terminal regions are relatively less conserved (only 33% for residues 1 to 283 in *E. coli*), and it is this region that contributes important contacts to the allosteric effector site in the mammalian phosphorylase. Gln 71, Gln 72, and Tyr 75 are changed, there are considerable variations in the whole of the loop 40 to 50, but arginines 242, 309, and 310 are conserved. These features indicate that the phosphate binding site may be intact although the adenosine-ribose recognition site is not present. There are considerable changes in the N-terminal tail around the phosphorylatable serine 14. In the potato enzyme, the serine is not conserved and the sequence is more acidic than in the rabbit muscle phosphorylase. In *E. coli* the sequence is missing and the chain starts at residue 18. At the glycogen storage site, the important residues Tyr 404 and Glu 433 are not conserved, and in the potato enzyme, the insertion after residue 412 is likely to interfere with this site. At the nucleoside inhibition site, one residue, Tyr 613, is conserved but the other residue, Phe 285, has been substituted by a nonaromatic residue. Thus, these changes provide a ready rationalization for the lack of binding at the control sites in the potato and the *E. coli* phosphorylases.

In the rabbit muscle enzyme, the initial changes from a low affinity to high affinity catalytic site are likely to involve the displacement of residue Asp 283 by Arg 569 and a movement of the glycine loop, residues 130 to 137. These residues are conserved on the nonregulatory enzymes, but it is possible that they are held in the high affinity conformation by the rest of the structure. It is interesting that Asn 284 is replaced by Glu in the potato enzyme, a change which would favor a more exposed conformation of this loop.

Examination of the residues involved in the subunit-subunit contact (Section II.C) shows very little conservation for the cap/α-2 interactions that form part of the allosteric site. The tower-tower interactions show a greater degree of similarity but there are some puzzling changes. Both the Asn 270 to Asn 274′ and the Val 266 to Val 278′ interactions that form such an essential part of the contact in the rabbit muscle structure are not conserved in *E. coli* or potato enzymes. The subunit-subunit interactions have been reported to be stronger in potato phosphorylase than in rabbit muscle phosphorylase which suggests there may be more substantial changes than a simple substitution of one amino acid for another in a conserved structure.

III. CATALYTIC MECHANISM

An understanding of the allosteric control mechanism of phosphorylase *b* requires an understanding of the catalytic mechanism. Here we review recent advances in this area and interpret the results in terms of the X-ray structure.

A. Substrate Specificity

Rabbit muscle phosphorylase can utilize either glycogen or oligosaccharide with similar catalytic rates at saturating substrate concentrations, but the relative affinity for glycogen is much higher than with linear oligosaccharides. From kinetic studies on a series of semi-synthetic branched saccharides, Hu and Gold[185] proposed a model in which two chain termini from the same saccharide molecule could bind to the phosphorylase molecule simultaneously. Following the discovery of the glycogen storage site, a more likely explanation of the apparent high specificity for branched substrates seemed to lie in masking of the true dissociation constant of the saccharide at the active site by the tight binding to the storage site.[139] Immobilization of the enzyme on the glycogen particle effectively produces a high local concentration of glycogen nonreducing ends at the active site. Philip et al.[143] have questioned this interpretation in studies in which cyanide bromide activated oligosaccharides were covalently linked to phosphorylase. The active modified enzyme had K_m values for glycogen and maltoheptaose which still differed by nearly two orders of magnitude. In the case of a covalently bound glycogen complex, however, the enzyme did exhibit activity on its own carrier.[186] It appears that the storage site does play a role, but not an exclusive one, in determining substrate specificity. Neither of the two nonregulatory phosphorylases from potato and *E. coli* exhibit a glycogen storage site, and these enzymes have a preference for linear polysaccharide substrates (see Section II.J for references).

Phosphorylase will digest up to four sugar residues from an $\alpha(1$ to $6)$-branch point and the smaller oligosaccharide that will serve as a good acceptor is maltotetraose.[23] These observations are in accord with the observation from the X-ray crystallographic studies which showed the narrow channel leading to the active site to be approximately 15- to 20-Å long, about the length of a tetrasaccharide. Maltotriose and maltose will also serve as acceptors but with activities only 0.1% of that observed with glycogen.[243]

The specificities for glucose-1-phosphate and glucosamine-1-phosphate and heptenitol are discussed in Section II.F and the use of glucosylic substrates glucal and glucosyl-fluoride in Section III.E.

B. Kinetics

The kinetic mechanism has been investigated by initial velocity experiments, inhibition studies, and isotopic exchange reaction at equilibrium.[180-184] The general conclusion is that the reaction proceeds through a random equilibrium Bi-Bi kinetic mechanism in which the rate limiting step is the interconversion of the ternary enzyme-substrate complex. Examination of the catalytic site in the crystal structure of phosphorylase *b* shows the major route of access to be a narrow channel. It is hard to see how phosphate or glucose-1-phosphate could have access if the channel were filled with glycogen. On the other hand, there is a second channel which is more restricted and takes a longer route via the pyridoxal moiety to the acidic region around residues 123. This second channel might provide access for the smaller substrates.

The notion of a single ternary-enzyme complex structure is, of course, an over simplification. Recent ^{19}F NMR experiments by Chang et al.[187,188] with 6-fluoropyridoxal phosphate reconstituted phosphorylase have shown that the two ternary complexes (i.e., phosphorylase-AMP-maltopentaose-Glc-1-P and phosphorylase-AMP-maltopentaose-phosphate) result in different chemical shifts for the ^{19}F nucleus. They propose that two changes occur during turnover: (1) a conformational interconversion of protein structure, and (2) chemical steps of bond making and bond breaking. The activation energy of catalysis (estimated to be about 15 to 18 kcal mol^{-1},)[24,189] may well be influenced by the energy barrier of the conformational change.

C. Glucosyl Intermediates

The reaction proceeds with breaking of α-Cl-01 glycosidic bond with retention of configuration.[23] A double displacement mechanism with the formation of either a β-glucosyl enzyme covalent intermediate or a carbonium-oxonium ion stabilized by electrostatic groups on the protein has therefore been invoked to account for retention of configuration. In the absence of a second substrate, phosphorylase will not catalyze either molecular exchange between Glc-1-P and phosphate[190] or positional isotope exchange between the peripheral and ester oxygens of Glc-1-P.[191] Thus, no observable bond breakage occurs in the absence of the second substrate. However, in the potato enzyme, where it was possible to use cyclodextrin as a pseudo second substrate, Kokesh and Kakuda[192] showed that positional isotope exchange did occur with rates similar to those observed for catalysis. This experiment provides the major evidence for the formation of a glucosyl intermediate.

Studies by Tu et al.[193] on the secondary isotope effect for phosphorylase gave a ratio of $k_h/k_d = 1.1$ consistent with a carbonium ion intermediate but in the work of Firsov et al.[194] a ratio of 1.0 with Glc-1-P, C-1[H] and C-1[^{2}H] was obtained which is more consistent with a covalent intermediate. Klein et al.[225] have provided preliminary evidence for a covalent glucosyl intermediate in the reaction between potato phosphorylase and glucal and have suggested a carboxyl group might be involved. Continued efforts to isolate such a complex have indicated that the putative intermediate is unstable in the presence of phosphate.[195] Chemical modification studies with carbodiimide reagents have indicated that modification of a single carboxyl group results in inactivation,[196,197] but the residue number has not been identified.

If the reaction does proceed through a carbonium ion intermediate then it might be anticipated that those compounds which resemble this intermediate by a half chair conformation of the glucopyranose ring might be potent inhibitors. D-gluconolactone is found to be a reasonable inhibitor (K_i = 1 mM) of glycogen phosphorylase.[193,198] The binding is tighter than for the substrate Glc-1-P, but is not as tight as usually envisaged for a transition-state analog.

Heptulose-2-phosphate (K_i = 14 μM) and glucose-1,2-phosphate (K_i = 0.5 mM) are potential transition-state analogs for the phosphate moiety as described in Section II.F.

D. pH Rate Studies

The pH rate profile studies by Kasvinsky and Meyer[199] have indicated two groups involved in catalysis with pK values for the ES complex of pK_a = 6.12 and pK_b = 7.0 at 30°. At 15°C, pK_a is very much less than 6.1, and it was suggested that this group could be the cofactor phosphate. The group with a pK near 7 had a heat of ionization of 8 kcal/mol and suggested that a histidine might be involved. Two histidines, His 377 and His 571, are present at the catalytic site although neither is optimally placed for catalysis. Measurements of Km as a function of pH gave a pK value of 6.56 which could be associated with either the phosphate substrate or protein. Withers et al.[200] investigated the pH profile of the pyridoxal reconstituted enzyme in the presence of phosphite or fluorophosphate and showed that the alkali limb of the activity/pH profiles corresponded to that of the native enzyme (pK for the enzyme substrate complex ∼7.2) but that the pK_m/pH curves were substantially different. Tagaya and Fukui,[201] in studies with pyridoxal (5′)disphospho(1)-α-D-glucose, interpreted their results in terms of a histidine residue associated with the acid limb and two or more basic groups associated with the alkali limb. Thus, interpretation of pH rate studies is confusing, although it is clear that there are at least three ionizable groups involved.

E. The Role of the Pyridoxal Phosphate

The precise role of the pyridoxal phosphate has dominated the design of experiments on the catalytic mechanism; almost to the exclusion of possible role of other groups on the enzyme. A variety of experiments have shown that the 5′-phosphate of the cofactor plays an obligatory role in catalysis. The evidence has been well reviewed.[23,28,29] One important experiment in the development of these ideas was the work of Parrish et al.[202] in 1977, who showed that pyridoxal reconstituted phosphorylase was inactive but some activity (19% maximum) could be regained on addition of a noncovalently bound dianion such as phosphate (or phosphite). Pyrophosphate was found to be a potent competitive inhibitor of phosphate activation and glucose-1-phosphate binding, and provided the first evidence, soon confirmed by X-ray studies on both phosphorylase a and phosphorylase b, of the likely proximity of the cofactor and substrate phosphates.

High resolution ^{31}P NMR studies in Professor Helmreich's laboratory[28] established that in inactive glycogen phosphorylase b, the state of ionization of the 5′-phosphate is a mono-anion, but in all activated states (either with AMPS and arsenate or by phosphorylation with ATP-γ-S, Mg^{2+}, and phosphorylase kinase or in the nonregulatory phosphorylases) and 5′-phosphate is a dianion.[154,203-205] In potato and rabbit muscle phosphorylase, the dianionic form is pH independent over the range 6 to 7.8, the range over which phosphorylase is active, but in *E. coli* phosphorylase, the 5′-phosphate group does titrate (around pH 6) indicating that in this form the cofactor phosphate is less shielded from the solvent. Cofactor analogs in which the 5′-phosphate cannot take on a dianionic state such as pyridoxal-phosphate monomethyl ester[206] or pyridoxal-fluorophosphate[195] are inactive, providing support for the notion that the dianionic form is important for activity. Inhibitors which from other evidence are known to stabilize the T-state (inactive) form of phosphorylase such as glucose and caffeine[77] or glucosyl fluoride[127] also produce a shift in the ^{31}P resonance to that characteristic of the monoanion. An anomaly has been observed with pyridoxal-5′-methylene phosphate reconstituted phosphorylase *b*. This enzyme has 25% activity of native phosphorylase and binds AMP ten times more tightly, suggesting that it is in the R conformation.[207] It is the only cofactor analog modified at the 5′ position that is able to confer activity on apophosphorylase. The pH dependent chemical shift of this analog is in the opposite direction to that observed for pyridoxal phosphate.[208] Observations on the ^{31}P resonances in the reconsituted enzyme have led to the conclusion that in inactive phosphorylase *b* the 5′-phosphonate is a dianion, while in activated phosphorylases it is a monoanion. This puzzling reversal of the states of ionization from that observed with native phosphorylase is not understood, but it may arise from different environment or geometry around the CH_2 of the phosphonate.

The striking observation that the 5′-phosphate dianion is a characteristic of active phosphorylases led to the first stereochemical proposals for the mechanism of action based on X-ray crystallographic studies on the binding of Glc-1-P with phosphorylase *b*.[119] The mechanism invoked the dianionic 5′-phosphate as an electrostatic stabilizing group for the carbonium ion intermediate. It was based on the assumption that the binding of Glc-1-P in the crystal was nonproductive. The proposals for a major reorientation of the sugar turned out to be wrong. All substrates and substrate analogs studied since have their glucosyl moiety in essentially the same orientation, including those where turnover is observed in the crystal. Further, Takagi et al.[209] showed that pyridoxal 5′-phospho-β-D-glucose, a possible intermediate of the reaction if the electrostatic interaction should collapse to a covalent intermediate, did not transfer glucose to glycogen.

Subsequent NMR studies have shown that there are further changes in the state of ionization or geometry of the dianion on formation of the ternary enzyme complex. In the potato enzyme, Klein and Helmreich[28,205] observed a chemical shift to higher field on binding glucose or maltoheptaose, a result which was interpreted as arising from a partial protonation of the 5′-phosphate group. With maltoheptaose and arsenate, there was a fast equilibrium between the protonated and deprotonated forms. This led to proposals for a role of the 5′-phosphate as a proton donor in the catalytic mechanism, a mechanism that had already been anticipated by Helmreich in 1968,[210] and considered by Jenkins et al.[99] in 1981.

In this mechanism[121-124] (Figure 11A), the 5′-phosphate becomes protonated on formation of the ternary complex, and interacts directly via a hydrogen bond with the substrate phosphate to promote general acid attack on the glycosidic bond. The resulting carbonium ion intermediate is then stabilized by the phosphate dianion which can complete the reaction by nucleophilic attack. The mechanism has the advantage that it explains why the enzyme transfers the glycosyl carbonium ion to phosphate rather than water, since phosphate is optimally situated for attack and is an obligatory part of the reaction.

The proposals have been supported by experiments with glucosylic substrates glucal, glucosyl fluoride, and heptenitol. D-glucal, which contains a reactive double bond between the C1 and C2 atoms, can replace glucose-1-phosphate as the glucosyl donor in phosphorylase catalyzed glucosyl transfer to oligosaccharide acceptor to yield 2-deoxy-α-D-glucosyl(glucose)$_n$.[195] On the basis of proton NMR experiments in D_2O, it was shown that the proton (deuterium) had been transferred to the equatorial position of the 2-deoxyglucose in a stereospecific manner. The fact that, both with glucal and with glucosyl fluoride,[211] phosphate is required for protonation and catalysis but is not incorporated into the final

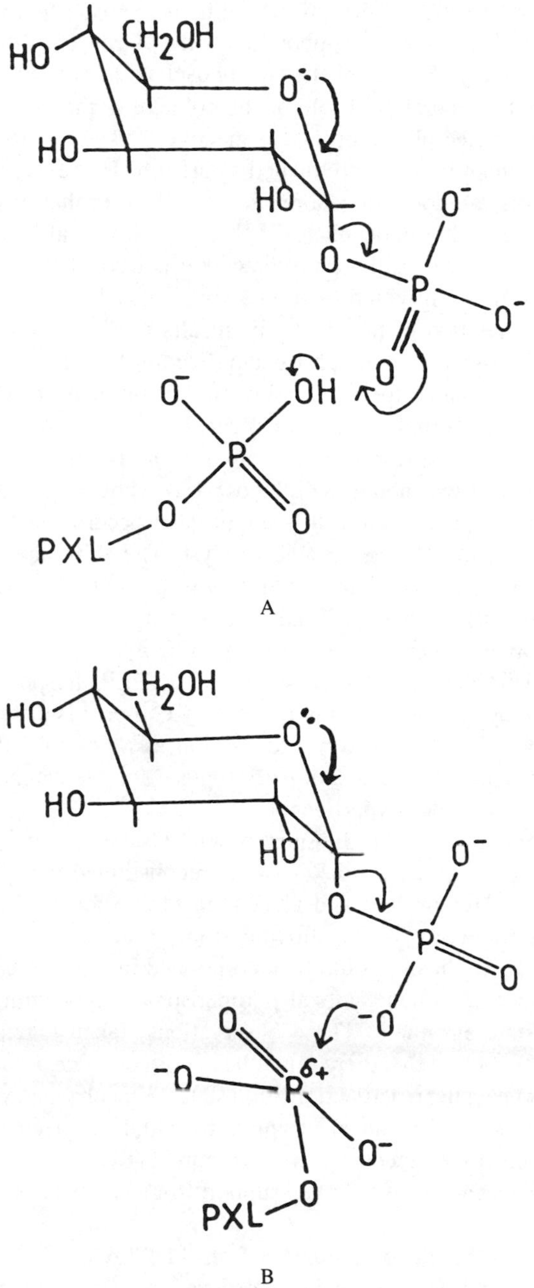

FIGURE 11. Possible mechanisms for phosphorylase catalysis. (A) The pyridoxal phosphate 5′ phosphate group acts as a general acid-base. (B) The phosphorus atom of the pyridoxal phosphate acts as an electrophile.

product, argues for a role of the phosphate in the protonation step of these already partially activated substrates. In the work with heptenitol[122-124] (Section II.F), the ^{31}P NMR results indicated that the dianionic 5′-phosphate of the natural cofactor becomes partially protonated on binding heptulose-2-phosphate in the potato enzyme. With the monoprotonated 5′-phosphonate group of partially active 5′-deoxy-pyridoxal-5′-methylene phosphonate phosphorylase *b*, there was a sharing of the proton with the phosphate moiety of heptulose-2-phosphate. These conclusions were supported by the X-ray crystallographic studies that showed the close proximity (5.1 Å) of the two phosphate groups.[120,121]

Also in support of the general acid role of the cofactor is the observation that pyridoxal fluorophosphate reconstituted phosphorylase is inactive.[195] The modified cofactor phosphate has a pK $\sim$ 4.8 and cannot become protonated at pH 6.8. However, it is surprising to find the pyridoxal phosphorylase does show some activity (13% of that observed with the native enzyme) in the presence of fluorophosphate.[123,200,202,212] Klein et al.[123] have argued that some of the steps in catalysis in which the phosphates are not covalently linked to the cofactor may be different from those in which there is a covalent link.

An alternative interpretation of the ^{31}P NMR results has been given by Withers et al.[155] These authors used glucose-1,2-cyclic phosphate (Section II.F) as a substrate analog whose phosphorus resonance is some 7 to 10 ppm downfield from that of pyridoxal phosphate. In the presence of the ternary complex with glucose-1,2-cyclic phosphate, maltopentaose, and AMPS, the cofactor phosphate resonates at 0 ppm in the position of the monoanion but is considerably broadened. It was noted that glucose-1,2-cyclic phosphate did not decompose under these conditions, a phenomenon that might have occurred if the cofactor phosphate had become protonated and acted as an acid catalyst. The shift was, therefore, interpreted as representing a tighter binding of the dianion with possible distortion of the tetrahedral geometry at the phosphorus. Klein et al.[123] have pointed out that glucose-1,2-cyclic phosphate cannot accept a proton and, therefore, is unlikely to be hydrolyzed under these conditions. Moreover Chang et al.[188] in their ^{19}F NMR studies on 6-FPLP phosphorylase have concluded that the binding of the cyclic compound in the ternary complex is different from that of Glc-1-P. However, the notion of the constrained dianion as a possible alternative explanation for the NMR observations had led to a novel proposal for the catalytic mechanism which has received support from other experiments.

In 1981, a combination of the Edmonton and Osaka groups[213] showed that pyridoxal(5′)diphospho(1)-α-D-glucose (PLPP-α-Glc) reconstituted phosphorylase yielded pyridoxal-pyrophosphate phosphorylase and glucose in the presence of maltopentaose. Later, Tagaya and his colleagues[201,214] showed that radioactive glucose from pyridoxal(5′)diphospho(1)α-D-glucose could be incorporated into the nonreducing group of glycogen with a $T_{1/2}$ about 13 min. Pyridoxal pyrophosphate reconstituted phosphorylase has properties of the R-state enzyme.[85] These observations, taken together with those of the constrained dianion, suggested that there might be a more intimate association of the cofactor and phosphate substrates in the reactive transition state. A mechanism was proposed in which the cofactor phosphorus acted as an electrophile to withdraw electrons from the substrate phosphate and destabilize the glycosidic bond (Figure 11B).

The electrophilic mechanism has received support from a number of different experiments:

1.   Kinetic studies on the glucosyl transfer from PLPP-α-Glc[201,214] showed the reaction to be similar to the normal reaction in almost every respect. Nevertheless, it is surprising to find that the linkage of the two adjacent phosphates to the pyridoxal, which is an important part of the arguments used to show the compatibility of the PLPP-α-Glc reaction with the electrophilic mechanism is not essential. Pyridoxal reconstituted phosphorylase can react with α-D-glucose-1-diphosphate to transfer a glucosyl residue to oligosaccharide and release pyrophosphate with a rate which is faster ($T_{1/2}$ = 4 min) than that observed with PLPP-α-Glc.[123]

2.   The close proximity of the cofactor and substrate phosphate sites has been supported by the observation[134] that methylenediphosphonate (P-P separation = 3.0 Å) is a more effective inhibitor of the pyridoxal enzyme than ethylenediphosphonate (P-P = 5 Å) or propylenediphosphonate (P-P = 6Å), but not as good as pyrophosphate (P-P =

2.9 Å). However, since the phosphate analogs are not covalently bound, one of the phosphate sites cannot be identical to that of the cofactor 5'-phosphate but is likely to be displaced, perhaps as much as 2 Å.

3. If the pentavalent-like distorted phosphorus proposed in the electrophilic mechanism is a structure on the catalytic pathway, then it is expected that compounds which resemble this structure might be potent inhibitors of the pyridoxal reconstituted enzyme. Chang et al.[212] showed that molybdate, which is capable of binding as chelates in a trigonal bipyramidal conformation, competes with phosphite for activation of the enzyme, although its inhibition constant ($\sim$0.1 mM) is not especially tight for a transition-state analog.

4. Oligoanions such as vanadate, molybdate, and tungstate are potent inhibitors of the pyridoxal reconstituted reaction, competing both with substrate Glc-1-P and phosphite with K_i values of the order of a few μM.[215] Studies of ^{51}V NMR suggested that arginine residues were important for this tight binding. The results of the two papers[212,215] were interpreted as providing general support for the electrophilic mechanism. In addition, however, the authors suggest that the 5'-phosphate of the cofactor may also have an important role in holding other groups in the correct orientation for catalysis.

Recently, the role for the rest of the cofactor has been probed with apophosphorylase reconstituted with 6-fluoropyridoxal phosphate.[111] Experiments with ^{19}F NMR indicate conformational changes during activation and catalysis, and the results have led to the suggestion that the ring nitrogen (N1) may interact with important groups on the protein. The role of the ring nitrogen has always been difficult to ascertain because N1-methyl PLP does not bind to the apoenzyme,[216] and pyridoxal phosphate-oxide binds but is gradually converted to pyridoxal phosphate,[217] leading to uncertainty in activity measurements. Observations on the crystal structure show that position six of the pyridine ring is in Van der Waals contact with Gly 134 and two water molecules. A fluorine atom can easily be accommodated here,[248] and this would be a position sensitive to substrate binding and conformational change. The N1 atom of the pyridoxal is linked via a water to Tyr 90 (OH), Asn 133 (CO) (Figure 5). Arg 569 (NH1) is also linked to Asn 133 (CO) and is a key residue in the allosteric response (Section III.F). Finally, we note that a flash excited state of the enol form of the Schiff base appears sensitive to the state of ionization of the 5'-phosphate and its environment in the protein.[218]

F. Summary of Proposals for the Catalytic Mechanism

In mechanism A of Figure 11, the 5'-phosphate group serves as an acid. It acquires a proton on formation of the ternary-enzyme-substrate complex and then promotes a proton transfer to the substrate phosphate. In the case of the degradative pathway, the substrate phosphate protonates the glycosidic bond that links the nonreducing terminal sugar to the oligosaccharide or glycogen chain resulting in the formation of a carbonium ion intermediate. The carbonium ion could be stabilized by the nearby presence of the phosphate dianion. The reaction is completed by nucleophilic attack of the phosphate dianion on the C1 atom to yield α-D-glucose-1-phosphate. In the course of the reaction, the phosphate must occupy two positions. First, it is in the noncovalent attacking position and second, in the covalent glucose-1-phosphate. The general flexibility of the phosphate recognition site for substrates and inhibitors as observed in the crystal structure (Section II.F) is in support of this notion. The origin of the proton is not specified. It could come from other groups in the vicinity or from water. The frontside attack by phosphate is in accordance with retention of configuration and with the results for positional isotope exchange. The proton transfer mechanism between the two phosphates has some analogies with the proton switch conduction mechanism proposed to account for the high electrical conductivity of phosphoric acid.[219]

In mechanism of B Figure 11, it is proposed that the formation of the ternary enzyme complex leads to a tight coordination of the coenzyme phosphate by basic amino acid side chains and distortion of the geometry at the phosphorus atom toward a trigonal bipyramid with the empty apical position oriented toward the phosphate of the substrate. Such a distorted 5' phosphate would be electrophilic and could effect a partial withdrawal of electron density from the substrate phosphate in the attacking position leading to a pseudo-pyrophosphate intermediate. As Madsen and Withers have pointed out,[29,220] there is an attractive similarity between the active site complex proposed (and that actually present in the reaction with PLPP Glc-phosphorylase) and uridine diphosphoglucose, the natural substrate of glycogen synthase, thus suggesting a common structural requirement in the phosphorylysis and biosynthesis of glycogen.

An attempt to assess the likelihood of these mechanisms in the light of the X-ray structural evidence is met with several difficulties. First, the conformational changes that accompany the substrate binding have not been analyzed in detail. The description of the contacts to the phosphate moiety of the substrate in particular may be misleading in the static structure as this is where the most significant changes take place. Second, no binding of the oligosaccharide substrate has been observed at the catalytic site in the crystal structures. Model building studies show that both the loops 133 to 137 and 283 to 284 need to move in order to allow access. Definition of the oligosaccharide component is important in order to test current ideas on the route of attack of the phosphate in the degradative pathway or the geometry of the glycan as an acceptor in the synthetic pathway. Third, the present crystal structures represent a structure close to the T-state of the enzyme. Further, more radical structural changes may take place on formation of the R-state enzyme, especially with those groups that govern the affinity of the enzyme for the substrate. Nevertheless, the X-ray results do provide a framework in which to test proposals.

The environment about the C1 and O5 ring oxygen substrate atoms in phosphorylase provides no electrostatic group in the vicinity capable of stabilizing a carbonium ion or glucosyl intermediate. Indeed, the ring oxygen participates as a hydrogen bond acceptor. This situation is quite different from that in lysozyme where Asp 52 plays an important role. Thus, the notion that the phosphate ion itself acts as an electrostatic stabilizing group and nucleophile, is an attractive feature of both mechanisms.

Direct observation of the pseudopyrophosphate complex, which is an obligatory intermediate of the "electrophilic" mechanism is unlikely in X-ray crystal structure analysis. The complex is likely to be unstable and very high resolution studies would be required in order to detect significant distortion of the tetrahedral phosphate geometry. On the other hand, the results with heptulose-2-phosphate (Section II.F) show a direct cofactor substrate phosphate-phosphate interaction that is consistent with the "general acid" hypothesis. The constellation of groups suggests that the proton may come from the phosphate or from other amino acid side chains after a conformational change or from a water. The structural results for native phosphorylase *b* have shown the importance of solvation of the cofactor 5'-phosphate. Presumably, some of this water would need to be displaced in the tightly coordinated constrained dianion envisaged in the electrophilic mechanism. Thus, the crystallographic results do not exclude the electrophilic mechanism, but they do provide support for the general acid mechanism, but this may be because the steric relationships proposed for this mechanism are more amenable to direct observation by crystallography.

The concerted movements of Arg 569 and Asp 283 appear to be the key to the activation response. In the phosphorylase *b*-heptulose-2-phosphate complex, the whole of the loop Pro 281 to Gly 288 is displaced, and Asp 283 is removed from the catalytic site. The side chain of Arg 569 swings in and is in a position to make a direct contact to the substrate phosphate. In the native structure, Arg 569 is at the center of a complex network of hydrogen bonds (Figure 12). The concerted movements would allow

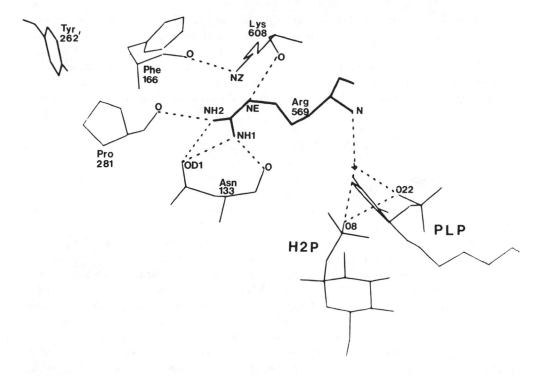

FIGURE 12. The hydrogen bond network for Arg 569 in the native structure of glycogen phosphorylase *b*. When heptulose-2-phosphate is bound, the arginine side chain changes conformation to contact with the phosphate of heptulose-2-phosphate (see Figure 7).

1. A possible explanation for the change in state of ionization of the cofactor phosphate on activation. The presence of Asp 283 in the active site would tend to promote the monoanionic state while replacement by Arg 569 could promote the dianionic state.

2. The formation of the substrate phosphate recognition site. Asp 283 partly occupies the substrate site and would tend to repel phosphate. Arg 569 could provide a group that would not only stabilize the substrate, but also the transition state of the catalyzed reaction.

3. The formation of the oligosaccharide substrate site. In the crystals of phosphorylase *b*, no oligosaccharide binding has been observed at the catalytic site. This is because there is a major restriction in the active site channel between the regions 133 to 136 and 283 to 284. Arg 569 is hydrogen bonded to Asn 133. Thus, movement of this residue could also trigger a movement of the "glycine loop" (133 to 136) and open up the oligosaccharide site.

4. Information to be conveyed to the subunit-subunit interface that could be important for homotropic and heterotropic interaction: Arg 569 is linked both via Pro 281 and via Lys 608 to Phe 166 to Tyr 262′ from the top of the tower of the symmetry related subunit (Figures 12 and 4B). This interface appears to be important for allostery as described in Section IV. Pro 281 is the pivot point for the movement of the loop 281 to 288. Movement of this loop is critical for site I, and allows a possible communication to site N (Section IV) and to the glycogen storage site G (Section II.G).

The structural results are likely to become more definitive in the not too distant future as crystallographic refinement of the ternary enzyme complexes progresses. In particular, it will be important to establish the conformation of Arg 569 in the ternary enzyme complex. In addition, other groups that might be important in catalysis and stabilization of the transition state include His 377, His 571, Tyr 573, Lys 574, and Glu 672. The gene cloning experiments both with mammalian and bacterial phosphorylases (Sections II.B and II.J) offer exciting possibilities for site-directed mutagenesis experiments designed to probe the roles of specific groups.[252] These, together with the chemical and recent crystallographic experiments using novel inhibitors[253] and time-resolved studies on the 1-s time scale,[243] indicate that although the mechanism is not understood at present, more definitive data should be available before too long.

IV. ALLOSTERIC MECHANISM

A simple two-state[8,159] model of allosteric response in phosphorylase is an over-simplification, but has been a good working hypothesis in a number of experiments. A three-state model of activation was proposed to describe the activity of the enzyme with low concentrations of AMP.[221] To account for the different ways of activation by IMP or AMP, a four-state model was suggested,[47,151] and four-state models were proposed for the description of the heterotropic interactions between AMP and Glc-1-P[222] and AMP and Glc-6-P.[160] Madsen and co-workers[24] proposed a six-state model to interpret their fluorescence measurements. Evidence is gathering that the "T-" to "R-" state transition, i.e., gaining access to the active site and/or the rearrangement of the catalytic groups, can be achieved in more than one way and in more than one step in phosphorylase, just as inactive T conformations may be produced through several routes and states. This was also demonstrated by cross-linking experiments[170-171,223-224] which showed a multiplicity of conformational states as reflected in the change of reactivities and relative distances of lysyl ϵ-NH$_2$ groups on the surface of the protein. This notion is further supported by the fact that positive homotropic interactions exist between the respective N sites,[157,226,228-229] between the catalytic sites,[226,230] and between the I sites[164] of the dimer. Moreover, there are a number of heterotropic interactions between binding sites: positive heterotropic interactions exist between the C sites and the N sites,[231] between the C sites and the G sites,[139] and also between the G sites and the N sites.[75,232] Negative heterotropic interactions exist between the N sites and the I sites[164] and the C sites and the I sites.[164]

The picture is further complicated by the fact that rabbit muscle phosphorylase undergoes a reversible dimer to tetramer transition upon activation *in vitro* (reviewed in Reference 20). *In vivo*, however, only a small proportion of phosphorylase is in the tetrameric form even if fully activated, because binding to glycogen dissociates tetramers.[141,170,223] Occupation of the glycogen storage site induces a conformational change in the dimer enhancing activity.[139] *In vivo*, the enzyme is almost always bound to glycogen; therefore, the effect of glycogen must be superimposed on the effect of all other effectors and substrates in the living cell. These combined effects of ligands must have a dominant part in modulating the activity of the enzyme.

The amount of biochemical data on allosteric response in phosphorylase is vast. We shall try to amalgamate biochemical observations with the available X-ray crystallographic results. In this respect, we are somewhat limited because the crystallographic observations relate to only one conformational state — a conformation close to the T-state.

The basic allosteric control of phosphorylase is exercised on two different binding sites: the N site and the I site. Ligands which bind to the I site inhibit the enzyme and dissociate tetramers, whereas ligands to the N site either inhibit (e.g., Glc-6-P) or activate (e.g., AMP, IMP, P$_i$, or SO$_4^{2-}$) (Sections II.H and II.I). Binding of ligands to the N site can lead to tetramerization (with AMP), dimerization (Glc-6-P in the case of phosphorylase *a*), or to no effect (IMP). Communication between these sites and the catalytic site is likely to involve a global response of the enzyme that has still to be characterized. Central to this response, however, is the "rein" leading from the N site (helix α-8, Arg 309, Arg 310) down to the

I site (Phe 285) and then to the active site (Asn 284, Asp 283) before turning back toward the subunit-subunit interface (tower, helix α-7) (Figures 2 and 3). The loop 281 to 288, like a curtain, plays an important role in allowing access to the catalytic site (Section II.F). Pulling it down can be achieved by occupying site I where Phe 285 makes the major contact to aromatic ligands at this site (Section II.I). Pushing it up can be achieved by sugar phosphate binding at the catalytic site which leads to the displacement of Asp 283 and its likely replacement by Arg 569. The central role of Arg 569 in triggering changes here and elsewhere has been discussed in Section III.F. Pulling up the curtain could also be achieved by the binding of AMP to the N site where the phosphate moiety of the nucleotide is in close contact with Arg 310 and Arg 309 at the end of helix α-8 (one end of the rein). The precise position of the phosphate is determined by the ligand bound. With AMP, the adenosine component is localized by contacts to Tyr 75 and residues 42' to 45' from the symmetry related cap region of the other subunit, and shifts in positions of Arg 309 and Arg 310 and the cap are required to tailor the site to fit AMP. For phosphate and Glc-6-P, the existing site, some 2 Å from the observed for AMP in the T-state already has a high affinity for phosphate. It involves Arg 242 and Arg 310, although some changes in Arg 309 and Arg 310 are observed. The binding of the glucose component of Glc-6-P results in different relative movements of the cap, helix α-2, and helix α-8 from those observed with AMP (Section II.H).

The position of the allosteric effector, AMP, at the subunit-subunit interface and the need for relative movements of residues at this interface in order to generate a high affinity site provide an explanation for the homotropic effects of AMP (n = 1.4). These effects could either be transmitted by a concerted response in which the twofold symmetry of the dimer is conserved or they could be transmitted by a sequential response in which movement of the cap (42 to 45) promotes movements by the N-terminus of its own α-2-helix. In this connection, we note that the α-2-helix is partially unwound at residues 62 to 65 in the T-state structure of phosphorylase b.

Homotropic effects between sites C and heterotropic effects are likely to involve the other end of the rein in the region of Tyr 262 at the top of the tower. Tyr 262 is in Van der Waals contact with Phe 166' and this residue via Lys 608' is in contact with site C and site I of the other subunit. Cleavage by subtilisin between Gln 264 and Ala 265 inactivates the enzyme[233] and destroys the heterotropic interaction between AMP and glucose sites in phosphorylase a.[234] Binding of ligands to the N, I, and C sites moderates the rate of inactivation by subtilisin.[235]

The active site is located between the N-terminal and the C-terminal domains (residues 1 to 484 and 485 to 842, respectively). Access to this site is restricted at two points in tetragonal crystals of the enzyme: loops 281 to 286 (part of the rein) and 131 to 137 (the glycine loop) block the channel. The area covered between the N- and C-terminal domains is about 3000 Å² on each side. Both of the two loops contribute to the contacts between the domains. Opening and closing the cleft between the domains could provide an obvious explanation for some features of the allosteric transitions. There are experimental data which fall in line with this hypothesis:

1. All ligands that bind to the I site inhibit the enzyme. These ligands are aromatic and form a cross-link between the two domains by forming a three-layered sandwich with Phe 285 (N-terminal domain) and Tyr 613 (C-terminal domain) (Section II.I).
2. Site I is a hydrophobic site. Short chain aliphatic solvents prevent ligand binding at site I and activate the enzyme by stabilizing an active conformation.[39,40] Low temperatures also facilitate the T to R transition.[221,236-237]
3. Activation of the intact rabbit muscle enzyme is always accompanied by the formation of tetramers *in vitro*. Ligands of the I site dissociate all tetrameric forms of phosphorylase regardless of the way they were formed.

4.   Binding of glucose to the C site dissociates tetramers. Glucose provides a link between the two domains by connecting Leu 136 (glycine loop), Asn 284, His 377, and Asn 484 of the N-terminal domain to Tyr 573, Glu 672, and Glu 675 of the C-terminal half. This fixed conformation restricts access to the active site. Crystals of phosphorylase *a* were grown in the presence of glucose and access to the active site is restricted there. Glc-1-P, on the other hand, does not dissociate tetramers in spite of the fact that the sugar moiety provides similar contacts with the same residues as glucose does. However, the phosphate groups of Glc-1-P forces Asp 283 out of its previous position destroying contacts between Phe 285 (part of the I site) and the C-terminal domain. This could loosen up the tight lips of the active site cleft and lead to the formation of tetramers.

For the explanation of catalysis, the movements envisaged here are rather less extensive than those proposed by Madsen and Withers.[220] In our kinetic X-ray crystallographic results on catalysis with phosphate and the substrate analog heptenitol to form heptulose-2-P in tetragonal crystals of phosphorylase *b* (Section II.F), there is no indication of domain movement during catalysis, only local movements of residues 281 to 288, His 377, Glu 672, and Arg 569. This suggests that the chemical step in the reaction does not require a major rearrangement of the protein. However, our kinetic crystallographic data with long chain oligosaccharides gave very noisy maps suggesting that the binding of oligosaccharide at or around the active site in the crystals causes the enzyme molecule to move or rearrange. Binding of oligosaccharide at the storage site does not have this effect. This would be in line with the assumption that the opening up of the active site cleft is obligatory for access of larger oligo- or polysaccharides to the active center.

The crystallographic and solution studies indicate that phosphorylase acts by tailoring phosphate recognition sites. The structural changes that accompany to the T $\rightarrow$ R transition result in an increased affinity for AMP and a decreased affinity for Glc-6-P at the allosteric effector site, in an increase in affinity for P_i by nearly two orders of magnitude at the catalytic site, and in a change in the state of ionization (and perhaps geometry) of the cofactor 5'-phosphate. The X-ray results have shown that the latter phosphate is highly solvated despite its buried environment. The thermodynamics of water replacement by phosphate and the interactions of phosphate with water are likely to form a significant factor at these recognition sites, a factor which has not been fully explored. We have come some way in understanding the allosteric mechanism but the essential features await direct elucidation by structural studies on R-state crystals. This work is in progress.

ACKNOWLEDGMENTS

We are most grateful to those who sent us reprints and preprints in advance of publication: M. Cortijo, V. Dombradi, R. J. Fletterick, T. Fukui, D. J. Graves, J. R. Griffiths, E. J. M. Helmreich, H. Klein, P. J. Kasvinsky, J. W. Leadbetter, N. B. Madsen, D. Palm, and S. G. Withers. This work has been supported by the Medical Research Council and the Science and Engineering Research Council.

This manuscript was submitted in November 1986. Brief references to publications after that date have been made at the proof stage.

REFERENCES

1. **Cori, C. F. and Cori, G. T.,** Mechanism of formation of hexose monophosphate in muscle and isolation of a new phosphate ester, *Proc. Soc. Exp. Biol. Med.,* 34, 702, 1936.
2. **Cori, G. T. and Cori, C. F.,** The kinetics of the enzymic synthesis of glycogen from glucose-1-phosphate, *J. Biol. Chem.,* 135, 733, 1940.
3. **Cori, C. F., Cori, G. T., and Green, A. A.,** Crystalline muscle phosphorylase III kinetics, *J. Biol. Chem.,* 151, 39, 1943.
4. **Fischer, E. H., Graves, D. J., Crittenden, E. R. S., and Krebs, E. G.,** Structure of the site phosphorylated in the phosphorylase *b* to *a* reaction, *J. Biol. Chem.,* 234, 1696, 1959.
5. **Madsen, N. B. and Cori, C. F.,** Inhibition of muscle phosphorylase by p-chloromercuribenzoate, *Biochim. Biophys. Acta,* 18, 156, 1955.
6. **Helmreich, E. and Cori, C. F.,** The role of adenylic acid in the activity of phosphorylase, *Proc. Nat. Acad. Sci., U.S.A.,* 51, 131, 1964.
7. **Monod, J., Changeux, J.-P., and Jacob, F.,** Allosteric proteins and cellular control systems, *J. Mol. Biol.,* 6, 306, 1963.
8. **Monod, J., Wyman, J., and Changeux, J.-P.,** On the nature of allosteric transitions: a possible model, *J. Mol. Biol.,* 12, 88, 1965.
9. **Baronowsky, T., Illingworth, B., Brown, D. H., and Cori, C. F.,** The isolation of pyridoxal 5'-phosphate from crystalline muscle phosphorylase, *Biochim. Biophys. Acta,* 25, 16, 1957.
10. **Illingworth, B., Jansz, H. S., Brown, D. H., and Cori, C. F.,** Observations on the function of pyridoxal 5'-phosphate in phosphorylase, *Proc. Natl. Acad. Sci., U.S.A.,* 44, 1180, 1958.
11. **Porter, K. R. and Bruni, C.,** An electron microscope study of the early effects of 3'-Me-DAB on rat liver cells, *Cancer Res.,* 19, 997, 1959.
12. **Wanson, J.-C. and Drochmans, P.,** Rabbit skeletal muscle glycogen, *J. Cell. Biol.,* 38, 130, 1968.
13. **Meyer, F., Heilmeyer, L. M. G., Haschke, R. H., and Fischer, E. H.,** Control of phosphorylase activity in the glycogen particle. I. Isolation and characterisation of the protein glycogen complex, *J. Biol. Chem.,* 245, 6642, 1970.
14. **Heilmeyer, L. M. G., Meyer, F., Haschke, R. H., and Fischer, E. H.,** Control of phosphorylase activity in a muscle glycogen particle. II. Activation by calcium, *J. Biol. Chem.,* 245, 6649, 1970.
15. **Haschke, R. H., Heilmeyer, L. M. G., Meyer, F., and Fischer, E. H.,** Control of phosphorylase activity in the glycogen particle. III. Regulation of phosphorylase phosphatase, *J. Biol. Chem.,* 245, 6657, 1970.
16. **Wanson, J. C. and Drochmans, P.,** Role of the sarcoplasmic reticulum in glycogen metabolism. Binding of phosphorylase, phosphorylase kinase and primer complexes to sarcovesicles of rabbit skeletal muscle, *J. Cell. Biol.,* 54, 206, 1972.
17. **Haschke, R. H., Gratz, K. W., and Heilmeyer, L. M. G.,** The role of phosphorylase activity in a muscle glycogen particle, *J. Biol. Chem.,* 247, 5351, 1972.
18. **Busby, S. J. W. and Radda, G. K.,** Regulation of the glycogen phosphorylase system, *Curr. Top. Cell. Regul.,* 10, 89, 1976.
19. **Caudwell, B., Antoniw, J. F., and Cohen, P.,** Calsequestrin, myosin and the components of the protein-glycogen complex in rabbit skeletal muscle, *Eur. J. Biochem.,* 86, 511, 1978.
20. **Dombradi, V.,** Structural aspects of the catalytic and regulatory function of glycogen phosphorylase, *Int. J. Biochem.,* 13, 125, 1981.
21. **Hurd, S. S., Teller, D., and Fischer, E. H.,** Probable formation of partially phosphorylated intermediates in interconversion of phosphorylase *a* and *b, Biochem. Biophys. Res. Commun.,* 24, 79, 1966.
22. **Fischer, E. H., Pocker, A., and Saari, J. C.,** The structure, function and control of glycogen phosphorylase, *Essays Biochem.,* 6, 23, 1970.
23. **Graves, D. J. and Wang, J. H.,** α-glucan phosphorylases — chemical and physical basis of catalysis and regulation, *The Enzymes,* 3rd ed. Boyer, P., Ed., Academic Press, New York, 1972, 435.
24. **Madsen, N. B., Avramovic-Zikic, Lue, P. F., and Honikel, K. O.,** Studies in allosteric phenomena in glycogen phosphorylase *b, Mol. Cell. Biochem.,* 11, 35, 1976.
25. **Fletterick, R. J. and Madsen, N. B.,** The structures and related functions of phosphorylase *a, Annu. Rev. Biochem.,* 49, 31, 1980.
26. **Fletterick, R. J. and Sprang, S. R.,** Glycogen phosphorylase structure and function, *Acc. Chem. Res.,* 15, 361, 1982.
27. **Madsen, N. B.,** Glycogen phosphorylase, in *The Enzymes: Enzyme Control by Phosphorylation,* Boyer, P. D. and Krebs, E. G., Eds., Academic Press, New York, 17, 366, 1986.
28. **Helmreich, E. J. M. and Klein, H. W.,** The role of pyridoxal phosphate in the catalysis of glycogen phosphorylases, *Angew. Chem. Int. Ed. Engl.,* 19, 441, 1980.
29. **Madsen, N. B. and Withers, S. G.,** Glycogen phosphorylase, in *Coenzymes, and Cofactors: Pyridoxal Phosphate and Derivatives,* Vol 1B, Dolphin, D., Poulson, R., and Avramovic, O., Eds., John Wiley & Sons, New York, 1986, 355.

30. **Cohen, P.,** The role of protein phosphorylation in neural and hormonal control of cellular activity, *Nature (London),* 296, 613, 1982.

31. **Griffiths, J. R.,** Non-covalent control of glycogenolysis in muscle, in *Short Term Regulation of Liver Metabolism,* Hue L. and Van de Werve, G., Ed., Elsevier, Amsterdam, 1981, 77.

32. **Griffiths, J. R.,** A fresh look at glycogenolysis in skeletal muscle, *Bioscience Rep.,* 1, 595, 1981.

33. **Rahim, Z. H. A., Perrett, D., Lutaya, G., and Griffiths, J. R.,** Metabolic adaption in phosphorylase kinase deficiency, *Biochem. J.,* 186, 331, 1980.

34. **McArdle, B.,** Myopathy due to a defect in muscle glycogen breakdown, *Clin. Sci.,* 10, 13, 1951.

35. **Ross, B. D., Radda, G. K., Gadian, D. G., Rocker, G., Esivi, M., and Falconer-Smith, J.,** Examination of a case of suspected McArdles syndrome by ^{31}P nuclear magnetic resonance, *N. Engl. J. Med.,* 304, 1338, 1981.

36. **Wang, J. H., Humnisky, P. M., and Black, W. J.,** Effect of polyamines on glycogen phosphorylase — differential electrostatic interactions and enzyme properties, *Biochemistry,* 7, 2037, 1968.

37. **Ktenas, T. B., Oikonomakos, N. G., Sotiroudis, T. G., Nikolaropoulos, S., and Evangelopoulos, A. E.,** Interaction of aliphatic amines with glycogen phosphorylase, *J. Biochem.,* 92, 2029, 1982.

38. **Lutaya, G. and Griffiths, J. R.,** Rapid formation of spermine in skeletal muscle during tetanic stimulation, *FEBS Lett.,* 123, 186, 1981.

39. **Dreyfus, M., Vandenbunder, B., and Buc, H.,** Stabilization of a phosphorylase *b* active conformation by hydrophobic solvents, *FEBS Lett.,* 95, 185, 1978.

40. **Uhing, R. J., Janski, A. M., and Graves, D. J.,** The effect of solvents on nucleotide regulation of glycogen phosphorylase, *J. Biol. Chem.,* 254, 3169, 1979.

41. **Stura, E. A., Zanotti, G., Babu, Y. S., Sansom, M. S. P., Stuart, D. I., Wilson, K. S., Johnson, L. N., and Van der Werve, G.,** Comparison of AMP and NADH binding to glycogen phosphorylase *b,* *J. Mol. Biol.,* 170, 529, 1983.

42. **Dombradi, V., Vereb, G., and Bot, G.,** Interaction of ligands with glycogen phosphorylase as revealed by affinity chromatography, *Int. J. Biochem.,* 10, 905, 1979.

43. **Hers, H. G.,** The control of glycogen metabolism in the liver, *Annu. Rev. Biochem.,* 45, 167, 1976.

44. **Stalmans, W.,** The role of the liver in the homeostasis of blood glucose, *Curr. Top. in Cell Regul.,* 11, 51, 1976.

45. **Engers, H. D., Bridger, W. A., and Madsen, N. B.,** Kinetic mechanism of phosphorylase *b,* *J. Biol. Chem.,* 244, 5936, 1969.

46. **Michaelides, M. C., Sherman, R., and Helmreich, E.,** The interaction of muscle phosphorylase with soluble antibody fragments, *J. Biol. Chem.,* 239, 4171, 1964.

47. **Black, W. J. and Wang, J. H.,** Studies on the allosteric activation of glycogen phosphorylase *b* by nucleotides, *J. Biol. Chem.,* 243, 5892, 1968.

48. **Griffiths, J. R., Dwek, R. A., and Radda, G. K.,** Conformational changes in glycogen phosphorylase studied with a spin label probe, *Eur. J. Biochem.* 61, 237, 1976.

49. **Morgan, H. E. and Parmeggiani, A.,** Regulation of glycogenolysis in muscle, *J. Biol. Chem.,* 239, 2440, 1964.

50. **Wang, J. H., Tu, J. I., and Lo, F. M.,** Effect of glucose-6-phosphate on the nucleotide site of glycogen phosphorylase *b,* *J. Biol. Chem.,* 245, 3115, 1970.

51. **Battersby, M. K. and Radda, G. K.,** The stereospecificity of glucose-6-phosphate binding site of glycogen phosphorylase *b,* *FEBS Lett.,* 72, 319, 1976.

52. **Madsen, N. B.,** Inhibition of glycogen phosphorylase by uridine diphosphate glucose, *Biochem. Biophys. Res. Commun.,* 6, 310, 1961.

53. **Mateo, P. L., Baron, C., Lopez-Mayorga, Jimenez, J. S., and Cortijo, M.,** AMP and IMP binding to glycogen phosphorylase *b,* *J. Biol. Chem.,* 259, 9384, 1984.

54. **Melpidou, A. E. and Oikonomakos, N. G.,** Effect of glucose-6-P on the catalytic and structural properties of glycogen phosphorylase *a,* *FEBS Lett.,* 154, 105, 1983.

55. **Danforth, W. H. and Lyon, J. B.,** Glycogenolysis during tetanic contraction of isolated mouse muscles in the presence and absence of phosphorylase *a,* *J. Biol. Chem.,* 239, 4047, 1964.

56. **Bailey, I. A., Williams, S. R., Radda, G. K., and Gadian, D. G.,** Activity of phosphorylase in total global ischaemia in the rat heart, *Biochem. J.,* 196, 171, 1981.

57. **Lyon, J. B. and Porter, J.,** The relation of phosphorylase to glycogenolysis in skeletal muscle and heart of mice, *J. Biol. Chem.,* 238, 1, 1963.

58. **Rahim, Z. H. A., Perrett, D., and Griffiths, J. R.,** Skeletal muscle purine nucleotide levels in normal and phosphorylase kinase deficient mice, *FEBS Lett.,* 69, 203, 1976.

59. **Lutaya, G., Rahim, Z. H. A., Shuttlewood, R. J., Bashford, C. L., and Griffiths, J. R.,** Alkalinization of phosphorylase kinase deficient muscle during tetranic contraction, *Bioscience Rep.,* 1, 177, 1981.

60. **Dawson, M. J., Gadian, D. G., and Wilkie, D. R.,** Mechanical relaxation rate and metabolism studied in fatiguing muscle by phosphorus nuclear magnetic resonance, *J. Physiol.,* 299, 465, 1980.

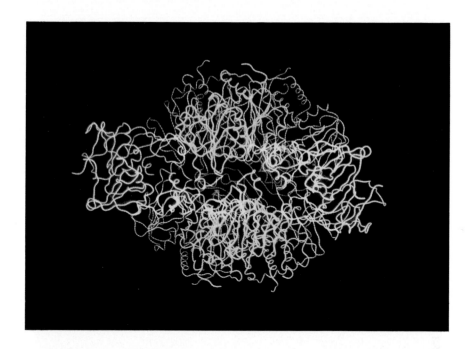

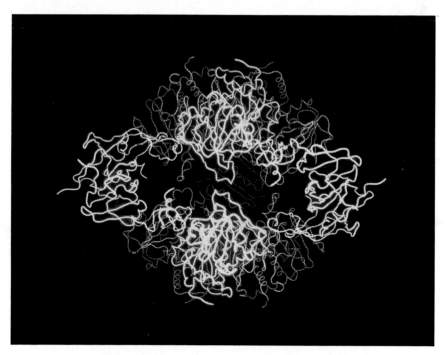

PLATE 1. The quaternary structure change in ATCase. α-Carbone backbone of ATCase in the T-state (top) and R-state (bottom). Note the repositioning of the 240 loop. Computer drawn by J. Cherfils and J. Janin (Université Paris-Sud) on the basis of the atom coordinates generously provided by W. Lipscomb.[13]

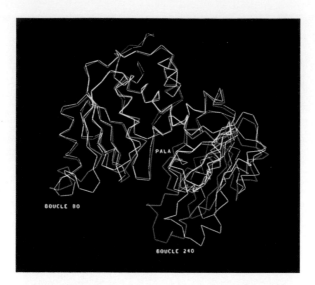

PLATE 2. Secondary structure of the catalytic chain in the T- and R-states of ATCase. Computer drawn by J. Cherfils and J. Janin (Université Paris-Sud) on the basis of the atom coordinates generously provided by W. Lipscomb.[13] Blue: T-state; red: R-state. PALA indicates the location of the catalytic site.

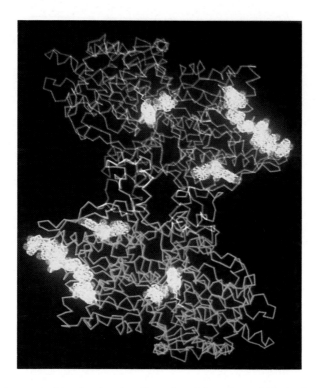

PLATE 3. An α-carbon trace of the phosphorylase dimer viewed down the twofold axis of symmetry: one subunit is red and the other blue. Ligands are represented by Van der Waals spheres. With reference to the blue subunit, oligosaccharide at the glycogen storage site is lower left, AMP is at the allosteric effector site near the subunit-subunit boundary, and Glc-l-P and pyridoxal phosphate is at the catalytic site in the center of the subunit.

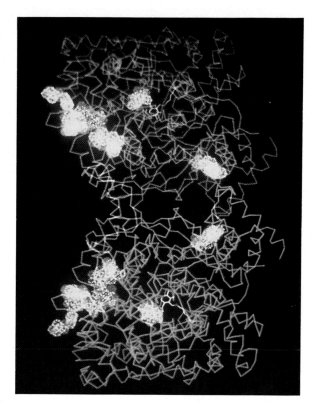

PLATE 4. A view of the α-carbon trace of the phosphorylase dimer normal to the twofold axis. The pyridoxal phosphate is shown (center of subunit) in skeletal form. Other ligands are shown as in Plate 3. The glycogen storage sites are on one side of the dimer (left) and the allosteric effector sites on the opposite side (right). The N-terminal helix for both the blue and red subunit is on the right-hand side of the dimer.

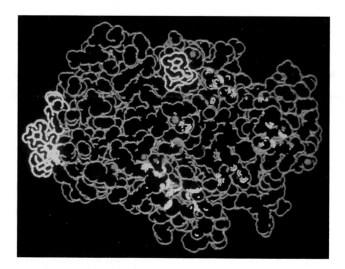

PLATE 5. A ''foot-print'' diagram showing the surface residues involved in the subunit-subunit contacts. The diagram shows a view of the molecule down the crystallographic z axis, normal to the molecular diad (i.e., the vertical direction in Plate 3). Oligosaccharide (to the left) and AMP (upper center) surfaces are shown pale orange. The nonpolar subunit-subunit contacts are shown white and the polar contacts red and blue. The tower contacts are lower center and the cap contacts are at the right. Only a small proportion of the surface residues are involved in subunit contacts (see text).

61. **Dawson, M. J., Gadian, D. G., and Wilkie, D. R.,** Studies of the biochemistry of contracting and relaxing muscle by the use of ^{31}P NMR in conjunction with other techniques, *Philos. Trans. R. Soc. London,* B289, 445, 1980.

62. **Wilkie, D. R.,** Discussion to a paper on phosphorylase: control and activity, *Philos. Trans. R. Soc. London,* B293, 40, 1981.

63. **Piras, R. and Staneloni, R.,** *In vivo* regulation of rat muscle glycogen synthetase activity, *Biochemistry,* 8, 2153, 1969.

64. **Eagles, P. A. M., Iqbal, M., Johnson, L. N., Mosley, J., and Wilson, K. S.,** A tetragonal crystal form of phosphorylase b, *J. Mol. Biol.,* 71, 803, 1972.

65. **Fischer, E. H. and Krebs, E. G.,** Muscle phosphorylase b, in *Methods in Enzymology,* Vol. 5, Academic Press, New York, 1962, 369.

66. **Johnson, L. N., Madsen, N. B., Mosley, J., and Wilson, K. S.,** The crystal structure of phosphorylase b at 6Å resolution, *J. Mol. Biol.,* 90, 703, 1974.

67. **Weber, I. T., Johnson, L. N., Wilson, K. S., Yeates, D. G. R., Wild, D. L., and Jenkins, J. A.,** Crystallographic studies on the activity of glycogen phosphorylase b, *Nature (London),* 274, 433, 1978.

68. **Sansom, M. S. P., Stuart, D. I., Acharya, K. R., Hajdu, J., McLaughlin, P. J., and Johnson, L. N.,** Glycogen phosphorylase b — the molecular anatomy of a large regulatory enzyme, *J. Mol. Struct.,* 123, 3, 1985.

69. **Stuart, D. I., Acharya, K. R., Sansom, M. S. P., Babu, Y. S., and Johnson, L. N.,** The Crystallographic Refinement of Glycogen Phosphorylase b, unpublished results.

70. **Fletterick, R. J., Sygusch, J., Madsen, N. B., and Johnson, L. N.,** Low-resolution structure of the glycogen phosphorylase a monomer and comparison with phosphorylase b, *J. Mol. Biol.,* 103, 1, 1976.

71. **Fletterick, R. J., Sygusch, J., Sample, H., and Madsen, N. B.,** Structure of glycogen phosphorylase a at 3.0Å resolution and its ligand binding sites at 6Å, *J. Biol. Chem.,* 251, 6142, 1976.

72. **Sprang, S. and Fletterick, R. J.,** The structure of glycogen phosphorylase a at 2.5Å resolution, *J. Mol. Biol.,* 131, 523, 1979.

73. **Fletterick, R. J.,** Glycogen phosphorylase: plasticity and specificity in ligand binding, *Proc. Robert A. Welch Found. Conf. Chem. Res.,* 27, 173, 1984.

74. **Johnson, L. N., Stura, E. A., Wilson, K. S., Sansom, M. S. P., and Weber, I. T.,** Nucleotide binding to glycogen phosphorylase b in the crystal, *J. Mol. Biol.,* 134, 639, 1979.

75. **Helmreich, E., Michaelides, M. C., and Cori, C. F.,** Effects of substrates and a substrate analog on the binding of 5'-adenylic acid to muscle phosphorylase a, *Biochemistry,* 6, 3695, 1967.

76. **Kasvinsky, P. J., Shechosky, S., and Fletterick, R. J.,** Synergistic regulation of phosphorylase a by glucose and caffeine, *J. Biol. Chem.,* 253, 9102, 1978.

77. **Withers, S. G., Sykes, B. D., Madsen, N. B., and Kasvinsky, P. J.,** Identical structural changes induced in glycogen phosphorylase by two non-exclusive allosteric inhibitors, *Biochemistry,* 18, 5342, 1979.

78. **Wang, J. H., Shonka, M. L., and Graves, D. J.,** The effect of glucose on the sedimentation and catalytic activity of glycogen phosphorylase, *Biochem. Biophys. Res. Commun.,* 18, 131, 1965.

79. **Kasvinsky, P. J. and Madsen, N. B.,** Activity of glycogen phosphorylase in the crystalline state, *J. Biol. Chem.,* 251, 6852, 1976.

80. **Wang, J. H. and Tu, J.-I.,** Modification of phosphorylase b by glutaraldehyde, *Biochemistry,* 8, 4403, 1969.

81. **Mathews, F. S.,** X-ray crystallographic study of glycogen phosphorylase, *Fed. Proc.,* 26, 831, 1967.

82. **Fasold, H., Ortanderl, F., Huber, R., Bartels, K., and Schwager, P.,** Crystallisation and crystallographic data of rabbit muscle phosphorylase a and b, *FEBS. Lett.,* 21, 229, 1972.

83. **Madsen, N. B., Honikel, K. O., and James, M. N. G.,** *Studies in Metabolic Interconversion of Enzymes,* Wieland, O., Helmreich, E., and Holzer, H., Eds., Springer-Verlag, Berlin, 1972, 448.

84. **Bartels, K. and Colman, P. M.,** Subunit symmetry of tetrameric phosphorylase a, *Biophys. Struct. Mech.,* 2, 43, 1976.

85. **Withers, S. G., Madsen, N. B., and Sykes, B. D.,** Covalently activated glycogen phosphorylase: a phosphorus-31 NMR and ultracentrifugation analysis, *Biochemistry,* 21, 6716, 1982.

86. **Sprang, S. and Fletterick, R. J.,** Subunit interactions and the allosteric response in phosphorylase, *Biophys. J.,* 32, 175, 1980.

87. **Oikonomakos, N. G., Melpidou, A. E., and Johnson, L. N.,** Crystallisation of pig skeletal phosphorylase b. Purification, physical and catalytic characterisation, *Biochim. Biophys. Acta,* 832, 248, 1985.

88. **Hwang, P., Stein, M., and Fletterick, R. J.,** Purification and crystallisation of bovine liver phosphorylase, *Biochim. Biophys. Acta,* 791, 252, 1984.

89. **Buehner, M. and Bender, C.,** Crystallisation and crystallographic data of *Escherichia Coli* maltodextrin phosphorylase, *FEBS Lett.,* 85, 91, 1978.

90. **Titani, K., Koide, A., Hermann, J., Ericsson, L. H., Kumar, S., Wade, R. D., Walsh, K. A., Neurath, H., and Fischer, E. H.,** Complete amino acid sequence of rabbit muscle glycogen phosphorylase, *Proc. Natl. Acad. Sci. U.S.A.,* 74, 4762, 1977.

91. **Koide, A., Titani, K., Ericsson, L. H., Kumar, S., Neurath, H., and Walsh, K. A.,** Sequence of amino terminal 349 residues of rabbit muscle glycogen phosphorylase including the sites of covalent and allosteric control, *Biochemistry,* 17, 5657, 1978.

92. **Hermann, J., Titani, K., Ericsson, L. H., Wade, R. D., Neurath, H., and Walsh, K. A.,** Amino acid sequence of two cyanogen bromide fragments of glycogen phosphorylase, *Biochemistry,* 17, 5672, 1978.

93. **Titani, K., Koide, A., Ericsson, L. H., Kumar, S., Hermann, J., Wade, R. D., Walsh, K. A., Neurath, H., and Fischer, E. H.,** Sequence of the carboxy terminal 492 residues of rabbit muscle glycogen phosphorylase including the pyridoxal 5'-phosphate binding site, *Biochemistry,* 17, 5680, 1978.

94. **Hwang, P. K., See, Y. P., Vincentini, A. M., Powers, M. A., Fletterick, R. J., and Crerar, M. M.,** Comparative sequence analysis of rat, rabbit and human muscle glycogen phosphorylase cDNAs, *Eur. J. Biochem.,* 152, 267, 1985.

95. **Lebo, R. V., Govin, F., Fletterick, R. J., Kao, F.-T., Cheung, M. C., Bruce, B. D., and Kan, Y. W.,** High resolution chromosome sorting and DNA spot-blot analysis assign McArdles syndrome to chromosome 11, *Science,* 225, 57, 1984.

96. **Cohen, P., Saari, J. C., and Fischer, E. H.,** Comparative study of dogfish and rabbit muscle phosphorylase, *Biochemistry,* 12, 5233, 1973.

97. **Lerch, K. and Fischer, E. H.,** Amino acid sequence of two functional sites in yeast glycogen phosphorylase, *Biochemistry,* 14, 2009, 1975.

98. **Sprang, S., Yang, D., and Fletterick, R. J.,** Solvent accessibility properties of complex proteins, *Nature (London),* 280, 333, 1979.

99. **Jenkins, J. A., Johnson, L. N., Stuart, D. I., Stura, E. A., Wilson, K. S., and Zanotti, G.,** Phosphorylase: control and activity, *Philos. Trans. R. Soc. London,* B293, 23, 1981.

100. **Rossmann, M. G., Moras, D., and Olsen, K. W.,** Chemical and biological evolution of a nucleotide-binding protein, *Nature (London),* 250, 194, 1974.

101. **Fletterick, R. J., Sprang, S., and Madsen, N. B.,** Analysis of the surface topography of glycogen phosphorylase *a:* implications for metabolic interconversion and regulatory mechanisms, *Can. J. Biochem.,* 57, 789, 1979.

102. **McLaughlin, P. J. and Stuart, D. I.,** unpublished results.

103. **Carty, T. J., Tu, J.-I., and Graves, D. J.,** Regulation of glycogen phosphorylase — role of the peptide region surrounding the phosphoserine residue in determining enzyme properties, *J. Biol. Chem.,* 250, 4980, 1975.

104. **Vogel, H. J. and Bridger, W. A.,** Phosphorus-31 NMR pH titration studies of the phosphoproteins tropomyosin and glycogen phosphorylase *a, Can. J. Biochem. Cell. Biol.,* 61, 363, 1983.

105. **Sealock, R. W. and Graves, D. J.,** Effect of salt solutions on glycogen phosphorylase, *Biochemistry,* 6, 201, 1967.

106. **Janski, A. M. and Graves, D. J.,** Use of an antibody probe to study regulation of glycogen phosphorylase by its NH$_2$-terminal region, *J. Biol. Chem.,* 254, 1644, 1979.

107. **Martensen, T. M., Brotherton, J. E., and Graves, D. J.,** Kinetic studies of the activation of muscle phosphorylase phosphatase, *J. Biol. Chem.,* 248, 8329, 1973.

108. **Graves, D. J., Mann, S. A. S., Philip, G., and Oliveira, R. J.,** A probe into the catalytic activity and subunit assembly of glycogen phosphorylase, *J. Biol. Chem.,* 248, 6090, 1968.

109. **Janski, A. M. and Graves, D. J.,** Complementation of glycogen phosphorylase b' with synthetic peptides, *J. Biol. Chem.,* 254, 4033, 1979.

110. **Kent, A. B., Krebs, E. G., and Fischer, E. H.,** Properties of crystalline phosphorylase *b, J. Biol. Chem.,* 232, 549, 1958.

111. **Chang, Y. C. and Graves, D. J.,** The use of 6-fluoro derivatives of pyridoxal and pyridoxal phosphate in the study of co-enzyme function in glycogen phosphorylase, *J. Biol. Chem.,* 260, 2709, 1985.

112. **Sansom, M. S. P., Babu, Y. S., Hajdu, J., Stuart, D. I., Stura, E. A., and Johnson, L. N.,** The role of pyridoxal phosphate in glycogen phosphorylase *b;* structure, environment and relationship to catalytic mechanism, in *Chemical and Biological Aspects of Vitamin B$_6$ Catalysis, Part A,* Evangelopoulos, A. E., Ed., Alan R. Liss, New York, 1984.

113. **Hol, W. G. J., Van Duijen, P. T., and Berendsen, H. J. C.,** *Nature (London),* 273, 443, 1978.

114. **Barrett, A. N. and Palmer, R. A.,** The crystal structure of pyridoxal phosphate oxime, *Acta Crystallogr.,* B25, 688, 1969.

115. **Shimomura, S. and Fukui, T.,** Characterisation of the pyridoxal phosphate site in glycogen phosphorylase *b* from rabbit muscle, *Biochemistry,* 17, 5359, 1978.

116. **Shimomura, S., Nakano, K., and Fukui, T.,** Affinity labelling of the cofactor site in glycogen phosphorylase *b* with a pyridoxal phosphate analog, *Biochem. Biophys. Res. Commun.,* 82, 462, 1978.

117. **Ford, G. C., Eichele, G., and Jansonius, J. N.,** Three dimensional structure of a pyridoxal phosphate dependent enzyme, mitochondrial aspartate aminotransferase, *Proc. Natl. Acad. Sci. U.S.A.,* 77, 2559, 1980.

118. **Kirsch, J. F., Eichele, G., Ford, G. C., Vincent, M. G., Jansonius, J. N., Gehring, H., and Christen, P.**, Mechanism of action of aspartate aminotransferase proposed on the basis of the spatial structure, *J. Mol. Biol.*, 174, 497, 1984.

119. **Johnson, L. N., Jenkins, J. A., Wilson, K. S., Stura, E. A., and Zanotti, G.**, Proposals for the catalytic mechanism of glycogen phosphorylase *b* prompted by crystallographic studies in glucose-1-phosphate binding, *J. Mol. Biol.*, 140, 565, 1980.

120. **Hajdu, J., Acharya, K. R., Stuart, D. I., McLaughlin, P. J., Barford, D., Klein, H., and Johnson, L. N.**, Catalysis in the crystal: synchrotron radiation studies with glycogen phosphorylase *b*, *EMBO J.*, 6, 539, 1987.

121. **McLaughlin, P. J., Stuart, D. I., Klein, H. W., Oikonomakos, N. G., and Johnson, L. N.**, Substrate cofactor interactions for glycogen phosphorylase *b*. A binding study in the crystal with heptenitol and heptulose-2-phosphate, *Biochemistry*, 23, 5862, 1984.

122. **Klein, H. W., Im, M. J., and Helmreich, E. J. M.**, The role of pyridoxal 5'-phosphate and orthophosphate in general acid-base catalysis by α-glucan phosphorylases, in *Chemical and Biological Aspects of Vitamin B_6 Catalysis*, Part A, Evangelopoulos, A. E., Ed., Alan R. Liss, New York, 1984, 147.

123. **Klein, H. W., Im, M. J., Palm, D., and Helmreich, E. J. M.**, Does pyridoxal 5'-phosphate function in glycogen phosphorylase as an electrophilic or a general acid catalyst?, *Biochemistry*, 23, 5853, 1984.

124. **Klein, H. W., Im, M. J., and Palm, D.**, Mechanism of the phosphorylase reaction: utilisation of D-gluco-hept-1-enitol in the absence of primer, *Eur. J. Biochem.*, 157, 107, 1986.

125. **Kokesh, F. C., Stephenson, R. K., and Kakuda, Y.**, Inhibition of potato starch phosphorylase by α-D-glucopyranose-1,2-cyclic phosphate, *Biochim. Biophys. Acta*, 483, 258, 1977.

126. **Hu, H.-Y. and Gold, A. M.**, Inhibition of rabbit muscle glycogen phosphorylase by α-D-glucopyranose 1,2-cyclic phosphate, *Biochim. Biophys. Acta*, 525, 55, 1978.

127. **Withers, S. G., Madsen, N. B., and Sykes, B. D.**, Active form of Pyridoxal phosphate in glycogen phosphorylase. Phosphorus 31 nuclear magnetic resonance investigation, *Biochemistry*, 20, 1748, 1981.

128. **Hajdu, J., Acharya, K. R., Stuart, D. I., McLaughlin, P. J., Barford, D., Klein, H., and Johnson, L. N.**, Time resolved structural studies on catalysis in the crystal with glycogen phosphorylase *b*, *Biochem. Soc. Trans.*, 14, 538, 1986.

129. **Ariki, M. and Fukui, T.**, Affinity of glucose analogs for α-glucan phosphorylases from rabbit muscle and potato tubers, *J. Biochem.*, 81, 1017, 1977.

130. **Kirkman, B. R. and Whelan, W. J.**, Glucosamine is a normal component of liver glycogen, *FEBS Lett.*, 194, 6, 1986.

131. **Romero, P. A., Smith, E. E., and Whelan, W. J.**, Glucosamine is a substitute for glucose in glycogen metabolism, *Biochem. Int.*, 1, 1, 1980.

132. **Najmudin, S.**, The Control and Activity of Phosphorylase: Crystallographic Binding Studies of Glucosamine-1-Phosphate and Adenosine 5'-Phosphate to Phosphorylase *b*, Biochemistry FHS Part II thesis, University of Oxford, 1984.

133. **Sprang, S. R., Goldsmith, E. J., Fletterick, R. J., Withers, S. G., and Madsen, N. B.**, Catalytic site of glycogen phosphorylase: structure of the T state and specificity for α-D-glucose, *Biochemistry*, 21, 5364, 1982.

134. **Withers, S. G., Madsen, N. B., Sprang, S. R., and Fletterick, R. J.**, Catalytic site of glycogen phosphorylase: structural changes during activation and mechanistic implications, *Biochemistry*, 21, 5372, 1982.

135. **Madsen, N. B., Kasvinsky, P. J., and Fletterick, R. J.**, Allosteric transitions of phosphorylase *a* and the regulation of glycogen metabolism, *J. Biol. Chem.*, 253, 9097, 1978.

136. **Dreyfus, M., Vandenbunder, B., and Buc, H.**, Mechanism of allosteric activation of glycogen phosphorylase probed by the reactivity of essential arginyl residues, *Biochemistry*, 19, 3634, 1980.

137. **Vandenbunder, B. and Buc, H.**, The reactivity of arginine residues interacting with glucose 1-phosphate in glycogen phosphorylase, *Eur. J. Biochem.*, 133, 509, 1983.

138. **Miller, J. F., Seybold, M. C., and Graves, D. J.**, Use of arginine compounds to examine the role of an essential arginine in the mechanism of glycogen phosphorylase, *Biochemistry*, 20, 4579, 1981.

139. **Kasvinsky, P. J., Madsen, N. B., Fletterick, R. J., and Sygusch, J.**, X-ray crystallographic and kinetic studies of oligosaccharide binding to phosphorylase, *J. Biol. Chem.*, 253, 1290, 1978.

140. **Johnson, L. N., Stura, E. A., Sansom, M. S. P., and Babu, Y. S.**, Oligosaccharide binding to glycogen phosphorylase *b*, *Biochem. Soc. Trans.*, 11, 142, 1983.

141. **Wang, J. H., Shonka, M. L., and Graves, D. J.**, Influence of carbohydrate on phosphorylase structure and activity. I. Activation by preincubation with glycogen, *Biochemistry*, 4, 2296, 1965.

142. **Metzger, B., Helmreich, E., and Glaser, L.**, The mechanism of activation of skeletal muscle phosphorylase *a* by glycogen, *Proc. Natl. Acad. Sci. U.S.A.*, 57, 994, 1967.

143. **Philip, G., Gringel, G., and Palm, D.**, Rabbit muscle phosphorylase derivatives with oligosaccharides covalently bound to the glycogen storage site, *Biochemistry*, 21, 3043, 1982.

144. **Goldsmith, E., Sprang, S., and Fletterick, R.,** Structure of maltoheptaose by difference Fourier methods and a model for glycogen, *J. Mol. Biol.,* 156, 411, 1982.

145. **McLaughlin, P. J.,** Crystallographic Studies on Glycogen Phosphorylase *b*, D. Phil. thesis, University of Oxford, 1985.

146. **Okazaki, T., Nakazawa, A., and Hayaishi, O.,** Studies on the interaction between regulatory enzymes and effectors, *J. Biol. Chem.,* 243, 5266, 1968.

147. **Black, W. J. and Wang, J. H.,** Studies on the allosteric activation of glycogen phosphorylase *b* by nucleotides, *Biochim. Biophys. Acta,* 212, 257, 1970.

148. **Lee, Y. M. and Benisek, W. F.,** Inactivation of phosphorylase *b* by potassium ferrate, *J. Biol. Chem.,* 253, 5460, 1978.

149. **Anderson, R. A., Parish, R. F., and Graves, D. J.,** Affinity labeling of AMP site in phosphorylase, *Biochemistry,* 12, 1901, 1973.

150. **Hulla, F. W. and Fasold, H.,** Synthesis of an adenosine 5'-monophosphate analog and its use for affinity labelling muscle phosphorylase *b*, *Biochemistry,* 11, 1056, 1982.

150a. **Mott, D. M. and Bieber, A. L.,** Structural specificity of the adenosine 5'-phosphate site on glycogen phosphorylase *b*, *J. Biol. Chem.,* 245, 4058, 1970.

151. **Morange, M., Garcia Blanco, F., Vandenbunder, B., and Buc, H.,** AMP analogs: their function in the activation of glycogen phosphorylase *b*, *Eur. J. Biochem.,* 65, 553, 1976.

152. **Madsen, N. B. and Withers, S. G.,** Nucleotide activation of glycogen phosphorylase *b* occurs only when the nucleotide is in a dianionic form, *Biochem. Biophys. Res. Commun.,* 97, 53, 1980.

153. **Murray, A. W. and Atkinson, R. R.,** Adenine 5'-phosphorothioate as an AMP analog, *Biochemistry,* 7, 4023, 1968.

154. **Feldman, K. and Hull, W. E.,** ^{31}P nuclear magnetic resonance studies of glycogen phosphorylase for rabbit muscle. Ionisation states of pyridoxal 5'-phosphate, *Proc. Natl. Acad. Sci. U.S.A.,* 74, 856, 1977.

155. **Withers, S. G., Madsen, N. B., and Sykes, B. D.,** Active form of pyridoxal phosphate in glycogen phosphorylase. Phosphorus-31 nuclear magnetic resonance investigation, *Biochemistry,* 20, 1748, 1981.

156. **Buc, H.,** On the allosteric interaction between 5'-AMP and orthophosphate on phosphorylase *b*, *Biochem. Biophys. Res. Commun.,* 28, 59, 1967.

157. **Engers, H. D. and Madsen, N. B.,** The effects of anions on the activity of phosphorylase *b*, *Biochem. Biophys. Res. Commun.,* 33, 49, 1968.

158. **Madsen, N. B. and Sheckosky, S.,** Allosteric properties of phosphorylase *b*, *J. Biol. Chem.,* 242, 3301, 1967.

159. **Buc-Caron, M. H. and Buc, H.,** Quaternary changes of rabbit muscle glycogen phosphorylase *b* at low temperatures: relaxation studies and titration of sulphydryl groups, *Eur. J. Biochem.,* 52, 575.

160. **Battersby, M. K. and Radda, G. K.,** Intersubunit transmission of ligand effects in the glycogen phosphorylase *b* dimer, *Biochemistry,* 18, 3774, 1979.

161. **Lorek, A., Wilson, K. S., Sansom, M. S. P., Stuart, D. I., Stura, E. A., Jenkins, J. A., Zanotti, G., Hajdu, J., and Johnson, L. N.,** Allosteric interaction of glycogen phosphorylase *b*, *Biochem. J.,* 218, 45, 1984.

162. **Wang, J. H., Kwok, S.-C., Wirch, E., and Susuki, I.,** Distinct AMP sites in glycogen phosphorylase *b* as revealed by calorimetric studies, *Biochem. Biophys. Res. Commun.,* 40, 1340, 1970.

163. **Ho, H. C. and Wang, J. H.,** Calorimetric studies of the interaction between phosphorylase *b* and its nucleotide activators, *Biochemistry,* 12, 4750, 1973.

164. **Kasvinsky, P. J., Madsen, N. B., Sygusch, J., and Fletterick, R. J.,** Regulation of glycogen phosphorylase *a* by nucleotide derivatives, *J. Biol. Chem.,* 253, 3343, 1978.

165. **Sprang, S., Fletterick, R., Stern, M., Yang, D., Madsen, N. B., and Sturtevant, J.,** Analysis of an allosteric binding site: the nucleoside inhibitor site of phosphorylase *a*, *Biochemistry,* 21, 2036, 1982.

166. **Kasvinsky, P. J., Fletterick, R. J., and Madsen, N. B.,** Regulation of the dephosphorylation of glycogen phosphorylase *a* and synthase b by glucose and caffeine in isolated hepatocytes, *Can.J. Biochem.,* 59, 387, 1981.

167. **Oikonomakos, N. G., Sotiroudis, T. G., and Evangelopoulos, A. E.,** Interactions with phosphorylase *b* with Eosin, *Biochem. J.,* 181, 309, 1979.

168. **Steiner, R. F., Greer, L., Bhat, R., and Oton, J.,** Structural changes induced in glycogen phosphorylase *b* by the binding of glucose and caffeine, *Biochim. Biophys. Acta,* 611, 269, 1980.

169. **Soman, G. and Philip, G.,** The nature of the binding site for aromatic compounds in glycogen phosphorylase *b*, *Biochem. J.,* 147, 369, 1971.

170. **Hajdu, J., Dombradi, V., Bot, G., and Friedrich, P.,** Structural changes in glycogen phosphorylase as revealed by cross-linking with bifunctional diimidates: phosphorylase *b*, *Biochemistry,* 18, 4037, 1979.

171. **Dombradi, V., Toth, B., Bot, G., Hajdu, J., and Friedrich, P.,** Interactions of ligands in phosphorylase *a* as monitored by cross-linking and enzymic modifications; synergism of glucose and caffeine manifested in the exposure of the N-terminal segment, *Int. J. Biochem.,* 14, 277, 1982.

172. **Nakano, K. and Fukui, T.**, The complete amino acid sequence of potato α-glucan phosphorylase, *J. Biol. Chem.*, 261, 8230, 1986.

173. **Palm, D., Goerl, R., and Burger, K. J.**, Evolution of catalytic and regulatory sites in phosphorylases, *Nature (London)*, 313, 500, 1985.

174. **Palm, D., Schaechtele, K. H., Schiltz, E., Fischer, B., Zeier, R., and Goerl, R.**, *E. Coli* maltodextin phosphorylase: primary structure and deletion mapping of the C-terminal site, *Z. Naturforsch.*, 42c, 394, 1987.

175. **Fukui, T., Shimomura, S., and Nakano, K.**, Potato and rabbit muscle phosphorylases: comparative studies on the structure, function and regulation of regulatory and non-regulatory enzymes, *Mol. Cell. Biochem.*, 42, 129, 1982.

176. **Hatfield, D., Hofnung, M., and Schartz, M.**, Genetic analysis of the maltose A region in *Escherishia Coli*, *J. Bacteriol.*, 98, 559, 1969.

177. **Chapon, C.**, Role of the catabolite activator protein in the maltose regulon of *Escherichia Coli*, *J. Bacteriol.*, 150, 722, 1982.

178. **Schwartz, M. and Hofnung, M.**, La maltodextrine phosphorylase d' *Escherichia Coli*, *Eur. J. Biochem.*, 2, 132, 1967.

179. **Schiltz, E., Palm, D., and Klein, H. W.**, N-terminal sequences of *Escherichia Coli* and potato phosphorylase, *FEBS Lett.*, 109, 59, 1980.

180. **Maddaiah, V. T. and Madsen, N. B.**, Kinetics of purified liver phosphorylase, *J. Biol. Chem.*, 241, 3873, 1966.

181. **Chao, J., Johnson, G. F., and Graves, D. J.**, Kinetic mechanism of maltodextrin phosphorylase, *Biochemistry*, 8, 1459, 1969.

182. **Engers, H. D., Shechosky, S., and Madsen, N. B.**, Kinetic mechanism of phosphorylase *a*, *Can. J. Biochem.*, 48, 746, 1970.

183. **Engers, H. D., Bridger, W. A., and Madsen, N. B.**, Kinetic mechanism of phosphorylase *a*, *Can. J. Biochem.*, 48, 755, 1970.

184. **Gold, A. M., Johnson, R. M., and Tseng, J. K.**, Kinetic mechanism of rabbit muscle glycogen phosphorylase *a*, *J. Biol. Chem.*, 245, 2564, 1970.

185. **Hu, H. Y. and Gold, A. G.**, Kinetics of glycogen phosphorylase *a* with a series of semisynthetic, branched saccharides. A model for binding of polysaccharide substrates, *Biochemistry*, 14, 224, 1975.

186. **Sotiroudis, T. G., Oikonomakos, N. G., and Evangelopoulos, A. E.**, Phosphorylase *b* covalently bound to glycogen. Properties of the complex, *Eur. J. Biochem.*, 88, 573, 1978.

187. **Chang, Y. C. and Graves, D. J.**, The use of 6-fluoroderivatives of pyridoxal and pyridoxal phosphate in the study of coenzyme function in glycogen phosphorylase, *J. Biol. Chem.*, 260, 2709, 1985.

188. **Chang, Y. C., Scott, R. D., and Graves, D. J.**, Function of PLP in glycogen phosphorylase: ^{19}F NMR and kinetic studies of phosphorylase reconstituted with 6-fluoropyridoxal and 6-fluoropyridoxal phosphate, *Biochemistry*, 25, 1932, 1986.

189. **Fukui, T., Tagaya, M., Takagi, M., and Shimomura, S.**, Role of pyridoxal 5'-phosphate in the catalytic mechanism of glycogen phosphorylase, in *Chemical and Biological Aspects of Vitamin B$_6$ Catalysis*, Evangelopoulos, A. E., Ed., Alan R. Liss, New York, 1984.

190. **Cohn, M. and Cori, G. T.**, On the mechanism of action of muscle and potato phosphorylase, *J. Biol. Chem.*, 175, 89, 1948.

191. **Gold, A. M. and Osber, M. P.**, On the mechanism of glycogen phosphorylase, *Arch. Biochem. Biophys.*, 153, 784, 1972.

192. **Kokesh, F. C. and Kakuda, Y.**, Evidence for intermediate formation in the mechanism of potato starch phosphorylase from exchange of the ester and phosphoryl oxygens of α-D-glucopyranosyl 1-phosphate, *Biochemistry*, 16, 2467, 1977.

193. **Tu, J. I., Jacobson, G. R., and Graves, D. J.**, Isotopic effects and inhibition of polysaccharide phosphorylase by 1,5-gluconolactone. Relationship to the catalytic mechanism, *Biochemistry*, 10, 1229, 1971.

194. **Firsov, L. M., Bogacheva, T. I., and Bressler, S. E.**, Secondary isotope effect in the phosphorylase reaction, *Eur. J. Biochem.*, 42, 605, 1974.

195. **Klein, H. W., Palm, D., and Helmreich, E. J. M.**, General acid base catalysis of α-glucan phosphorylases: stereospecific glucosyl transfer from D-glucal is a pyridoxal-5'-phosphate and orthophosphate (arsenate) dependent reaction, *Biochemistry*, 21, 6675, 1982.

196. **Avramovic-Zikic, O., Brildenbach, W. C., and Madsen, N. B.**, Evidence for an essential carboxyl group in glycogen phosphorylase, *Can. J. Biochem.*, 52, 146, 1974.

197. **Ariki, M. and Fukui, T.**, Modification of rabbit muscle phosphorylase *b* by a water soluble carbodiimide, *J. Biochem.*, 83, 183, 1980.

198. **Gold, A. M., Legrand, E., and Sanchez, G.**, Inhibition of muscle phosphorylase *a* by 5-gluconolactone, *J. Biol. Chem.*, 246, 5700, 1971.

199. **Kasvinsky, P. J. and Meyer, W. L.**, The effect of pH and temperature on the kinetics of native and altered glycogen phosphorylase, *Arch. Biochem. Biophys.*, 181, 616, 1977.

200. **Withers, S. G., Shechosky, S., and Madsen, N. B.,** Pyridoxal phosphate is not the acid catalyst in the glycogen phosphorylase catalytic mechanism, *Biochem. Biophys. Res. Commun.,* 108, 322, 1982.
201. **Tagaya, M. and Fukui, T.,** Catalytic reaction of glycogen phosphorylase reconstituted with a coenzyme substrate conjugate, *J. Biol. Chem.,* 259, 4860, 1984.
202. **Parrish, R. F., Uhing, R. J., and Graves, D. J.,** Effect of phosphate analogues on the activity of pyridoxal reconstituted glycogen phosphorylase, *Biochemistry,* 16, 4824, 1977.
203. **Horl, M., Feldman, K., Schnackerz, K. D., and Helmreich, E. J. M.,** Ionisation of pyridoxal-5'-P and the interactions of AMPS and thiophosphoseryl residues in native and succinylated rabbit muscle phosphorylase *b* and *a* as inferred from ^{31}P-NMR spectra, *Biochemistry,* 18, 2457, 1979.
204. **Palm, D., Schachtele, Feldmann, K., and Helmreich, E. J. M.,** Is the action form of pyridoxal-P in glycogen phosphorylases a 5'-phosphate dianion?, *FEBS Lett.,* 101, 403, 1979.
205. **Klein, H. W. and Helmreich, E. J. M.,** A proton donor acceptor function of the 5'-phosphate group of pyridoxal-P in potato phosphorylase inferred from ^{31}P NMR spectra, *FEBS Lett.,* 108, 209, 1979.
206. **Weisshaar, H. D. and Palm, D.,** Role of pyridoxal 5'-phosphate in glycogen phosphorylase, *Biochemistry,* 11, 2146, 1972.
207. **Vidgoff, J. M., Pocker, A., Hullar, T. L., and Fischer, E. H.,** Interaction of muscle glycogen phosphorylase with pyridoxal 5'-methylenephosphonate, *Biochem. Biophys. Res. Commun.,* 57, 1166, 1974.
208. **Schnackerz, K. D. and Feldman, K.,** Pyridoxal-5'-deoxymethylenephosphonate reconstituted D-Serine hydratase: a ^{31}P NMR study, *Biochem. Biophys. Res. Commun.,* 95, 1832, 1980.
209. **Takagi, M., Shimomura, S., and Fukui, T.,** Function of the phosphate group of pyridoxal 5'-phosphate in the glycogen phosphorylase reaction, *J. Biol. Chem.,* 256, 728, 1981.
210. **Kastenschmidt, L. L., Kastenschmidt, J., and Helmreich, E. J. M.,** The effect of temperature on the allosteric transitions of rabbit muscle phosphorylase *b,* *Biochemistry,* 7, 4543, 1968.
211. **Palm, D., Blumenauer, G., Klein, H. W., Blanc-Muesser, M.,** α-glucan phosphorylases catalyse the glucosyl transfer from α-D-glucosyl fluoride to oligosaccharides, *Biochem. Biophys. Res. Commun.,* 111, 530, 1983.
212. **Chang, Y. C., McCalmont, T., and Graves, D. J.,** Functions of the 5'-phosphoryl group of pyridoxal 5'-phosphate in phosphorylase: a study using pyridoxal reconstituted enzyme as a model system, *Biochemistry,* 22, 4987, 1983.
213. **Withers, S. G., Madsen, N. B., Sykes, B. D., Takagi, M., Shimomura, S., and Fukui, T.,** Evidence for direct phosphate phosphate interaction between pyridoxal phosphate and substrate in the glycogen phosphorylase mechanism, *J. Biol. Chem.,* 256, 10759, 1981.
214. **Takagi, M., Fukui, T., and Shimomura, S.,** Catalytic mechanism of glycogen phosphorylase: pyridoxal (5')diphospho(1)-α-D-glucose as a transition state analogue, *Proc. Nat. Acad. Sci. U.S.A.,* 79, 3716, 1982.
215. **Soman, G. M., Chang, Y. C., and Graves, D. J.,** The effect of oxyanions of the early transition metals on rabbit skeletal muscle phosphorylase, *Biochemistry,* 22, 4994, 1983.
216. **Pocker, A. and Fischer, E. H.,** Synthesis of analogs of pyridoxal 5'-phosphate, *Biochemistry,* 8, 518, 1969.
217. **Pfeuffer, T., Ehrlich, J., and Helmreich, E.,** Role of pyridoxal 5'-phosphate in glycogen phosphorylase-II, *Biochemistry,* 11, 2125, 1972.
218. **Cornish, T. J. and Ledbetter, J. W.,** Interactions at the active site of glycogen phosphorylase *b:* a new laser probe, *Eur. J. Biochem.,* 143, 63, 1984.
219. **Grenwood, N. N. and Earnshaw, A.,** *Chemistry of the Elements,* Pergamon Press, Oxford, 1985.
220. **Madsen, N. B. and Withers, S. G.,** The catalytic mechanism of phosphorylase: novel role of the coenzyme, in *Chemical and Biological Aspects of Vitamin B$_6$ Catalysis,* Evangelopoulos, A. E., Ed., Alan R. Liss, New York, 1984, 117.
221. **Kastenschmidt, L. L., Kastenschmidt, J., and Helmreich, E.,** The effect of temperature on the allosteric transitions of rabbit skeletal muscle phosphorylase *b,* *Biochemistry,* 7, 4543, 1968.
222. **Birkett, D. J., Dwek, R. A., Radda, G. K., Richards, R. E., and Salmon, A. G.,** Probes for the conformational transitions of phosphorylase *b,* *Eur. J. Biochem.,* 20, 494, 1971.
223. **Dombradi, V., Hajdu, J., Bot, G., and Friedrich, P.,** Structural changes in glycogen phosphorylase as revealed by cross-linking with bifunctional diimidates: phospho-dephospho hybride and phosphorylase *a,* *Biochemistry,* 19, 2295, 1980.
224. **Gusev, N. B., Hajdu, J., and Friedrich, P.,** Motility of the N-terminal tail of phosphorylase *b* as revealed by cross-linking, *Biochem. Biophys. Res. Commun.,* 90, 70, 1979.
225. **Klein, H. W., Schiltz, E., and Helmreich, E. J. M.,** A catalytic role of the dianionic 5'-phosphate of pyridoxal phosphate in glycogen phosphorylases: formation of a covalent glycosyl intermediate, *Protein Phosphorylation,* 8, 305, 1981.
226. **Ullman, A., Vagelos, P. R., and Monod, J.,** The effect of 5'-adenylic acid upon the association between bromthymol blue and muscle phosphorylase *b,* *Biochem. Biophys. Res. Commun.,* 17, 86, 1964.
227. **Pathy, L.,** *Chemical Modification of Arginine Residues in Protein Structure and Evolution,* Fox, J. L., Deyl, Z., and Blacej, A., Eds., Marcel Dekker, New York, 1976, 91.
228. **Avramovic, O. and Madsen, N. B.,** Allosteric properties of phosphorylase *b.* III. Inactivation by cyanate and binding studies, *J. Biol. Chem.,* 243, 1656, 1968.

229. **Remy, P. and Buc, H.,** Evidence for a slow transition leading to co-operative binding of 5'-AMP onto rabbit muscle phosphorylase *b* at 5°, *FEBS Lett.,* 9, 152, 1970.

230. **Madsen, N. B. and Schechosky, S.,** Allosteric properties of phosphorylase *b*. II. Comparison with a kinetic model, *J. Biol. Chem.,* 242, 3301, 1967.

231. **Madsen, N. B.,** Allosteric properties of phosphorylase *b, Biochem. Biophys. Res. Commun.,* 15, 390, 1964.

232. **Merino C. G., Garcia-Blanco, I., and Layner, J.,** The effects of glycogen on phosphorylase *b* interactions with 5'AMP and phosphate, *FEBS Lett.,* 73, 97,

233. **Raibaud, O. and Goldberg, M. E.,** Characterization of two complementary polypeptide chains obtained by proteolysis of rabbit muscle phosphorylase, Biochemistry, 12, 5154, 1973.

234. **Gergely, P., Toth, B., Dombradi, V., Matko, J., and Bot, G.,** Heterotropic interactions of AMP and glucose binding sites in phosphorylase *a* are destroyed by limited proteolysis, *Biochem. Biophys. Res. Commun.,* 113, 825, 1983.

235. **Dombradi, V., Toth, B., Gergely, P., and Bot, G.,** Limited proetolysis of glycogen phosphorylase *a* by subtilisin BPN, *Int. J. Biochem.,* 15, 1329, 1983.

236. **Shimomura, S. and Fukui, T.,** Circular dichroism studies on glycogen phosphorylase from rabbit muscle. Interaction with the allosteric activator AMP, *Biochemistry,* 15, 4438, 1976.

237. **Vanderbunder, B., Dreyfus, M., and Buc, H.,** Conformational changes and local events at the AMP site of glycogen phosphorylase *b*. A fluorescene temperature jump relaxation study, *Biochemistry,* 17, 4153, 1978.

238. **Barford, D.,** unpublished results.

239. **Buehner, M.,** private communication.

240. **Sprang, S.,** private communication.

241. **Nakano, K., Hwang, P., and Fleterick, R. J.,** Complete cDNA sequence for rabbit muscle glycogen phosphorylase, *FEBS Lett.,* 204, 283, 1987.

242. **Oikonomakos, N. G., Acharya, K. R., Stuart, D. I., Melpidou, A. E., Mclaughlin, P. J., and Johnson, L. N.,** Uridine(5')diphospho(1)-D-glucose: a binding study to phosphorylase b in the crystal, *Eur. J. Biochem.,* 173, 569, 1988.

243. **Hadju, J., Machin, P. A., Campbell, J., Reenhough, T. J., Clifton, I. J., Zurek, S., Gover, S., Johnson, L. N., and Elder, M.,** Millisecond X-ray diffraction and the first electron density map from Laue photographs of a protein crystal, *Nature (London),* 329, 178, 1987.

244. **Hwang, P. K. and Fletterick, R. J.,** Convergent and divergent evolution of regulatory sites in eukaryotic phosphorylases, *Nature (London),* 324, 80, 1986.

245. **Newgard, C. B., Nakano, K., Hwang, P. K., and Fletterick, J.,** Sequence analysis of the cDNA encoding human liver glycogen phosphorylase reveals tissue specific codon usage, *Proc. Natl. Acad. Sci. U.S.A.,* 83, 8132, 1986.

246. **Burke, J., Hwang, P. K., Gorin, F., Lebo, R. V., and Fletterick, R. J.,** Intron/exon structure of the human gene for the muscle isozyme of glycogen phosphorylase, *Proteins,* 2, 177, 1987.

247. **Sprang, S. R., Acharya, K. R., Goldsmith, E. J., Stuart, D. I., Varvill, K., Fletterick, R. J., Madsen, N. B., and Johnson, L. N.,** Structural changes in glycogen phosphorylase induced by phosphorylation, *Nature London,* in press, 1988.

248. **Oikonomakos, N. G., Johnson, L. N., Acharya, K. R., Stuart, D. I., Barford, D., Hajdu, J., Varvill, K. M., Melpidou, A. E., Papageorgiou, T., Graves, D. J., and Palm, D.,** Pyridoxal phosphate site in glycogen phosphorylase b structure in native enzyme and in three derivatives with modified cofactors, *Biochemistry,* 26, 8381, 1987.

249. **Street, I. P., Armstrong, C. R., and Withers, S. G.,** Hydrogen bonding and specificity: fluoro sugars as probes of hydrogen bonding in the glycogen phosphorylase glucose complex, *Biochemistry,* 25, 6021, 1986.

250. **Johnson, L. N., Cheetham, J., Mclaughlin, P. J., Acharya, K. R., Barford, D., and Phillips, D. C.,** Protein-oligosaccharide interactions: lysozyme, phosphorylase and amylases, Curr. Top. Microbiol. Immunol., 139, 81, 1988.

251. **Sprang, S. R., Goldsmith, E. J., and Fletterick, R. J.,** Structure of the nucleotide activation switch in glycogen phosphorylase a, *Science,* 237, 1012, 1987.

252. **Palm, D., Schinzel, R., Zeier, R., and Klein, H.,** Site specific mutations in phosphorylase exerting different effects on substrate binding and catalysis, in *Biochemistry of Vitamin B6,* Proceedings of 7th International Congress on Chemical and Biological Aspects of Vitamin B6 Catalysis, Korpela, T. and Christen, P., Eds., Birkhaser, Basel, 1987.

253. **Barford, D., Schwabe, J. W. R., Oikonomakos, N. G., Acharya, K. R., Hajdu, J., Papageorgiou, A. C., Martin, J. L., Knott, J. C. A., Vasella, A., and Johnson, L. N.,** Channels at the catalytic site of glycogen phosphorylase b: binding and kinetic studies with the β-glycosidase inhibitor D-gluconohydroximo-1,5-lactone-N-phenylurethane, *Biochemistry,* 27, 6733, 1988.

254. **Kabsch, W. and Sander, C.,** Dictionary of protein structure: pattern recognition of hydrogen bonded and geometrical features, *Biopolymers,* 22, 2577, 1983.

Chapter 5

REGULATION OF BACTERIAL GLUTAMINE PHOSPHORIBOSYLPYROPHOSPHATE AMIDOTRANSFERASE

Robert L. Switzer

TABLE OF CONTENTS

I. INTRODUCTION

Even before the formal enunciation of the concept of allostery by Monod, Changeux, and Jacob[1] in 1963, regulation of glutamine phosphoribosylpyrophosphate (PRPP) amidotransferase by end product inhibition had been described.[2] This enzyme catalyzes the reaction

$$\text{Glutamine } + \text{ PRPP} \rightarrow \text{5-Phosphoribosyl-1-amine } + \text{ Glutamate } + \text{ PPi} \qquad (1)$$

which is the first committed step of purine nucleotide biosynthesis *de novo*.[3,4] Thus, it is an appropriate target for metabolic regulation. The initial studies by Wyngaarden's group were with the PRPP amidotransferase from avian liver,[2,5,6] but allosteric regulation of the enzyme from every source studied — bacteria,[7-10] yeast,[11,12] plants,[13] rats,[14] and humans[15,16] — has been observed. It is the purpose of this chapter to review the experimental evidence concerning the regulation of bacterial PRPP amidotransferases. Primary emphasis will be placed on the enzymes from *Bacillus subtilis* and *Escherichia coli* because they have been studied in the most detail. They also present a striking biochemical contrast: the *B. subtilis* enzyme contains an essential [4Fe-4S] cluster, whereas the *E. coli* enzyme does not.

As with many biosynthetic enzymes in bacteria, end product regulation of PRPP amidotransferase is exerted both by repression of enzyme synthesis and allosteric inhibition. This review will focus chiefly on the latter. Until recently, little was known concerning the biochemical details of repression of PRPP amidotransferase. Nierlich and Magasanik[17] reported that full repression of the *Klebsiella aerogenes* PRPP amidotransferase required the presence of both adenine and guanine. Adenosine and guanosine act to repress the enzyme in *B. subtilis*.[18,19] The two nucleosides appear to act independently and with equal effectiveness.[18,19] Recent studies in Zalkin's laboratory[19a,19b] provide considerable insight into mechanisms governing repression of *purF*, the gene encoding PRPP amidotransferase. The *E. coli purF* is the second gene in a bicistronic operon; the first gene encodes a 17.9 kDa polypeptide of unknown function.[19a] The expression of *E. coli purF* is regulated at the transcriptional level and is repressed by adenine. The promoter-regulatory region for the *purF* operon has been identified by sequencing, mapping of transcripts, and deletion analysis.[19a] A sequence that overlaps the promoter has been proposed to provide an operator site for the binding of a repressor protein. Preliminary evidence that *purF* transcription is repressed via the stringent response was also obtained.[19a] In *B. subtilis purF* lies in the middle of a polycistronic operon, which encodes all nine enzymes involved in biosynthesis of IMP.[19b] Mapping of the transcripts formed from this operon *in vivo* demonstrated two modes of regulation.[19b] Adenosine decreases the amount of *pur* mRNA, which indicates that it suppresses initiation of transcription, possibly via a regulatory protein. Guanosine, on the other hand, does not repress transcription initiation, but leads to elevated levels of prematurely terminated transcripts, which are encoded by an untranslated 5′ leader sequence in the operon. The organization of the *B. subtilis pur* operon into three clusters of overlapping cistrons punctuated by short untranslated segments may indicate that additional termination-antitermination mechanisms govern the relative abundance of gene products formed from the operon.[19b]

A third mode of regulation of PRPP amidotransferase in *B. subtilis* has been discovered and characterized in my laboratory. The enzyme is inactivated[20] and degraded[21] in nutrient-starved cells. These processes are believed to be part of a group of metabolic adaptations that commit the starving and sporulating cell to synthesis of new nucleic acid and proteins at the expense of turnover of already synthesized polymers.[22] The disappearance of the PRPP amidotransferase has been dissected into two steps:[21] (1) inactivation as a consequence of oxidation of the Fe-S cluster of the enzyme, and (2) degradation of the inactive enzyme to peptides and amino acids. These processes, especially the former, will also be discussed in this chapter.

II. COVALENT AND QUATERNARY STRUCTURE

A. Primary Structure

The primary structures of *E. coli*[23] and *B. subtilis*[24] PRPP amidotransferases have been determined in Zalkin's laboratory by cloning and sequencing of their structural genes. In each case, the amino acid sequence deduced from the DNA sequence is well supported by studies of the purified proteins.[24,25] Figure 1 shows the structures of the two enzymes, aligned by computer methods to give maximal homology.[24] Overall, the sequences of the two enzymes are about 40% conserved, with more highly conserved sequences in what are believed to be functional regions. Comparison of the coding region of the gene for the *B. subtilis* enzyme to the N-terminal sequence of the purified protein shows that the protein must undergo posttranslational processing to remove 11 N-terminal amino acid residues to form the mature enzyme, which has cysteine at the N-terminus. Only an N-terminal methionyl residue is removed from the *E. coli* enzyme. The *E. coli* enzyme (503 residues) is somewhat larger than the *B. subtilis* enzyme (465 residues in the mature protein) because of a C-terminal extension.

The gene for *B. subtilis* PRPP amidotransferase was cloned in an *E. coli purF* host.[24] The PRPP amidotransferase produced in such cells was isolated and shown to be identical to the enzyme produced by *B. subtilis*. In particular, the enzyme had a normal content of Fe-S clusters and had undergone N-terminal processing exactly as in *B. subtilis*.[24] Thus, even though *E. coli* cells do not insert Fe-S clusters into or cleave the 11 N-terminal amino acid residues from the *E. coli* PRPP amidotransferase, these processes occurred normally with the *B. subtilis* amidotransferase protein in an *E. coli* host. This indicates that specific *B. subtilis* enzymes are not required to catalyze these maturation events and that they may occur spontaneously or autocatalytically.

The N-terminal regions of both PRPP amidotransferases, which display some homology, have been clearly implicated in glutamine utilization.[25,26] The N-terminal cysteinyl residues of both mature enzymes react stoichiometrically with [^{14}C]6-diazo-5-oxo-L-norleucine to yield modified enzymes that are incapable of glutamine utilization, but which function normally with ammonia as the substrate. An elegant confirmation of the role of this residue in catalysis was obtained by Mäntsälä and Zalkin,[27] who replaced the cysteinyl residue with phenylalanine in the *B. subtilis* enzyme by site-directed mutagenesis. The mutant enzyme was functional *in vitro* and *in vivo* only with ammonia. It had also failed to undergo removal of the N-terminal 11 residues from the primary translation product, which demonstrates that the translation initiation codon that functions *in vivo* has been correctly identified. Failure to remove the N-terminal leader did not affect insertion of the Fe-S cluster or the allosteric properties of the enzyme.[27]

Further insight into the relation between the structure of the N-terminal segment of *B. subtilis* PRPP amidotransferase and N-terminal processing, Fe-S cluster insertion, and enzyme activity has been obtained from a series of point and deletion mutants prepared by site-directed mutagenesis.[27a] Correct processing of the proenzyme in the heterologous *E. coli* host was prevented by substitution of either of the two Glu residues that precede the active site Cys, as well as by substitution of the Cys residue itself. Correct processing is less dependent on other residues in the propeptide, but the propeptide must be longer than three residues. The enzyme has no glutamine-dependent activity if the proenzyme is not correctly processed. Surprisingly, the NH$_3$-dependent activity is also severely reduced in most unprocessed forms of the enzyme. Unprocessed forms of the enzyme and a mutant from which the entire propeptide was deleted were synthesized in *E. coli* with normal [4Fe-4S] clusters. The mutant enzyme from which the propeptide was deleted had activity equal to the wild type enzyme; its initiator formyl-Met was removed to uncover the N-terminal Cys. These results establish that the propeptide is not essential for N-terminal maturation or Fe-S cluster insertion. The physiological function of N-terminal processing remains unclear.

```
              1                    20                   40
E. coli       M C G I V G I A G V M P V N Q S I Y D A L T V L Q H R G Q D A A G I T I D A N N C F R
B. subtilis   M L A E I K G L N E E C G G V F G I W G H E E A P Q I T Y Y G L H S L Q H R G Q E G A G I V A T D G E K L T
              1                    20                   40

                        60                   80                   100
E. coli       S L K A N A L V S D V F E A R H M Q R L Q G N M G I G H V R Y P T A G S S S A S E A Q P F Y V N S P Y G I T L A H N G
B. subtilis   A H K G Q G L I T E V F Q N G E L S K V V K G K G A I G H V R Y A T A G G G G Y E N V Q P L L F R S Q N N G S L A L A H N G
                        60                   80                   100

                        120                  140                  160
E. coli       N L T N A H E L R K K L F E E K R R H I N T T S D S E I L L N I F A S E L D N F R H Y P L E A D N I F A A I A A T N R L I
B. subtilis   N L V N A T Q L K Q Q L E N Q G S I F Q T S S D T E V L A H L I K R S G H F T L K D Q I K N S L S M L
                        120                  140                  160

                        180                  200                  220
E. coli       R G A Y A C V A M I I G H G M V A F R D P N G I R P L V L G K R D I D E N R T E Y M V A S E S V G S I R W A L I S C V T S
B. subtilis   K G A Y A F L I M T E T E M I V A L D P N G L R P L S I G M G D A Y V V A S E T C A F D V V G A T Y L R E V
                        180                  200                  220

                        240                  260                  280
E. coli       R R A R I Y T E E G Q L F T R Q C A D N P V S N P C L F E Y V Y F A R P D S F I D K I S V Y S A R A N M G T K V G E K I
B. subtilis   E P G E M L I N D E G M K S E R F S M N I N R S I C S M E Y I Y F S R P D S N I D G I N V H S A R K N L G K M L
                        240                  260                  280

                        300                  320                  340
E. coli       A R E W E D L D I D V V I P I P F T S C I A L F I A R I L G K P Y R Q G F V K N R Y V G R T F I M P G Q L R R K S V R
B. subtilis   A Q E S A V E A D V V T G V P D S S I S A A I G Y A E A T G I P Y E L G L I K N R R Y V G R T F I Q P S Q A L R E Q G V R
                        300                  320                  340

                        360                  380                  400
E. coli       R K L N A N R A E F R D K N V L L V D D S I V R G T T S E Q L I E M A R E A G A K K V Y L A S A A P E I R F P N V Y G I D
B. subtilis   M K L S A V R G V V E G K R V V M V D D S I V R G T T S R R I V T M L R E A G A T E V H V K I S S P P I A H P C F Y G I D
                        360                  380
```

```
                    420                         440
M P S A T E L I A H G R E V D E I R Q I I G A D G L I F Q D  L N D L L I D A V R A E N P D I Q Q F E C S V F N G V V V T K
T S T H E E L I A S S H S V G E I R Q E I G A D T L S F L S V E G L L K G I G R K Y D D S N C G Q C L A C F T G K Y P T E
  400                                  420                         440                              460

D V D Q G Y L D F L D T L R N D D A K A V Q R Q S E V E N L E M H N E G
I Y Q D T V L P H V K E A V L T K
  460                        480                        500
```

FIGURE 1. Amino acid sequence of *E. coli* and *B. subtilis* PRPP amidotransferases. The sequences have been aligned to give maximal homology. Identical amino acid residues are underlined and overlined. Numbering refers to the sequence of the primary translation products deduced from the gene sequence; both enzymes undergo N-terminal processing to form the mature proteins (see text). A, Ala; C, Cys; D, Asp; E, Glu; F, Phe; G, Gly; H, His; I, Ile; K, Lys; L, Leu; M, Met; N, Asn; P, Pro; Q, Gln; R, Arg; S, Ser; T, Thr; V, Val; W, Trp; Y, Tyr. (From Makaroff, C. A. et al., *J. Biol. Chem.*, 258, 10586, 1983. With permission.)

Cysteinyl residues have been implicated in glutamine utilization for a number of amidotransferases. One of these amidotransferases, glucosamine 6-phosphate synthetase (encoded by *glmS*) from *E. coli* shares considerable homology in its N-terminal 180 residues with the two PRPP amidotransferases.[28] After removal of *N*-formylmethionine, the protein encoded by *glmS* would have a N-terminal cysteinyl residue. It is likely that this residue is involved in glutamine hydrolysis, as has been shown for the PRPP amidotransferases. It is very interesting that cysteinyl residues involved in glutamine hydrolysis by a variety of other amidotransferases also lie in homologous sequences, but these cysteinyl residues are internal, rather than N-terminal, and occur in sequences which are not homologous to the putative glutamine-utilizing sequences of *glmS* or PRPP amidotransferase.[28,29] It appears that two different families of homologous sequences encoding glutamine utilizing sites in amidotransferases have evolved.

The C-terminal regions of the two PRPP amidotransferases have little homology; it is in this region that the binding site for the Fe-S cluster is found in the *B. subtilis* enzyme. By comparison to sequences of the cluster binding sites of [4Fe-4S] ferredoxins, the sequence -Cys(445)-Gly-Gln-Cys(448)-Leu-Ala-Cys(451)- was proposed to contain three of the four cysteinyl ligands of the Fe-S cluster.[24] The fourth ligand was proposed to be Cys-393, which lies in the sequence -Pro-Cys-Phe-Tyr-. No cysteinyl residues are found in homologous positions or in similar arrays in the *E. coli* enzyme, which contains no Fe-S cluster.[23] Further evidence that these cysteine residues are involved in assembly of the Fe-S cluster and in correct folding to yield an active enzyme has come from studies of the proteins formed when these residues are substituted by *in vitro* mutagenesis.[29a] Substitution of either Cys-448 or Cys-451 by Ser leads to formation of a mutant protein which contains no Fe-S clusters,[29b] is catalytically inactive, and is not processed at the N-terminus. Failure of these mutant proteins to incorporate Fe-S clusters provides direct evidence that Cys-448 and Cys-451 are liganded to the Fe-S cluster in the wild type enzyme. Another mutant protein in which Phe-394, which lies adjacent to the putative fourth cysteinyl ligand, was replaced by Val was capable of Fe-S cluster insertion and N-terminal processing.[29a] This enzyme had only about 15% of wild type catalytic activity, however, which suggests that the Phe to Val substitution perturbs the catalytic site. Since an intact [4Fe-4S] cluster is essential for activity, the reduced activity of the mutant is consistent with the proposal that Cys-393 is ligated to the Fe-S cluster and that the protein sequences adjacent to Cys-393 communicate directly with the active site. A double mutant in which Asp-442 is replaced by Cys and Cys-451 is substituted by Ser, i.e., in which the Cys-X-X-Cys-X-X-Cys liganding triad is shifted by three residues toward the N-terminus, was capable of forming an active, processed enzyme with an Fe-S cluster inserted.[29a] Not surprisingly, this mutant enzyme has only about 25% of wild type activity, which further documents the sensitivity of the catalytic site to perturbations in the Fe-S cluster and its ligand sequences.

Two homologous regions of the PRPP amidotransferase sequences have been suggested to encode functional regions by comparison to primary and secondary structures of several enzymes.[24,30] These are residues 234 to 353 in the *E. coli* sequence (229 to 344 in *B. subtilis*), which are predicted to encode a nucleotide binding site, and residues 354 to 450 in the *E. coli* sequence (345 to 440 in *B. subtilis*), which are homologous to the active sites of other phosphoribosyltransferases. These sequences are also conserved in the yeast PRPP amidotransferase structural gene.[31]

B. Quaternary Structure

The molecular weights of the mature *E. coli* and *B. subtilis* polypeptide chains are 56,263 and 50,397, respectively. These values are in excellent agreement with other estimates of the subunit molecular weights of these proteins.[8,32] The pure, native *E. coli* enzyme appears to exist entirely as a tetramer.[8] The quaternary structure of the *B. subtilis* PRPP amidotrans-

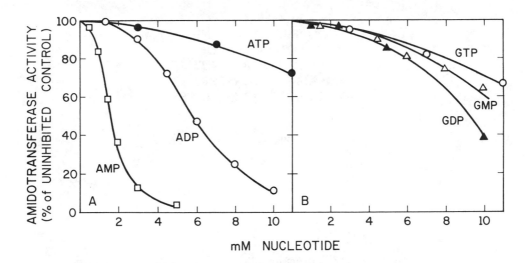

FIGURE 2.   Inhibition of *B. subtilis* PRPP amidotransferase by purine ribonucleotides. Standard assay conditions (2.5 m*M* PRPP, 20 m*M* glutamine, 5 m*M* MgCl$_2$, 2 m*M* EGTA, 50 m*M* Tris/HCl, pH 7.9, 37°C) were used. (From Meyer, E. and Switzer, R. L., *J. Biol. Chem.*, 254, 5397, 1979. With permission.)

ferase is more complex; this enzyme exists in an equilibrium among tetrameric, dimeric, and possibly monomeric forms, depending on enzyme concentration and the presence of ligands.[32] The tetrameric form predominates at higher protein concentrations; the dimeric form is seen upon dilution. The allosteric inhibitors AMP and GMP appear to stabilize the dimeric form, but another inhibitor, GDP, stabilizes the tetramer.[32] There is no obvious correlation between quaternary structure and binding of allosteric inhibitors, as has been suggested for the human enzyme.[33]

III. ALLOSTERIC REGULATION

A. Kinetic Patterns of Inhibition by End Products

The bacterial PRPP amidotransferases are especially suitable for the study of allosteric regulation of this reaction, because, unlike the avian liver enzymes,[5,6,34] their sensitivity to inhibitors is not lost or altered during purification. Kinetic studies of the end product inhibition of the pure enzymes from *E. coli*[8] and *B. subtilis*[10] are available. These are summarized here. Emphasis is placed on the results with *B. subtilis* PRPP amidotransferase, with which the author is more familiar and somewhat more detailed studies have been published, but important contrasts between the enzymes from both sources will be noted.

PRPP amidotransferase is specifically inhibited by purine ribonucleotides. Pyrimidine nucleotides and purine 2'-deoxyribonucleotides are not inhibitory. The only effective inhibitors of *B. subtilis* PRPP amidotransferase were adenine and guanine nucleotides (Figure 2). AMP was by far the most effective inhibitor, followed in order of effectiveness by ADP, GDP, GMP, GTP, and ATP. IMP and XMP were not inhibitory. Inhibition at the AMP site is extremely specific: 3'-AMP, 2'-dAMP, 2'3'-cAMP, 3'5'-cAMP, adenosine, adenine, formycin monophosphate, and 3-iso-AMP are not inhibitors. The specificity of the *E. coli* PRPP amidotransferase is quite different. GMP is the most effective inhibitor, followed in order by GDP, IMP, GTP, and AMP.

Concentration dependence curves of nucleotide inhibition exhibit positive cooperativity, which is quite strong in many cases (Table 1). Hill coefficients in the range of 3.3 to 3.8 are observed with the *B. subtilis* enzyme. In general, the most effective inhibitors exhibit the strongest cooperativity. Inhibition of the *E. coli* enzyme by GMP and AMP is also strongly cooperative (Figure 3).

Table 1
COOPERATIVITY AND
EFFECTIVENESS OF
INHIBITION OF *B. SUBTILIS*
PRPP AMIDOTRANSFERASE
BY PURINE NUCLEOTIDES[a]

Inhibitor	Hill coefficient	$I_{0.5}$[b] (mM)
AMP	3.3	1.6
ADP	3.8	5.8
ATP	1.8	18
GMP	1.2	12
GDP	3.3	9
GTP	1.7	15

[a]   Data from Reference 10.
[b]   Concentration giving 50% inhibition under standard assay conditions (2.5 mM PRPP, 20 mM glutamine, 5 mM MgCl$_2$, 2 mM EGTA, 50 mM Tris/HCl, pH 7.9, 37°C). These values are very sensitive to divalent cation and PRPP concentration and pH.

From Meyer, E. and Switzer, R. L., *J. Biol. Chem.*, 254, 5397, 1979. With permission.

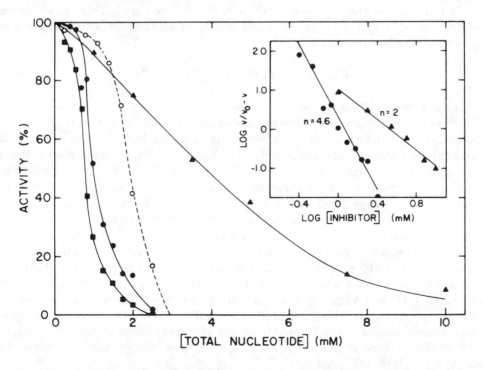

FIGURE 3.   Inhibition of *E. coli* PRPP amidotransferase by AMP and GMP. Assay conditions were 1 mM PRPP, 5 mM glutamine, 10 mM MgCl$_2$, pH 7.4, 37°C. AMP (▲), GMP (●), or equimolar mixtures of both (■) were present as indicated. The curve calculated for additive inhibition by equimolar AMP and GMP (○) is also shown. The insert shows that data for AMP (▲) and GMP (●) plotted according to the Hill equation for determination of Hill coefficients (n). (From Messenger, L. J. and Zalkin, H., *J. Biol. Chem.*, 254, 3382, 1979. With permission.)

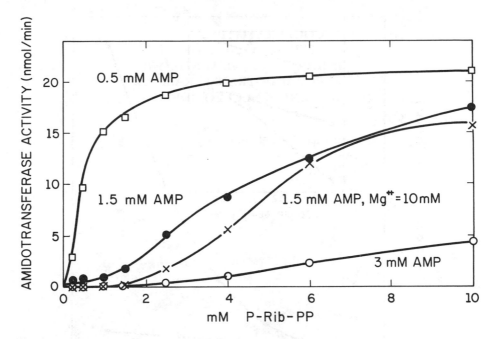

FIGURE 4. Interactions between AMP inhibition and PRPP saturation of *B. subtilis* PRPP amidotransferase. Standard assay conditions (Figure 2) were used except that the PRPP and AMP concentrations shown were added and the total MgCl$_2$ concentration was equal to PRPP + AMP + 3 mM. In one experiment (×) the total MgCl$_2$ was held constant at 10 m*M*. (From Meyer, E. and Switzer, R. L., *J. Biol. Chem.*, 254, 5397, 1979. With permission.)

The bacterial PRPP amidotransferases resemble this enzyme from other sources in displaying a strong antagonism between saturation with PRPP and nucleotide inhibition. For both the *B. subtilis* and *E. coli* enzymes, PRPP saturation is hyperbolic in the absence of inhibitors and has a K$_m$ value of about 70 μ*M*. In the presence of a purine nucleotide inhibitor, the PRPP saturation curve becomes sigmoid (Hill coefficient approaching 2), and the half-saturating concentration of PRPP is much higher. This is illustrated for AMP inhibition of the *B. subtilis* enzyme in Figure 4. AMP inhibition is reversed by saturation with PRPP, but the *B. subtilis* enzyme does not exhibit pure "K-system"[1] kinetics, because the V$_{max}$ is also reduced at high inhibitor levels. The degree of saturation by glutamine has relatively little effect on sensitivity to inhibitors. The *E. coli* PRPP amidotransferase presents an interesting contrast to the *B. subtilis* enzyme (Figure 5). Inhibition by GMP is competitive with PRPP and converts the PRPP saturation curve to sigmoid form as just described. However, AMP inhibition, while also competitive with PRPP saturation, does not induce cooperativity of PRPP saturation.

An interesting property of PRPP amidotransferase from most sources is their capacity for synergistic inhibition by pairs of nucleotides, i.e., mixtures of nucleotides give much stronger inhibition than would be expected from the inhibition observed with equivalent concentrations of each nucleotide acting alone. This forms the basis of a very efficient control mechanism, because it requires that both adenine and guanine nucleotide pools be elevated before *de novo* purine nucleotide synthesis is strongly inhibited. This phenomenon is illustrated for the *E. coli* enzyme in Figure 3. A thorough study of the *B. subtilis* enzyme demonstrated that three pairs of nucleotides were strongly synergistic inhibitors: ADP + GMP, ADP + GDP, and GMP + GDP (Table 2). Other pairs of nucleotides showed either weak or no synergism. AMP, which is a very potent inhibitor by itself, was not a partner in synergistic inhibition.

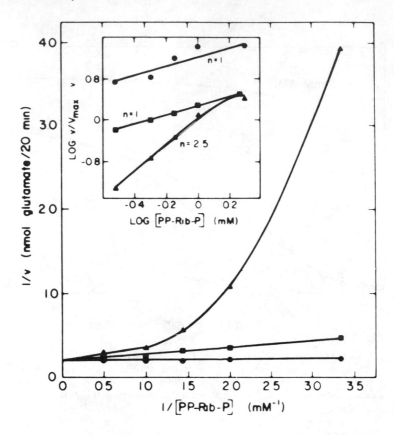

FIGURE 5.   Interations between AMP and GMP inhibition and PRPP saturation of *E. coli* PRPP amidotransferase. Conditions were as in Figure 3 except that the PRPP concentrations indicated were used with no inhibitor (●), 0.95 m*M* GMP (▲), or 2 m*M* AMP (■). The inset shows that data plotted according to the Hill equation for determination of Hill coefficients (n). (From Messenger, L. J. and Zalkin, H., *J. Biol. Chem.*, 254, 3382, 1979. With permission.)

Table 2
SYNERGISTIC INHIBITION OF *B. SUBTILIS* PRPP AMIDOTRANSFERASE BY PAIRS OF PURINE NUCLEOTIDES[a]

Inhibitors		Observed $I_{0.5}$[b] (m*M*)	Calculated $I_{0.5}$[c] (m*M*)	Synergistic inhibition?
AMP	+ ADP	2.0	3.5	Weak
AMP	+ GMP	3.5	3.5	None
AMP	+ GDP	3.8	3.5	None
ADP	+ GMP	1.1	11	Strong
ADP	+ GDP	3.2	11	Strong
ADP	+ ATP	10	11	None
GMP	+ GDP	5	15	Strong
GDP	+ GTP	13	15	None
ATP	+ GTP	14	19	Weak
IDP	+ GMP	7	11	Weak

[a]   Modified from Reference 10.
[b] $I_{0.5}$ is defined and determined as described under Table 1.
[c]   Calculated according to Reference 5.

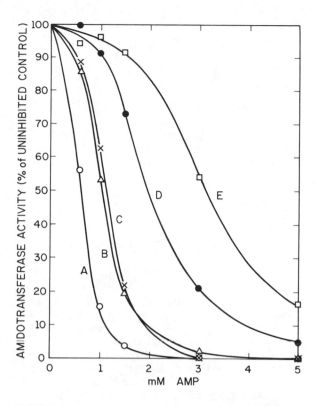

FIGURE 6. Effects of Mg^{2+} concentration on the inhibition of *B. subtilis* PRPP amidotransferase. Standard assay conditions (Figure 2) were used except as follows: A (○) 8.5 m*M* $MgCl_2$; B (△) 5.0 m*M* $MgCl_2$; C (X) total $MgCl_2$ concentration equaled the sum of PRPP + AMP + 1 m*M*; D (●) 2.5 m*M* $MgCl_2$; E (□) 1.25 mM $MgCl_2$. No EGTA was included in the assay mixture. (From Meyer, E. and Switzer, R. L., *J. Biol. Chem.*, 254, 5397, 1979. With permission.)

Two sets of kinetic observations strongly support the proposal that nucleotide inhibitors of the *B. subtilis* enzyme act at allosteric sites, rather than at the catalytic site. First, the enzyme exhibits essentially constant activity in the pH range from 6.0 to 8.4. At pH values of 6 and below, the enzyme is completely insensitive to inhibition by 1.5 m*M* AMP.[10] The sensitivity to AMP inhibition increases sharply as the pH is raised to 8.0 or above. Second, the *B. subtilis* enzyme requires divalent cations for activity and for sensitivity to allosteric inhibitors, but the concentration dependence and cation specificity of the two requirements are so different as to indicate that two separate cation binding sites are involved.[10] These points are documented in Figures 6 and 7, respectively. In Figure 6, it is seen that an increase in the total Mg^{2+} concentration from 1.25 to 8.5 m*M* markedly increases the sensitivity of the amidotransferase to inhibition by AMP. The same is true for ADP and GMP inhibition. Yet the activity of the enzyme in the absence of nucleotides is constant and maximal throughout this range of Mg^{2+} concentrations. This suggests that a divalent cation site required for catalysis was saturated and that a second, weaker cation site required for allosteric inhibition was not. The results shown in Figure 7 are also readily explained by postulating two cation sites. Co^{2+} and Mn^{2+} support activity nearly as well as Mg^{2+}, but the enzyme is virtually insensitive to AMP inhibition when Co^{2+} is the activating cation. In contrast, the Mn^{2+} activated enzyme is even more sensitive to AMP inhibition than when Mg^{2+} is present.

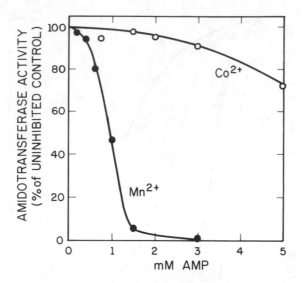

FIGURE 7.   Dependence of AMP inhibition of *B. subtilis* PRPP amidotransferase on the divalent cation present. Standard assay conditions (Figure 2) were used except EGTA was omitted and 5 mM CoCl$_2$ (○) or 5 mM MnCl$_2$ (●) were used in place of MgCl$_2$ and the buffer was 50 mM Tris/Maleate, pH 6.5. Activities without AMP were with MgCl$_2$, 100%; with CoCl$_2$, 93%; with MnCl$_2$, 90% under these conditions. (From Meyer, E. and Switzer, R. L., *J. Biol. Chem.*, 254, 5397, 1979. With permission.)

It is most likely that the high affinity cation site involves tight, direct binding of the cation to the PRPP amidotransferase because the K$_a$ for Mg^{2+} activation of the enzyme (0.2 mM) is independent of PRPP concentration and smaller than the dissociation constant for the Mg-PRPP complex (0.6 mM).[10] In the case of the lower affinity cation site, two possibilities are consistent with the data. The actual inhibitory species may be the divalent cation-nucleotide complex (e.g., Mg-AMP), or the cation may bind directly to the enzyme.

B. Number and Nature of Allosteric Sites

The kinetic studies described above do not fully define the number and specificity of nucleotide binding sites on the bacterial PRPP amidotransferases. To do so would require direct binding studies by equilibrium dialysis or other techniques. Nonetheless, the kinetic studies permit some reasonable suggestions about the nature of the allosteric sites. In the case of the *B. subtilis* PRPP amidotransferase, the marked effects of nucleotide inhibitors on the rate of reaction of the enzyme with oxygen, which leads to irreversible inactivation, provides an independent assessment of nucleotide binding.[35] These results, which are described in the next section, influence my interpretation of the kinetic results.

In the case of *E. coli* PRPP amidotransferase, the results[8] suggest the existence of two separate nucleotide sites. This is most strongly indicated by the potent synergistic inhibition by GMP plus AMP. Both nucleotides must bind simultaneously to have such effects. Simultaneous binding requires either that each subunit has two nucleotide sites with different specificity, or, less likely, that nucleotide sites on adjacent subunits in the native tetramer can interact more strongly when occupied by different nucleotides than when occupied by the same nucleotide. Differences in the kinetic consequences of AMP and GMP binding also suggest that these nucleotides bind at different sites. GMP induces cooperativity in PRPP binding; AMP does not. GMP inhibition exhibits a Hill coefficient of 4.6; the Hill coefficient for AMP inhibition is 2. This latter argument is not airtight. According to the London-Schmidt[36] model, which has received strong experimental confirmation, CTP and ATP bind in different orientations to the same site on *E. coli* aspartate transcarbamylase and cause quite different kinetic effects.

Because the *E. coli* and *B. subtilis* PRPP amidotransferases are homologous (Figure 1), it is likely that these two enzymes have the same number of nucleotide binding sites on each subunit, even though the specificity and interactions of these sites with the active site differ substantially. This constraint influences our conclusions concerning the number of allosteric sites on each enzyme.

The evidence that nucleotides bind at allosteric sites on the *B. subtilis* enzyme as opposed to the active site — in spite of the kinetic antagonism between PRPP and nucleotides — is very strong.[10] First, the specificity of the inhibitory site is very marked, especially for AMP binding. Furthermore, the sensitivity to allosteric inhibition can be abolished without loss of catalytic activity by simply conducting assays at pH 6, at low Mg^{2+}, or with Co^{2+} as the activating cation. Finally, separate regions encoding a nucleotide binding site, a phosphoribosyltransferase catalytic site, and the glutamine utilizing site have been identified in the amino acid sequences of the PRPP amidotransferases (see Section II).

The kinetic data and observations of the effects of nucleotides on the rate of reaction of *B. subtilis* PRPP amidotransferase with O_2 suggest that this enzyme has two types of allosteric sites.[35] One site binds 5′-AMP with high affinity and specificity. AMP binding at this site causes strong stabilization of the enzyme against O_2 in a positively cooperative manner (see Figure 9), except at low concentrations (25 to 100 μM) where destabilization is consistently observed. Interestingly, the same concentrations of AMP cause a 10% activation of the amidotransferase under some assay conditions.[10] Both inhibition by AMP and stabilization against O_2 are strongly antagonized by PRPP. AMP, when bound at this site, does not act as a partner in synergistic inhibition. GMP, GDP, and possibly ADP, appear to bind at another, somewhat less specific site. Binding of GMP and GDP leads to destabilization of the PRPP amidotransferase to O_2 (Figure 10). High concentrations of these nucleotides are required to inhibit the enzyme (Table 2), but their effects on inactivation are maximal at 1 mM. This apparent contradiction is probably resolved by the observation that the inhibitory effects are antagonized by PRPP, whereas the effects of reaction with O_2 are not (Figure 10). Nucleotides bound to this site appear to be effective partners in synergistic inhibition.

Altogether, the observations on both *E. coli* and *B. subtilis* PRPP amidotransferases fit well with the hypothesis that these enzymes have two allosteric nucleotide binding sites with different specificities on each subunit. It is possible, however, to argue that the *B. subtilis* enzyme has three nucleotide binding sites. This view arises from the observation that AMP is not a partner in synergistic inhibition. How do ADP and GMP or GMP and GDP yield synergistic inhibition if both bind at the same site? The existence of a third nucleotide binding site would provide a solution. An alternative explanation would be that occupancy of the GMP/GDP site alters the specificity of the AMP site such that other nucleotides can bind there and act synergistically. When AMP is present, it binds so effectively to the AMP site that such effects are not detectable. The effects of synergistic pairs on the reaction of the enzyme with O_2 tend to argue against this alternative. Occupancy of the AMP site strongly stabilizes, but all of the synergistic pairs destabilize even more strongly than the individual nucleotides.[35] Another way in which synergistic inhibition might occur is through strongly synergistic interactions between GMP/GDP/ADP sites on adjacent subunits in the tetrameric enzyme.

Can all of the observations on the *B. subtilis* PRPP amidotransferase be accommodated by the model in which there is only one allosteric nucleotide site per subunit? If one allows the assumption just stated that synergistic inhibition occurs by interactions between subunits, the answer is yes. This assumption requires that occupancy of identical sites by different nucleotides gives much stronger inhibition than when such sites are occupied by the same nucleotide, which seems somewhat unlikely. The differences between AMP effects and those of the other nucleotides can be accommodated in a one site model if one proposes that the nucleotides bind to the same site, but in different, highly specific conformations, as in the London-Schmidt[36] model for aspartate transcarbamylase. Clearly, a definitive resolution of these questions will require direct measurement of the number of nucleotide binding sites on the enzyme and determination of whether different nucleotides are bound simultaneously.

As with the *E. coli* enzyme, the *B. subtilis* enzyme displays strong positive cooperativity in nucleotide inhibition (Hill coefficients approaching 4) and PRPP saturation (Hill coefficients up to 2). This suggests that the nucleotide sites on adjacent subunits are strongly interactive and that the enzyme exists as a tetramer under assay conditions, even though sedimentation velocity and gel filtration studies of the enzyme indicated that the dimeric form predominates at high dilution.[32] PRPP promotes aggregation of the dilute enzyme,[32] so a tetramer may form during catalysis.

While the substrates and allosteric inhibitors of *B. subtilis* PRPP amidotransferase clearly induce conformational changes in the enzyme, I do not believe that the properties of this enzyme can be satisfactorily fit to a two-state model such as described by Monod et al.[37] The effects of nucleotides and PRPP on the state of aggregation,[32] catalytic activity,[10] and the rate of oxidative inactivation[35] (see Section IV.C) cannot be readily rationalized by postulating the existence of only two conformation states (i.e., R- and T-states). There is no simple correlation among the ligands that stabilize the dimeric vs. tetrameric forms of the enzyme and those that stabilize or destabilize the enzyme to O_2 or act as inhibitors or substrates. The results indicate that the enzyme must exist in a variety of conformational states.

IV. OXIDATIVE INACTIVATION OF *B. SUBTILIS* PRPP AMIDOTRANSFERASE

A. Physiological Significance

PRPP amidotransferase is one of a number of enzymes of nucleotide and amino acid biosynthesis that are inactivated during sporulation of *B. subtilis* cells.[22,38] Such inactivation processes contribute to the cellular economy by preventing wasteful biosynthesis of metabolites in starving cells. Culture conditions that led to inactivation of PRPP amidotransferase include starvation for a carbon source[20] or a nitrogen source,[39] or starvation of an auxotroph for an amino acid.[20,21] The addition of many antibiotics also leads to inactivation, although at widely differing rates.[21,39]

The inactivation of PRPP amidotransferase *in vivo* specifically requires O_2 and can be blocked, even after it has begun, by sparging the culture with an inert gas.[20] The requirement is for O_2 itself, not for O_2 supported metabolism or metabolic energy.[20] The enzyme is stable in exponentially growing cells,[20,21] however, so the disappearance of the enzyme from starving cells results from a change in stability rather than from cessation of synthesis combined with continuous inactivation.

B. Chemical Nature of the Inactivation

The role of O_2 in inactivation of *B. subtilis* PRPP amidotransferase became clear when it was discovered that the enzyme is inactivated by O_2 *in vitro,* both in crude extracts[40] and as the purified, homogeneous protein.[41] This inactivation is not prevented or reversed by reducing agents or thiols.[35,40] Rather, the exclusive site of reaction with O_2 appears to be a [4Fe-4S] cluster[41] in the enzyme whose integrity is essential for enzyme activity.[32,35,41] The active form of the oxidant is O_2 rather than other activated or reduced oxygen species such as singlet oxygen, peroxide superoxide, or hydroxy radical.[35] There is no evidence for an

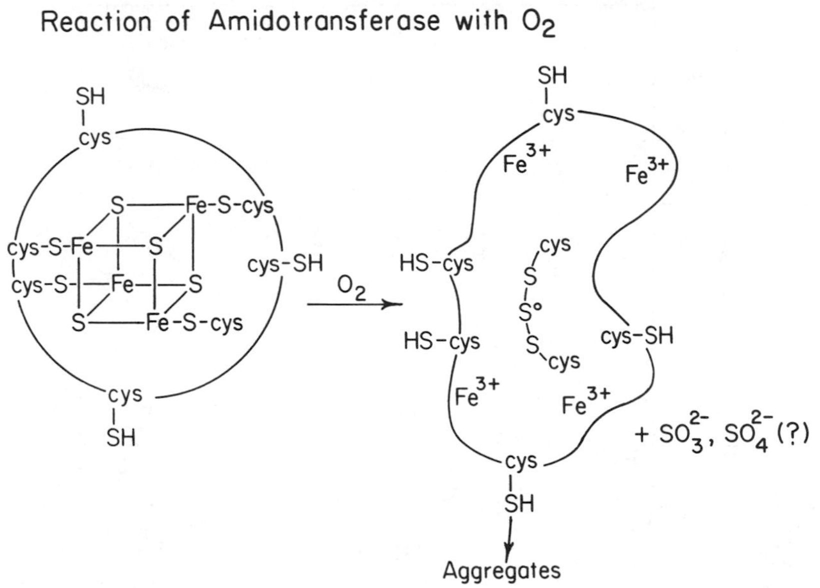

Reaction of Amidotransferase with O_2

FIGURE 8.   Schematic summary of the reaction of *B. subtilis* PRPP amidotransferase with oxygen. (From Bernlohr, D. A. and Switzer, R. L., *Biochemistry*, 20, 5675, 1981. With permission.)

inactivating enzyme in the oxidation process. It is important to note that the PRPP amido-transferases of *E. coli* and *S. cerevisiae,* which do not contain Fe-S clusters, are relatively stable in starving cells.[8,31]

The chemistry of the reaction of PRPP amidotransferase with O_2 was studied in detail with the pure enzyme.[35] The results are summarized in schematic form in Figure 8. Oxidation results in the total dissolution of the [4Fe-4S] cluster, conversion of the Fe atoms to high spin Fe^{3+}, trapping of 1 mol of S^{2-} as zero valent sulfur in a thiocystine linkage in the protein, and release of the rest of the S^{2-} as unidentified oxidized species. No stable Fe-S intermediates are detectable during the oxidation, but it is probable that a [3Fe-4S] "broken cube" structure is a transient intermediate.[43] Of the four sulfhydryl groups that are ligated to the Fe-S cluster in the native enzyme, two are released as free sulfhydryl groups and the other two are converted to the thiocystine residue. Three other sulfhydryl groups in each subunit remain unchanged. On prolonged exposure to O_2 the inactive enzyme aggregates and disulfide bonds are formed.

Oxidation of PRPP amidotransferase is accompanied by dramatic physical changes in the protein that are akin to heat denaturation.[35] The enzyme becomes much less soluble, es-pecially in the presence of divalent cations. Changes in the circular dichroism spectrum indicate not only dissolution of the Fe-S cluster, but changes in polypeptide folding as well. The oxidized enzyme is much more susceptible to proteolysis than is the native enzyme. We believe that these changes demonstrate a crucial role of the Fe-S cluster in maintaining the native tertiary structure of this enzyme. We also have proposed that similar changes accompany oxidation *in vivo* and that these changes lead to the proteolytic degradation of the enzyme in starving cells.[21]

C. Regulation of the Inactivation by Substrates and Nucleotides *In Vitro*

From studies conducted with crude extracts, it was already evident that substrates and allosteric ligands had marked effects on the rate of inactivation of PRPP amidotransferase by O_2 and might serve to regulate the inactivation *in vivo*.[40] Subsequently, these effects were fully characterized with the pure enzyme under carefully controlled conditions.[35]

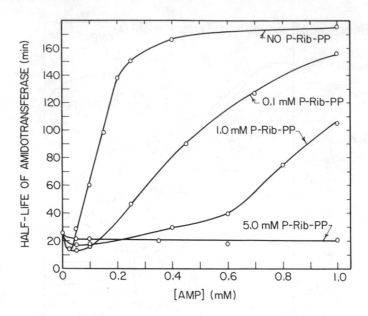

FIGURE 9. Effects of AMP and PRPP on the rate of oxidative inactivation of *B. subtilis* PRPP amidotransferase. Conditions were as described under Table 3. (From Bernlohr, D. A. and Switzer, R. L., *Biochemistry,* 20, 5675, 1981. With permission.)

The substrates glutamine or PRPP by themselves have little effect on the rate of inactivation by O_2. It is difficult to determine accurately the effect of the two substrates together because they are rapidly consumed by irreversible conversion to products. The best estimate from an indirect determination is that the two substrates decrease the rate of inactivation by O_2 by about threefold.[44]

As noted above, 5'-AMP is a highly specific and effective stabilizer (six- to sevenfold) of the enzyme against O_2. The concentration curve for AMP stabilization shows strong positive cooperativity, and AMP stabilization is antagonized by PRPP (Figure 9). All other allosteric inhibitors are destabilizers of PRPP amidotransferase toward O_2. GMP and GDP, for example, reduce the half-life of the enzyme by about half (Figure 10). These nucleotides act cooperatively and are effective at quite low concentrations. Their effects on stability are not reversed by PRPP. Pairs of nucleotides that act as synergistic inhibitors are also very effective destabilizers. The results of a series of studies of the effects of nucleotides and nucleotide pairs on oxidative inactivation of PRPP amidotransferase are summarized in Table 3. The effects of stabilizing and destabilizing nucleotides were antagonistic in mixtures, so that the rate of inactivation of the enzyme by O_2 could be varied over a 30-fold range by including appropriate nucleotide mixtures.

D. Regulation of the Inactivation *In Vivo*

PRPP amidotransferase is stable in exponentially growing *B. subtilis* cells, even though the medium is vigorously aerated.[20,21] The enzyme is inactivated in aerated nongrowing cells with a half-life ranging from 40 to 200 min depending on the culture conditions.[21,39] What factors account for these changes in stability? The pronounced effects of substrates and allosteric inhibitors on the rate of oxidative inactivation of the enzyme *in vitro* seem to provide a reasonable explanation.[35,40] According to this hypothesis, the PRPP amidotransferase is stabilized by the simultaneous presence of high levels of glutamine and PRPP in growing cells. When the cells starve for glucose or nitrogen, the concentration of one or both of the substrates decreases, and the enzyme becomes unstable. The rate of inactivation

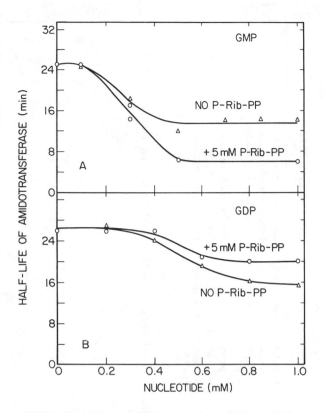

FIGURE 10. Effects of GMP (panel A) and GDP (panel B) on the oxidative inactivation of *B. subtilis* PRPP amidotransferase. Reaction conditions were as described under Table 3 without (△) and with (○) 5 m*M* PRPP. (From Bernlohr, D. A. and Switzer, R. L., *Biochemistry,* 20, 5675, 1981. With permission.)

would also be modulated by the ratio of stabilizing (AMP) to destabilizing (GMP, GDP, ADP) nucleotides, especially when substrate levels are subsaturating.

A careful study of the correlation between the intracellular concentrations of the substrates and effectors of PRPP amidotransferase in growing and nongrowing *B. subtilis* cells and the stability of the enzyme *in vivo* forced Bernlohr and Switzer[39] to conclude that this model was inadequate. PRPP and glutamine levels dropped sharply in glucose-starved cells concomitantly with the inactivation of PRPP amidotransferase, as predicted. However, intracellular levels of these substrates that were associated with stability of the enzyme in growing cells did not confer stability under other conditions, such as ammonia starvation or refeeding of glucose-starved cells. There was also a generally poor correlation between the ratio of stabilizing to destabilizing nucleotides and the rate of PRPP amidotransferase inactivation in starved cells or in cells to which antibiotic inhibitors of purine biosynthesis were added.

These observations indicate that additional physiological components or regulatory mechanisms are involved in controlling the stability of *B. subtilis* PRPP amidotransferase *in vivo*. What might these elements be? On the basis of apparent differences in the stability of newly synthesized PRPP amidotransferase in refed glucose-starved cells and enzyme that was present during starvation, Bernlohr and Switzer[39] suggested that inactivation might be regulated by rapid covalent modification of the enzyme prior to reaction with oxygen. Subsequent studies[45] have shown that the stability of PRPP amidotransferase in starved and refed cells is very dependent on the cell density and growth rate of the cells. In refed cells that are growing very rapidly at low cell density, all of the PRPP amidotransferase in the cells

Table 3

EFFECTS OF NUCLEOTIDES ON THE RATE OF INACTIVATION OF AMIDOTRANSFERASE BY REACTION WITH OXYGEN[a]

Inhibitor(s)	Effect on inactivation at saturation[b]	Concentration giving half-maximal effect (mM)	Shape of concentration dependence curve	Effect of 5 mM PRPP
AMP	Stabilizes to 170 min	0.15	Complex, sigmoid	Antagonizes stabilization
GMP	Destabilizes to 12 min	0.26	Sigmoid	Enhances destabilization
GDP	Destabilizes to 16 min	0.6	Sigmoid	Antagonizes destabilization
ADP	None		None	
ADP + GMP (equimolar)	Destabilizes to 8 min	0.05	Hyperbolic	Antagonizes destabilization
ADP + GDP (equimolar)	Destabilizes to 6 min	0.2	Sigmoid	None
GMP + GDP (equimolar)	Destabilizes to 12 min	0.3	Sigmoid (hyperbolic with PRPP)	Enhances destabilization

[a]   Reprinted with minor modifications from Reference 35.
[b]   Times given are half-lifes of PRPP amidotransferase at saturating concentrations of the effector at 37°C in O_2 saturated 50 mM Tris-HCl buffer, pH 7.9 containing 10 mM $MgCl_2$, and 2.5 mM EGTA.

becomes stable. These results do not exclude the possibility that reaction of PRPP amido-transferase is regulated by rapid covalent modification during starvation, but they require that such a modification be readily reversible, and they remove the experimental support previously advanced for this suggestion.

It is possible that other metabolites whose levels shift rapidly during starvation also alter the stability of PRPP amidotransferase to oxygen. No such metabolites were identified in a broad survey.[44] The fact the reactivity of the enzyme toward oxygen is very similar with the pure enzyme[35] and in crude extracts[40] suggests that other *B. subtilis* protein or cell components do not interact directly with the enzyme to modulate its stability.

A variable that deserves further attention is the role of intracellular oxygen levels in controlling the rate of PRPP amidotransferase inactivation. The rate of PRPP amidotrans-ferase inactivation *in vitro* is first order in oxygen concentration.[35] The simple model discussed above assumed that the intracellular oxygen level was essentially constant in vigorously aerated cultures. If intracellular oxygen levels are appreciably lower in growing cells than in starved cells (or in cells that are inhibited by antibiotics), inactivation of amidotransferase would necessarily be slower during growth. Are intracellular oxygen levels higher in ammonia-starved or amino acid-starved cells, in which PRPP amidotransferase inactivation is very rapid, than in glucose-starved cells, in which inactivation occurs half as fast? Could intracellular oxygen levels be sufficiently different in decoyinine-treated cells vs. hadicidin-treated cells to account for the fivefold more rapid inactivation of PRPP amidotransferase in decoyinine-treated cells?[39] The answers are not yet available. It is difficult to manipulate and measure intracellular (as opposed to extracellular) oxygen levels in bacterial cultures, but further studies are likely to be valuable.

V. DEGRADATION OF *B. SUBTILIS* PRPP AMIDOTRANSFERASE *IN VIVO*

Oxidative inactivation of PRPP amidotransferase in *B. subtilis* cells is followed by rapid degradation of the protein to immunochemically unrecognizable form.[21] The inactive, oxidized enzyme or proteolytic fragments normally do not accumulate *in vivo*, but inactive enzyme of native subunit molecular weight is accumulated in cells in which degradation is selectively inhibited.[21] The steps in the degradation of PRPP amidotransferase, the nature of the proteolytic apparatus and its physiological regulation present fascinating and important biochemical problems. A detailed review of these issues is beyond the scope of this chapter. A recent description of the current status of this area has been published.[46] The most important observations are summarized below.

1. Degradation of PRPP amidotransferase has not been detected in exponentially growing cells.[21]
2. All of the available evidence indicates that degradation follows oxidative inactivation of the enzyme *in vivo*. It is probable, but not conclusively proven, that inactivation is an essential first step in the turnover of this enzyme in *B. subtilis*.
3. The degradation requires metabolic energy; that is, it is inhibited if the energy metabolism of the cells is severely inhibited.[21]
4. Degradation of amidotransferase is strongly inhibited by addition of antibiotics that block protein synthesis prior to the normal onset of degradation.[21] Once degradation is under way, it is much less sensitive to these antibiotics.
5. Degradation is significantly inhibited, but not totally blocked, in *relA* and *relC* mutants.[21,47] This suggests that the stringent response plays an important role in regulation of PRPP amidotransferase degradation. Measurements of the levels of the nucleotide mediators of the stringent response, ppGpp and pppGpp, showed, however, that the rate of degradation is not directly correlated with the levels of these metabolites.[46,47]

6.   Degradation of PRPP amidotransferase appears to involve a regulated proteolytic system that is also required for degradation of other *B. subtilis* proteins, including the enzyme aspartate transcarbamylase.[46] The biochemical dissection and reconstruction *in vitro* of this system presents an important challenge for future research.

VI. CONCLUDING COMMENTS

It is evident that the bacterial PRPP amidotransferases are favorable objects for the study of regulatory mechanisms. Rapid progress has been made in moving studies of these enzymes from kinetic characterization toward a detailed biochemical description of structure-function relationships. The advent of convenient technology for cloning, sequencing, and site-directed mutagenesis[23,24,27,29a] have contributed greatly to these advances. The unexpected discovery of an essential [4Fe-4S] cluster in the *B. subtilis* PRPP amidotransferase has added unique elements to the study of this enzyme. It seems appropriate to close this chapter by listing some important unanswered questions about the biochemistry and regulation of the PRPP amidotransferases, which define avenues of future research.

1.   How many and what kind of allosteric nucleotide binding sites exist on PRPP amidotransferase? The kinetic studies cited above are most easily accommodated by postulating the existence of two fully separate sites per subunit, but the results can also be rationalized by models with only a single site or as many as three sites per subunit. Direct determination of the number of nucleotide sites per subunit and determination of whether the sites can be occupied simultaneously by different nucleotides would define the allosteric regulation of this enzyme much more clearly.

2.   What is the role of the Fe-S cluster in the *B. subtilis* PRPP amidotransferase? The view currently put forward is that the cluster plays no direct role in catalysis.[24,26,43] If so, it is the only known instance of such a "passive" Fe-S cluster in enzymology. The view that the Fe-S cluster plays a role in maintaining the structural integrity of the enzyme and renders it susceptible to inactivation, followed by degradation, in starving cells has received support in studies with *B. subtilis* cells.[20,21] It also is consistent with the observation that the PRPP amidotransferases of *E. coli*[8] and *S. cerevisiae*,[31] which do not contain Fe-S clusters, are relatively stable in starving cells. The PRPP amidotransferases from avian liver[6] and human placenta[48] also contain iron and probably contain Fe-S clusters. Do the clusters play a role in governing the rate of turnover of these enzymes in these species?

3.   Is the oxidative inactivation of *B. subtilis* PRPP amidotransferase the unique, obligatory first step in the degradation of this enzyme *in vivo*? How is the rate of reaction of oxygen with the Fe-S cluster in this enzyme regulated *in vivo*? A complete picture of the regulation of PRPP amidotransferase activity in growing and starving cells requires an answer to these questions.

4.   What is the physiological function, if any, of the N-terminal processing of the *B. subtilis* PRPP amidotransferase in which 11 amino acids are removed from the primary translation product? The processing occurs normally in an *E. coli* cytoplasm, even though the *E. coli* enzyme is not synthesized with an N-terminal leader peptide.[24] Does the enzyme catalyze its own N-terminal processing as suggested by Mäntsälä and Zalkin?[27]

5.   What is the three-dimensional structure of PRPP amidotransferase? Only a detailed X-ray diffraction analysis of crystals can yield a well-defined picture of the structure of the active site, the allosteric site(s), and the structural relationships between these sites and the Fe-S cluster in the *B. subtilis* enzyme. Both *E. coli* and *B. subtilis* enzymes can be prepared in substantial quantities, and the *B. subtilis* enzyme can be

readily crystallized.[49] However, crystals of quality suitable for diffraction analysis have yet to be obtained. Only when a three-dimensional structure is available for the enzyme can analysis of structure-function relationships in catalysis and allosteric regulation of PRPP amidotransferase proceed at a refined level.

ACKNOWLEDGMENTS

Research on PRPP amidotransferase in the author's laboratory has been supported by Grant No. AI11121 from the National Institutes of Health and Grant No. PCM 80-13476 from the National Science Foundation. It is a pleasure to acknowledge the collaboration and helpful comments of Professor Howard Zalkin of Purdue University.

REFERENCES

1. **Monod, J., Changeux, J. P., and Jacob, F.,** Allosteric proteins and cellular control systems, *J. Mol. Biol.,* 6, 306, 1963.
2. **Wyngaarden, J. B. and Ashton, D. M.,** The regulation of activity of phosphoribosylpyrophosphate amidotransferase by purine ribonucleotides: a potential feedback control of purine biosynthesis, *J. Biol. Chem.,* 234, 1492, 1959.
3. **Goldthwait, D. A.,** 5-Phosphoribosylamine, a precursor of glycinamide ribotide, *J. Biol. Chem.,* 222, 1051, 1956.
4. **Hartman, S. C. and Buchanan, J. M.,** Biosynthesis of the purines. XXI. 5-Phosphoribosylpyrophosphate amidotransferase, *J. Biol. Chem.,* 233, 451, 1958.
5. **Caskey, C. T., Ashton, D. M., and Wyngaarden, J. B.,** The enzymology of feedback inhibition of glutamine phosphoribosylpyrophosphate amidotransferase by purine ribonucleotides, *J. Biol. Chem.,* 239, 2570, 1964.
6. **Wyngaarden, J. B.,** Glutamine phosphoribosylpyrophosphate amidotransferase, in *The Enzymes of Glutamine Metabolism,* Prusiner, S. and Stadtman, E. R., Eds., Academic Press, New York, 1973, 365.
7. **Nierlich, D. P. and Magasanik, B.,** Regulation of purine ribonucleotide systhesis by end product inhibition. The effect of adenine and guanine ribonucleotides on the 5'-phosphoribosylpyrophosphate amidotransferase of *Aerobacter aerogenes, J. Biol. Chem.,* 240, 358, 1965.
8. **Messenger, L. J. and Zalkin, H.,** Glutamine phosphoribosylpyrophosphate amidotransferase from *Escherichia coli.* Purification and properties, *J. Biol. Chem.,* 254, 3382, 1979.
9. **Shiio, I. and Ishii, K.,** Regulation of purine ribonucleotide synthesis by end product inhibition. II. Effect of purine nucleotides on phosphoribosylpyrophosphate amidotransferase of *Bacillus subtilis, J. Biochem. (Tokyo),* 66, 175, 1969.
10. **Meyer, E. and Switzer, R. L.,** Regulation of *Bacillus subtilis* glutamine phosphoribosylpyrophosphate amidotransferase activity by end products, *J. Biol. Chem.,* 254, 5397, 1979.
11. **Nagy, M.,** Regulation of the biosynthesis of purine nucleotides in *Schizosaccharomyces pombe.* I. Properties of the phosphoribosylpyrophosphate:glutamine amidotransferase of the wild strain and of a mutant desensitized towards feedback modifiers, *Biochim. Biophys. Acta,* 198, 471, 1970.
12. **Satyanarayana, T. and Kaplan, J. G.,** Regulation of the purine pathway in baker's yeast: activity and feedback inhibition of phosphoribosylpyrophosphate amidotransferase, *Arch. Biochem. Biophys.,* 142, 40, 1971.
13. **Reynolds, P. H. S., Blevins, D. G., and Randall, D. D.,** 5-Phosphoribosylpyrophosphate amidotransferase from soybean root nodules: kinetic and regulatory properties, *Arch. Biochem. Biophys.,* 229, 632, 1984.
14. **Tsuda, M., Katunuma, N., and Weber, G.,** Rat liver glutamine 5-phosphoribosyl-1-pyrophosphate amidotransferase [EC 2.4.2.14]. Purification and properties, *J. Biochem. (Tokyo),* 85, 1347, 1979.
15. **Wood, A. W. and Seegmiller, J. E.,** Properties of 5-phosphoribosyl-1-pyrophosphate amidotransferase from human lymphoblasts, *J. Biol. Chem.,* 248, 138, 1973.
16. **Holmes, E. W., McDonald, J. A., McCord, J. M., Wyngaarden, J. B., and Kelley, W. N.,** Human glutamine phosphoribosylpyrophosphate amidotransferase. Kinetic and regulatory properties, *J. Biol. Chem.,* 248, 144, 1973.
17. **Nierlich, D. P. and Magasanik, B.,** Control by feedback repression of the enzymes of purine biosynthesis in *Aerobacter aerogenes, Biochim. Biophys. Acta,* 230, 349, 1971.

18. **Momose, H., Nishikawa, H., and Shiio, I.,** Regulation of purine nucleotide synthesis in *Bacillus subtilis.* I. Enzyme repression by purine derivatives, *J. Biochem. (Tokyo),* 59, 325, 1966.

19. **Nishikawa, H., Momose, H., and Shiio, I.,** Regulation of purine nucleotide synthesis in *Bacillus subtilis.* II. Specificity of purine derivatives for enzyme repression. *J. Biochem. (Tokyo),* 62, 92, 1967.

19a. **Makaroff, C. A. and Zalkin, H.,** Regulation of *Escherichia coli purF.* Analysis of the control region of a *pur* regulon gene, *J. Biol. Chem.,* 260, 10378, 1985.

19b. **Ebbole, D. J. and Zalkin, H.,** Cloning and characterization of a 12-gene cluster from *Bacillus subtilis* encoding nine enzymes for de novo purine nucleotide biosynthesis, *J. Biol. Chem.,* 262, 8274, 1987.

20. **Turnbough, C. L. and Switzer, R. L.,** Oxygen-dependent inactivation of glutamine phosphoribosylpyrophosphate amidotransferase in stationary-phase cultures of *Bacillus subtilis, J. Bacteriol.,* 121, 108, 1975.

21. **Ruppen, M. E. and Switzer, R. L.,** Degradation of *Bacillus subtilis* glutamine phosphoribosylpyrophosphate amidotransferase *in vivo, J. Biol. Chem.,* 258, 2843, 1983.

22. **Switzer, R. L.,** The inactivation of microbial enzymes *in vivo, Annu. Rev. Microbiol.,* 31, 135, 1977.

23. **Tso, J. Y., Zalkin, H., van Cleemput, M., Yanofsky, C., and Smith, J. M.,** Nucleotide sequence of *Escherichia coli purF* and deduced amino acid sequence of glutamine phosphoribosylpyrophosphate amidotransferase, *J. Biol. Chem.,* 257, 3525, 1982.

24. **Makaroff, C. A., Zalkin, H., Switzer, R. L., and Vollmer, S. J.,** Cloning of the *Bacillus subtilis* glutamine phosphoribosylpyrophosphate amidotransferase gene in *Escherichia coli.* Nucleotide sequence determination and properties of the plasmid-encoded enzyme, *J. Biol. Chem.,* 258, 10586, 1983.

25. **Tso, J. Y., Hermodson, M. A., and Zalkin, H.,** Glutamine phosphoribosylpyrophosphate amidotransferase from cloned *Escherichia coli purF.* NH$_2$-terminal amino acid sequence, identification of the glutamine site, and trace metal analysis, *J. Biol. Chem.,* 257, 3532, 1982.

26. **Vollmer, S. J., Switzer, R. L., Hermodson, M. A., Bower, S. G., and Zalkin, H.,** The glutamine-utilizing site of *Bacillus subtilis* glutamine phosphoribosylpyrophosphate amidotransferase, *J. Biol. Chem.,* 258, 10582, 1983.

27. **Mäntsälä, P. and Zalkin, H.,** Glutamine amidotransferase function. Replacement of the active-site cysteine in glutamine phosphoribosylpyrophosphate amidotransferase by site-directed mutagenesis, *J. Biol. Chem.,* 259, 14230, 1984.

27a. **Souciet, J.-L., Hermodson, M. A., and Zalkin, H.,** Mutational analysis of the glutamine phosphoribosylpyrophosphate amidotransferase pro-peptide, *J. Biol. Chem.,* 263, 3323, 1988.

28. **Walker, J. E., Gay, N. J., Saraste, M., and Eberle, A. N.,** DNA sequence around the *Escherichia coli unc* operon. Completion of the sequence of a 17 kilobase segment containing *AsnA, oriC, unc, glmS,* and *phoS, Biochem. J.,* 224, 799, 1984.

29. **Nyunoya, H. and Lusty, C. J.,** Sequence of the small subunit of yeast carbamyl phosphate synthetase and identification of its catalytic domain, *J. Biol. Chem.,* 259, 9790, 1984.

29a. **Makaroff, C. A., Paluh, J. L., and Zalkin, H.,** Mutagenesis of ligands to the [4Fe-4S] center of *Bacillus subtilis* glutamine phosphoribosylpyrophosphate amidotransferase, *J. Biol. Chem.,* 261, 11416, 1986.

29b. **Grandoni, J. A., Switzer, R. L., Makaroff, C. A., and Zalkin, H.,** unpublished results, 1987.

30. **Argos, P., Hanei, M., Wilson, J. M., and Kelley, W. N.,** A possible nucleotide-binding domain in the tertiary fold of phosphoribosyltransferases, *J. Biol. Chem.,* 258, 6450, 1983.

31. **Mäntsälä, P. and Zalkin, H.,** Nucleotide sequence of *Saccharomyces cerevisiae ADE4* encoding phosphoribosylpyrophosphate amidotransferase, *J. Biol. Chem.,* 259, 8478, 1984.

32. **Wong, J. Y., Bernlohr, D. A., Turnbough, C. L., and Switzer, R. L.,** Purification and properties of glutamine phosphoribosylpyrophosphate amidotransferase from *Bacillus subtilis, Biochemistry,* 20, 5669, 1981.

33. **Holmes, E. W., Wyngaarden, J. B., and Kelley, W. N.,** Human glutamine phosphoribosylpyrophosphate amidotransferase. Two molecular forms interconvertible by purine ribonucleotides and phosphoribosylpyrophosphate, *J. Biol. Chem.,* 248, 6035, 1973.

34. **Hartman, S. C.,** Phosphoribosylpyrophosphate amidotransferase. Purification and general catalytic properties, *J. Biol. Chem.,* 238, 3024, 1963.

35. **Bernlohr, D. A. and Switzer, R. L.,** Reaction of *Bacillus subtilis* glutamine phosphoribosylpyrophosphate amidotransferase with oxygen: chemistry and regulation by ligands, *Biochemistry,* 20, 5675, 1981.

36. **London, R. E. and Schmidt, P. G.,** A model for nucleotide regulation of aspartate transcarbamylase, *Biochemistry,* 11, 3136, 1972.

37. **Monod, J., Wyman, J., and Changeux, J. P.,** On the nature of allosteric transitions: a plausible model, *J. Mol. Biol.,* 12, 88, 1965.

38. **Maurizi, M. R. and Switzer, R. L.,** Proteolysis in bacterial sporulation, in *Current Topics in Cellular Regulation,* Vol. 16, Horecker, B. L. and Stadtman, E. R., Eds., Academic Press, New York, 1980, 163.

39. **Bernlohr, D. A. and Switzer, R. L.,** Regulation of *Bacillus subtilis* glutamine phosphoribosylpyrophosphate amidotransferase inactivation *in vivo, J. Bacteriol.,* 153, 937, 1983.

40. **Turnbough, C. L. and Switzer, R. L.,** Oxygen-dependent inactivation of glutamine phosphoribosylpyrophosphate amidotransferase *in vitro:* model for *in vivo* inactivation, *J. Bacteriol.,* 121, 115, 1975.

41. **Wong, J. Y., Meyer, E., and Switzer, R. L.,** Glutamine phosphoribosylpyrophosphate amidotransferase from *Bacillus subtilis.* A novel iron-sulfur protein, *J. Biol. Chem.,* 253, 7424, 1977.
42. **Averill, B. A., Dwivedi, A., Debrunner, P., Vollmer, S. J., Wong, J. Y., and Switzer, R. L.,** Evidence for a tetranuclear iron-sulfur center in glutamine phosphoribosylpyrophosphate amidotransferase from *Bacillus subtilis, J. Biol. Chem.,* 255, 6007, 1980.
43. **Vollmer, S. J., Switzer, R. L., and Debrunner, P. G.,** Oxidation-reduction properties of the iron-sulfur cluster in *Bacillus subtilis* glutamine phosphoribosylpyrophosphate amidotransferase, *J. Biol. Chem.,* 258, 14284, 1983.
44. **Bernlohr, D. A.,** Chemistry and Regulation of the Oxygen-Dependent Inactivation of Glutamine Phosphoribosylpyrophosphate Amidotransferase from *Bacillus subtilis,* Ph.D. thesis, University of Illinois, Urbana, 1982.
45. **Grandoni, J. A. and Switzer, R. L.,** unpublished data, 1985.
46. **Switzer, R. L., Bond, R. W., Ruppen, M. E., and Rosenzweig, S.,** Involvement of the stringent response in regulation of protein degradation in *Bacillus subtilis,* in *Current Topics in Cellular Regulation,* Vol. 27, Shaltiel, S. and Chock, P. B., Eds., Academic Press, New York, 1985, 373.
47. **Ruppen, M. E. and Switzer, R. L.,** Involvement of the stringent response in degradation of glutamine phosphoribosylpyrophosphate amidotransferase in *Bacillus subtilis, J. Bacteriol.,* 155, 56, 1983.
48. **Itakura, M. and Holmes, E. W.,** Human amidophosphoribosyltransferase. An oxygen-sensitive iron-sulfur protein, *J. Biol. Chem.,* 254, 333, 1979.
49. **Vollmer, S. J., White, S., Phillips, G., and Switzer, R. L.,** unpublished results, 1984.

Chapter 6

MAMMALIAN GLYCERALDEHYDE-3-PHOSPHATE DEHYDROGENASE AND ITS USE TO ELUCIDATE MOLECULAR MECHANISMS OF COOPERATIVITY

Yoav I. Henis and Alexander Levitzki

TABLE OF CONTENTS

I. INTRODUCTION*

Cooperativity is an essential feature of regulatory molecules, whether they represent soluble allosteric proteins or membrane-embedded enzymes and receptors. The most abundant form of cooperativity is positive cooperativity, whereas negative cooperativity is rarely observed. One of the few examples of a strong negatively cooperative system is the mammalian glyceraldehyde-3-phosphate dehydrogenase (GAPDH) which is discussed in this chapter. An attempt has been made to integrate the available information gathered, using a large variety of experimental techniques, and to present a complete picture of the evidence for negative cooperativity in coenzyme binding to GAPDH. The techniques employed in studying the negative cooperativity in NAD^+ binding to the enzyme include: (1) binding and displacement experiments of NAD^+ and many NAD^+ derivatives; (2) fluorescence spectroscopy and circular polarized luminescence (CPL) of NAD^+ analogs bound to GAPDH; (3) electron paramagnetic resonance (EPR) of spin-labeled NAD^+ analogs in their GAPDH-bound state; (4) fluorescence spectroscopy and CPL of ligand-bound GAPDH; and (5) X-ray crystallography of apo-GAPDH, $GAPDH(NAD^+)_1$, and $GAPDH(NAD^+)_4$. All of these techniques combined conclusively demonstrate that NAD^+ binding to GAPDH is negatively cooperative. The possible physiological significance of this negative cooperativity in GAPDH is discussed.

II. THE ORIGIN OF NEGATIVE COOPERATIVITY DEDUCED FROM LIGAND COMPETITION EXPERIMENTS

GAPDH displays strong negative cooperativity toward coenzyme (NAD^+) binding. The mechanisms of these cooperative interactions have been the subject of intensive studies in numerous laboratories for the past 15 years. Strong controversies arose mainly between two schools. One is represented by Levitzki and Koshland.[1,2] and argues that the apo-GAPDH is symmetric and that the negative cooperativity exhibited in NAD^+ binding is a result of sequential conformational changes, and another is represented by Seydoux et al.[3] and claims that GAPDH is asymmetric and that the negative cooperativity is a result of this preexistent asymmetry (PEA). The discrimination between genuine negative cooperativity which results from pure sequential conformational changes and *apparent* negative cooperativity which is due to the existence of heterogenous non-interacting binding sites is not a trivial problem.[1] A related problem is the frequent difficulty to definitively ascribe positive cooperativity to either the concerted Monod-Wyman-Changeux (MWC)[4] mechanism or to the Koshland-Nemethy-Filmer (KNF) sequential mechanism.[5,6] We felt that for both situations it was necessary to establish a rigorous method which would allow an unambiguous identification of the mechanism of cooperativity. We have indeed recently developed a novel method to determine the molecular mechanism of cooperativity. This method is based on the mode of competition between two ligands at the binding sites of the cooperative protein and is described below.

A. Basic Theoretical Features of the Competition Experiments

Cooperative ligand binding isotherms can usually be fitted by more than one molecular model. Among the three major cooperativity models, the sequential KNF model and the

---

* Abbreviations: KNF model, the sequential Koshland-Nemethy-Filmar model; MWC model, the Monod-Wyman-Changeux model; PEA model, the preexistent asymmetry model; GAPDH, glyceraldehyde-3-phosphate dehydrogenase (EC 1.2.1.12; the enzyme); n_H, Hill coefficient; ϵNAD^+, nicotinamide-1-N^6-ethenoadenine-dinucleotide; $APAD^+$, acetylpyridine adenine dinucleotide; ADP-Rib, adenosine diphosphoribose; CPL, circularly polarized luminescence; I-AEDANS, N-iodacetyl-N'-(5-sulfo-1-naphtyl)-ethylenediamine; Enzyme(alkyl)$_1$, product of the reaction of GAPDH with I-AEDANS at a 1:1 molar ratio: AAD^+, 3-aminopyridine adenine dinucleotide; CD, circular dichroism.

PEA model can give rise to negative cooperativity, while the KNF or the MWC models can generate positive cooperativity. However, examination of the effect of competition by one ligand on the mode of binding of a second ligand reveals differences between the predictions of the above models. It is relatively simple to demonstrate[7] that if the primary ligand binds with apparent negative cooperativity and the competing ligand binds in a noncooperative manner, any effect of the noncooperative competing ligand on the extent of negative co-operativity in the binding pattern of the primary ligand excludes the PEA model, and demonstrates the involvement of site-site interactions (as postulated in the sequential KNF model) in generating the negative cooperativity. The same approach can also be employed to distinguish between the KNF and the MWC models in cases of positive cooperativity.[7] Detailed mathematical analysis[7] demonstrates that the use of a noncooperative ligand as a competitor in the displacement experiments provides the simplest and most unambiguous way of determining the cooperativity-generating mechanism; however, if such a ligand is not available, a cooperative competing ligand may also be employed, although the distinction between the models may be more complex in such a case.[7,8]

The simplest case of the PEA model is that of a dimer or a tetramer possessing two classes of noninteracting sites. The Hill coefficient at 50% saturation by a ligand X, $n_H(X)$, is given under these circumstances by[1,2,9]

$$n_H(X) = \frac{4}{2 + (K'_x/K''_x)^{1/2} + (K''_x/K'_x)^{1/2}} \tag{1}$$

where K'_x, K''_x are the intrinsic association constants for the binding of the ligand X to the two classes of sites, respectively. In the presence of a competing ligand Z (which is added at a concentration well above that of the binding sites, so that the free Z concentration remains essentially constant during the titration with X), the Hill coefficient for the binding of X in the presence of Z, $n_H(X,Z)$, becomes[7]

$$n_H(X,Z) = \frac{4}{2 + \dfrac{K'_x + K''_x + (K'_x K''_z + K''_x K'_z)[Z]}{(K'_x K''_x)^{1/2} \{1 + (K'_z + K''_z)[Z] + K'_z K''_z [Z]^2\}^{1/2}}} \tag{2}$$

where K'_z and K''_z are the intrinsic association constants of Z to the two classes of sites. If Z binding is noncooperative, $K'_z = K''_z$ and Equation 2 reduces to Equation 1; i.e., a noncooperative competing ligand cannot affect the apparent negative cooperativity of X binding according to the PEA model $[n_H(X,Z) = n_H(X)]$. If Z binding itself reveals apparent negative cooperativity ($K'_z \neq K''_z$), Z will affect the cooperativity of X binding[7] (see Table 1).

Unlike the PEA model, the KNF model allows for a change in the cooperativity of X binding when the competing ligand binds noncooperatively. In the simplest case of the KNF model, the conformational change is limited to the ligand binding subunit, and the binding of X to a dimer or to a multimer of dimers yields.[1,2,9]

$$n_H(X) = \frac{2}{1 + (K_{AB}^2/K_{BB})^{1/2}} \tag{3}$$

where A and B are the conformations of unliganded and X liganded subunits, respectively. K_{AB} and K_{BB} are the subunit interaction constants between subunits in the A and B conformations (liganded and unliganded), and between two X liganded subunits, respectively.[5] In the presence of a high Z concentration, one obtains,[7]

Table 1
QUALITATIVE EFFECT OF A COMPETING LIGAND (Z) ON n_H FOR X BINDING TO AN OLIGOMER

Model	Effect of a cooperative Z on n_H for X	Effect of a noncooperative Z on n_H for X	
		Type 1	Type 2
PEA	$n_H(X)$ Changes, but always $n_H(X,Z) \le 1$	$n_H(X)$ does not change — $n_H(X,Z) = n_H(X)$.	
Simplest KNF	Any change in $n_H(X)$ is possible. $n_H(X,Z) > 1$ can be obtained even when $n_H(X) < 1$, and vice versa. $n_H(X,Z) \ne n_H(X)$.	Any change in $n_H(X)$ is possible: $n_H(X,Z) \ne n_H(X)$.	$n_H(X)$ does not change: $n_H(X,Z) = n_H(X)$.
General KNF	As in simplest KNF, but more variability possible. $n_H(X,Z) \ne n_H(X)$.	As in simplest KNF, but more variability is possible. $n_H(X,Z) \ne n_H(X)$.	Any change in $n_H(X)$ is possible-$n_H(X,Z) \ne n_H(X)$. Special case: if Z affinity to all conformations is identical, $n_H(X,Z) = n_H(X)$.
MWC*	$n_H(X)$ Changes, but always $n_H(X,Z) > 1$.	$n_H(X)$ Changes only to higher values: $n_H(X,Z) > n_H(X)$.	$n_H(X)$ does not change: $n_H(X,Z) = n_H(X)$.

Note: The definition of the properties of a noncooperative ligand (Z) depends on the model. In the PEA model, it is a ligand that binds with identical affinities to all sites. In the KNF model, a Type 1 noncooperative ligand is a ligand whose binding induces a conformation change (from A to C) in the ligand binding subunit, but this change has no effect on further Z binding to neighboring subunits ($K^2_{AC}/K_{CC} = 1$). A Type 2 noncooperative ligand in the KNF model does not induce a conformational change upon binding (a Type 1 ligand can behave as a Type 2, if the conformational change induced upon its binding does not affect X binding to the neighboring subunits). In the MWC model, a Type 1 noncooperative ligand is one that has a higher affinity for the T conformation in a system where L ≫ 1. A Type 2 ligand in the MWC model is a ligand which binds with identical affinities to the R and T conformations.

* The MWC model can account for positive cooperativity only. A detailed analysis of this model has been published elsewhere.[7]

$$n_H(X,Z) = \cfrac{2}{1 + \cfrac{K_{AB}^2 \left(1 + \cfrac{K_{BC}}{K_{AB}} K_{z_C} K_{t_{AC}} [Z]\right)^2}{K_{BB} \{1 + 2K_{AC} K_{z_C} K_{t_{AC}} [Z] + K_{CC}(K_{z_C} K_{t_{AC}})^2 [Z]^2\}}} \qquad (4)$$

where K_{AC} is the subunit interaction constant between a Z liganded subunit (conformation C) and an unliganded subunit (conformation A), while K_{BC} represents the interaction between an X liganded and a Z liganded subunit. $K_{t_{AC}}$ describes the free energy of transformation for converting a subunit from conformation A to C, and K_{z_C} is the intrinsic association contant of Z to a subunit in the C conformation. All these parameters are as originally defined by Koshland et al.[5,6]

Comparison of Equations 4 and 3 suggests that competition by Z can affect the co-operativity in X binding, depending on the relations between all the interaction constants. According to the KNF model, Z binds noncooperatively only if $K_{AC}^2/K_{CC} = 1$.[5] However, this does not prevent K_{BC}/K_{AB} in Equation 4 from obtaining any value. Thus, Equation 4 does not necessarily reduce to Equation 3 even where the competitor Z binds noncooperatively; namely a situation where $n_H(X,Z) \neq n_H(X)$ can still be obtained. Only in the special case, where $K_{BC}/K_{AB} = K_{AC}$ *in addition* to the noncooperative binding of Z, will $n_H(X,Z)$ equal $n_H(X)$. In the general KNF model, where conformational alterations may be transmitted to neighboring subunits,[6] a similar phenomenon is encountered, except that a higher degree of variability in the effects of Z is possible.[7] A qualitatively similar behavior is observed in the general case of a tetrameric protein, for which more involved algebraic manipulations have to be applied.[7] The conclusions derived from the theoretical analysis can also be extended to higher oligomers and suggest that the KNF model can account for altered cooperativity of X binding in the presence of a noncooperative competing ligand, while the PEA model cannot. Analogous treatment of the concerted MWC model for positive cooperativity[4] demonstrates differences between its predictions and those of the KNF model for the effects of a competing ligand on the cooperativity of the primary ligand.[7] The qualitative effects of competition on the cooperativity of the displacing (primary) ligand according to the three major cooperativity models are summarized in Table 1.

B. Ligand Competition Experiments with GAPDH

An accurate quantitative study of the mechanism of the negative cooperativity in GAPDH by competition experiments requires a primary ligand (an NAD^+ analog) whose binding can be measured easily and accurately, and competing NAD^+ analogs (Z ligands) that bind noncooperatively. The studies to be described are detailed in References 11 and 12.

A convenient primary ligand is the enzymatically active coenzyme analog nicotinamide-1-N^6-ethenoadenine dinucleotide (ϵNAD^+),[10] which binds to the enzyme with negative co-operativity.[11,12] This ligand is fluorescent, and its fluorescence is enhanced upon binding to the enzyme.[13,14] The enhancement (ΔF) is proportional to ϵNAD^+ binding.[11,13] The binding of ϵNAD^+ to the enzyme could therefore be computed from the expression:[11,12]

$$r = \frac{[\epsilon NAD^+]_B}{4[E]_t} = \frac{\Delta F}{\Delta F_\infty} \qquad (5)$$

where $[\epsilon NAD^+]_B$ is the concentration of bound ϵNAD^+, $[E]_t$ is the total enzyme concentration, 4 is the number of enzyme subunits, and ΔF is the maximal fluorescence enhancement (at saturation). The binding of ϵNAD^+ to the enzyme exhibited strong negative cooperativity (Figure 1, Table 2), as in the case of NAD^+ binding[15,16] (see Figure 1). The binding of ϵNAD^+ to the enzyme was analyzed using nonlinear regression curve-fitting procedures,[17] fitting the data to the general Adair equation for a tetramer:[18]

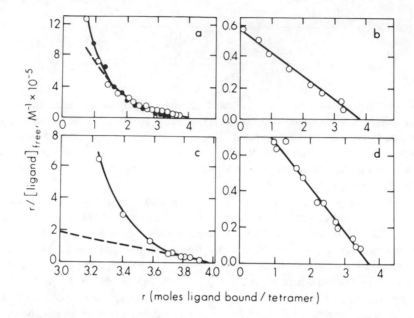

FIGURE 1.   Binding of coenzyme analogs to GAPDH. Data are presented as Scatchard plots. (a) εNAD$^+$ binding. ● and ○, two separate experiments. Solid line, best fit to Equation 6. Dashed line, best fit to the two-classes-of-sites (simplest case of PEA) model:

$$\frac{[X]_B}{[E]_t} = \frac{2[X]}{K' + [X]} + \frac{2[X]}{K'' + [X]}$$

where K$'$ and K$''$ are the intrinsic dissociation constants for X binding to the two classes of sites. The data are not sufficient to eliminate the latter model (correlation coefficients — 0.9997 and 0.9995 for the solid and dashed lines, respectively). [E]$_t$ was 0.84 × 10^{-5} M. (b) ATP binding. [E]$_t$ was 5.1 × 10^{-5} M. 5 mM phosphate were present to prevent ATP binding to the phosphate sites. Binding was determined by the dialysis rate method,[11] using an [α-^{32}P]-ATP concentration of 2.6 μM (7.7 Ci/mmol). (c) NAD$^+$ binding. [E]$_t$ was 1.0 × 10^{-5} M. Binding was determined by the dialysis rate method using a [^{14}C]-NAD$^+$ concentration of 7.4 μM (302 mCi/mmol). (d) APAD$^+$ binding. [E]$_t$ was 2.75 × 10^{-5} M. Binding was measured by following the difference spectrum formed upon APAD$^+$ binding to GAPDH at 350 nm, using previously described procedures.[21] The intensity of this band was found to be proportional to APAD$^+$ binding to the enzyme, as is the case with other NAD$^+$ analogs.[13,21] The extinction coefficient for bound APAD$^+$ was ε$_{350}$ = 1430 M^{-1} cm^{-1}.

Table 2
INTRINSIC DISSOCIATION CONSTANTS FOR εNAD$^+$ BINDING TO THE ENZYME

	Constants from Figure 1 (M × 10^5)	Constants from Reference 13 (M × 10^5)	Constants from Reference 14 (M × 10^5)
K$'_1$	0.15 ± 0.03	N.D.	N.D.
K$'_2$	0.63 ± 0.09	N.D.	N.D.
K$'_3$	1.3 ± 0.2	1.7	3.5
K$'_4$	1.5 ± 0.2	1.7	3.5

Note: In analyzing the data from Figure 1[11,12] curve fitting was performed using Equation 6, and the thermodynamic dissociation constants for the binding of the ith ligand molecule (K$_i$) were converted into the intrinsic dissociation constants (K$_i$ by the formula:[1,2] $K_i = \frac{(n - i + 1)K_i}{i}$, where n is the number of binding sites on the oligomer (4 in the case of GAPDH). The results are mean ± SD of three separate experiments. The constants from Reference 13 were obtained by reanalyzing the data with Equation 6. In Reference 14, enzyme with two bound NAD$^+$ molecules was used rather than apoenzyme; this may be the reason for the higher K$'_3$ and K$'_4$ values obtained. N.D. = not determined.

Table 3
INTRINSIC DISSOCIATION CONSTANTS FOR THE BINDING OF NONCOOPERATIVE NAD$^+$ ANALOGS

Ligand	Intrinsic dissociation constant $(M \times 10^5)$
ATP	6.7 ± 0.7
APAD$^+$	4.0 ± 0.4
ADP-Rib	4.5 ± 0.5

Note: The constants for ATP and APAD$^+$ (means $\pm$ SD of three separate experiments) were derived from the linear Scatchard plots as shown in Figure 1. The effect of ADP-Rib on ϵNAD$^+$ binding to the enzyme was purely competitive (Figure 2; see text), and thus the intrinsic dissociation constant for ADP-Rib (K$'$) could be derived from its competition with ϵNAD$^+$ given by:

$$[\epsilon NAD^+]_{0.5}^{obs} = [\epsilon NAD^+]_{0.5} \left(1 + \frac{[ADP - Rib]}{K'} \right)$$

$[\epsilon NAD^+]_{0.5}$ and $[\epsilon NAD^+]_{0.5}^{obs}$ are the free ϵNAD$^+$ concentrations yielding 50% saturation in the absence and presence of ADP-Rib, respectively.

$$\frac{[X]_B}{[E]_t} = \frac{\dfrac{[X]}{K_1} + \dfrac{2[X]^2}{K_1 K_2} + \dfrac{3[X]^3}{K_1 K_2 K_3} + \dfrac{4[X]^4}{K_1 K_2 K_3 K_4}}{1 + \dfrac{[X]}{K_1} + \dfrac{[X]^2}{K_1 K_2} + \dfrac{[X]^3}{K_1 K_2 K_3} + \dfrac{[X]^4}{K_1 K_2 K_3 K_4}} \tag{6}$$

where $[E]_t$ is the total tetramer concentration, $[X]_B$ and $[X]$ are the bound and free ligand concentrations, and K_1 through K_4 are the thermodynamic *dissociation* constants of the first through the fourth ligand molecules. This equation is model independent and fits ligand binding to any tetramer;[1,7,9] the inclusion of four dissociation constants makes it compatible with the KNF model. As can be seen (Figure 1), there is an excellent fit of the ϵNAD$^+$ binding data to this equation. However, the binding data cannot eliminate the PEA model, since even the simplest case of this model (two classes of sites) provides an excellent fit to the same data (Figure 1), although it has only two parameters (compared with four parameters in Equation 5).

In order to distinguish between the two models, one needs competitive ligands that bind noncooperatively. This is the case with acetylpyridine adenine dinucleotide (APAD$^+$), ATP, and adenosine diphosphoribose (ADP-Rib)[11,12] (see Figure 1 and Table 3). For ATP, the noncooperative binding (Figure 1, Table 3) was measured in the presence of 5 mM phosphate, in order to block ATP binding to the eight phosphate binding sites on the enzyme.[19] Such binding (with $K_d = 7 \times 10^{-4}\ M$) could be demonstrated when ATP binding to the rabbit muscle enzyme was measured in the presence of 5 mM NAD$^+$, under which conditions the four tighter ATP binding sites have disappeared.[11] These data indicate that the four tighter

ATP binding sites correspond to the NAD^+ sites on the enzyme, and that ATP also binds to the Pi binding sites.

The competition experiments[11,12] were performed primarily with ϵNAD^+ as the displacing ligand. The competing ligands were always introduced at a large molar excess over the binding sites and over their respective dissociation constants. This ensured that the concentration of free competing ligand would not change during titration with the displacing ligand, and that the competing ligand would initially saturate the binding sites, yielding the maximal effect on the cooperativity of the primary ligand.[7]

The results of the competition experiments with ϵNAD^{+}[11,12] are depicted in Figure 2 and Table 4. It is evident that competition by $APAD^+$ completely abolishes the negative cooperativity of ϵNAD^+ binding, and a somewhat weaker effect is exerted by ATP. The effect of the latter ligand is also due to competition for the NAD^+ sites, as evidenced by the insensitivity of the effect of ATP to the presence of 5 mM phosphate (phosphate by itself had no effect on ϵNAD^+ binding). ADP and AMP had a similar effect on ϵNAD^+ binding (Table 4). The reduced negative cooperativity is also not due to effects of the competing ligands on the aggregation state of the tetramer.[11] Unlike the marked effects of $APAD^+$ or ATP, ADP-Rib had no effect on the shape of the ϵNAD^+ binding curve, and only shifted it to higher ϵNAD^+ concentrations. This suggests that ADP-Rib does not only bind noncooperatively to the enzyme, but also that its binding is not affecting (and is not affected by) the binding of ϵNAD^+ to neighboring subunits.[7] This notion is in line with the finding of Eby and Kirtley[20] that ADP-Rib binds noncooperatively to GAPDH.

In order to demonstrate that the results obtained in the competition experiments using ϵNAD^+ are also valid for the natural coenzyme NAD^+, it was desired to study the effects of competition by noncooperative competitors on NAD^+ binding. Since the binding of the first three NAD^+ molecules to the enzyme is too tight to permit accurate determination of n_H for the binding of NAD^+ alone, the studies were restricted to comparison between the effects of different noncooperative competing analogs on NAD^+ binding.[11,12] The results are depicted in Figure 3 and Table 5. The effects of competition by ATP and ADP-Rib on the negative cooperativity of NAD^+ binding are clearly very different, ruling out the PEA model which predicts no effects on the cooperativity of the primary ligand for *any* noncooperative competing ligand.

The results of the competition experiments with either ϵNAD^+ or NAD^+ clearly demonstrate the involvement of site-site interactions in the negative cooperativity of coenzyme binding to GAPDH. The significant decrease in the extent of negative cooperativity of ϵNAD^+ in the presence of noncooperative ligands ($APAD^+$ and adenine nucleotides) eliminates the PEA model and establishes that the negatively cooperative coenzyme binding involves sequential conformational alterations which can be easily accounted for by the KNF sequential model.[1,2,5,6] Clearly, a mechanism which involves site heterogeneity but allows ligand-induced conformational changes is also compatible with these results.[7] This conclusion also holds for the natural coenzyme, NAD^+, in view of the highly different binding patterns observed for NAD^+ binding in the presence of ATP as compared with ADP-Rib. The analogous behavior of ϵNAD^+ and NAD^+ is further exemplified by the similarity of the values of their Hill coefficients in the presence of ADP-Rib or ATP.

Interestingly, the noncooperative NAD^+ analog ADP-Rib has no effect on the cooperativity of ϵNAD^+ binding (Figure 2, Table 4). This finding emphasizes the need for using more than a single noncooperative competing ligand in studies of cooperative mechanisms; different ligands may have varying effects on the cooperativity of the primary ligand depending on their specific interactions with the oligomeric protein. The failure of ADP-Rib, which lacks a pyridine moiety, to affect the cooperativity of ϵNAD^+ binding suggests that the pyridine ring of the coenzyme affects the mode of interaction between the enzyme and the coenzyme analog. This notion is strengthened by the fact that modifications in the pyridine

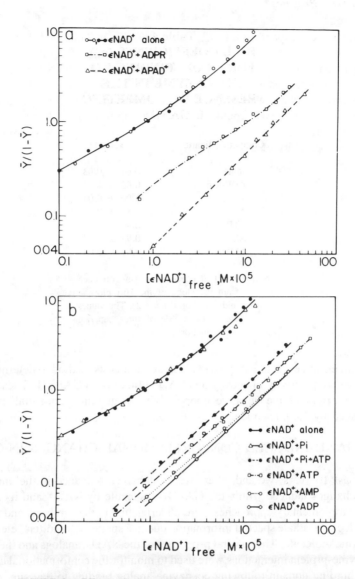

FIGURE 2. Effects of noncooperative competing ligands on ϵNAD^+ binding. Data are shown as Hill plots. $\bar{Y}$ is the fraction of binding sites occupied by ϵNAD^+. The enzyme concentration was 9 to 8 μM. (a) ϵNAD^+ binding alone (● and ○, two separate experiments) and in the presence of 0.41 mM APAD$^+$ (△--△) or 0.53 mM ADP-Rib (ADPR) (□—·—□). (b) ϵNAD^+ binding alone (●—●) and in the presence of 5 mM phosphate (△—△), 0.6 mM ATP (□—·—□), 0.65 mM ATP + 5 mM phosphate (●--●), 1.63 mM AMP (○....○), and 0.92 mM ADP (◇—◇). The concentrations of these ligands were chosen according to their intrinsic dissociation constants for the NAD$^+$ sites (8.1, 4.4, 3.8 × 10^{-5} M for AMP, ADP, and ATP, respectively — see Reference 11), to ensure similar saturation levels.

moiety significantly affect the cooperativity of coenzyme binding; for example, APAD$^+$ binds noncooperatively to the enzyme, although it differs from NAD$^+$ only in the pyridine side chain: APAD$^+$ possesses an acetyl group instead of a carboxamide group. The latter is capable of hydrogen bond formation, whereas the methyl group is devoid of such a capacity. It seems, therefore, that the pyridine moeity plays a role in propagating the cooperative interactions in GAPDH by the proper positioning of the coenzyme molecule within the binding site.

Table 4
HILL COEFFICIENTS (n_H)
FOR ϵNAD^+ BINDING TO
THE ENZYME IN THE
PRESENCE OF COMPETING
LIGANDS

Competing ligand	$n_H(\epsilon NAD^+)$
None	0.64 ± 0.03
ADP-Rib	0.65 ± 0.04
$APAD^+$	1.00 ± 0.04
ATP	0.89 ± 0.02
ATP + P_i	0.90 ± 0.03
ADP	0.93 ± 0.05
AMP	0.88 ± 0.05

Note: The Hill coefficients were obtained
from the slope of the Hill plots at
midsaturation (Figure 2). The values
are mean ± SD of three separate
determinations.

C. Conclusions

Ligand competition experiments provided an unambiguous method to determine the origin of the negative cooperativity in coenzyme (NAD^+) binding to GAPDH. These experiments show that the origin of the negative cooperativity is genuine sequential conformational changes induced by the nucleotide.

III. LIGAND-INDUCED CONFORMATIONAL CHANGES IN GAPDH

Over the past 10 years, we and others were intensively engaged in the analysis of conformational changes induced within the GAPDH molecule by NAD^+ and its analogs. Basically three complementary approaches were undertaken: (1) the synthesis and use of NAD^+ analogs which can irradiate spectral information due to a specific "reporter" element attached to the coenzyme molecule. The spectral properties of the NAD^+ analogs and their modulation by the coenzyme-protein interactions were used to monitor the conformational changes which occur at the binding domain following coenzyme analog binding to that site, and the transmission of these alterations to the remote sites on the neighboring subunits. (2) Examination of the spectral properties of the protein itself as a consequence of ligand binding. Special emphasis has been lent to the "Racker band" which is the charge transfer spectral band characteristic for the pyridinium moiety with the protein subsite. (3) Crystallographic data obtained for the apo-GAPDH, monobound holo-GAPDH $(NAD^+)_1$, and the fully saturated holoenzyme GAPDH $(NAD^+)_4$. Comparison between the three approaches sheds light on the mechanisms of coenzyme-induced conformational changes.

A. Data from NAD^+ and Its Analogs

The roles of nicotinamide and adenine subsites in the negative cooperativity of coenzyme binding to GAPDH were investigated employing two fluorescent derivatives of NAD^+- ϵNAD^+ ($1,N^6$-etheno-NAD^+) which has a fluorescent adenine derivative,[11,12] and 3-aminopyridine adenine dinucleotide (AAD^+) which contains a fluorescent pyridine derivative.[21] The ability to localize fluorescent groups within the adenine (ϵNAD^+) or nicotinamide (AAD^+) subsites enabled the detection of the conformational changes occurring at each subsite as a function of ligand occupancy.[13,21]

ϵNAD^+ was incubated with a high concentration (1.5 to $2 \times 10^{-4}\,M$) of the apoenzyme.[13]

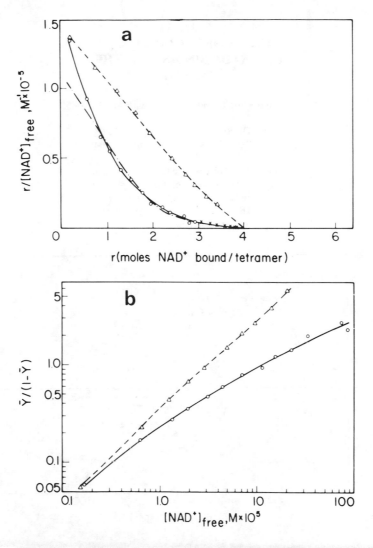

FIGURE 3. NAD$^+$ binding in the presence of ADP-Rib and ATP. The enzyme concentration was 2.7×10^{-5} M, and the [^{14}C]-NAD$^+$ concentration was 7.4 μM (302 mCi/mmol). ○—○, Titration in the presence of 10.45 mM ADP-Rib, best fit according to Equation 6; ○--○, same titration, best fit according to the two-classes-of-sites model (see caption of Figure 1a); △--△, titration in the presence of 11.85 mM ATP, best fit according to Equation 6. As in the case of εNAD$^+$, NAD$^+$ binding in the presence of ADP-Rib can be fitted well also to the PEA model. (a) Scatchard plots. x——x are points taken from NAD$^+$ binding without competing ligands (Figure 1c), and transferred to the conditions of 10.47 mM ADP-Rib assuming a purely competitive inhibition;[11] as can be seen, the fit is very satisfactory. (b) Hill plots. $\bar{Y}$ is the fraction of binding sites occupied by NAD$^+$.

Due to the rather high affinity of εNAD$^+$ to the enzyme (Table 2), stoichiometric binding of εNAD$^+$ occurred up to 3.6 mol per tetramer. The binding of εNAD$^+$ to the enzyme generated an absorption band with an extinction coefficient which was identical for the binding of the first through the fourth εNAD$^+$ molecules. The absorption band formed was optically active (Figure 4) and exhibited changes in the circular dichroism (CD) spectrum depending on the number of moles εNAD$^+$ bound per mole tetramer.

The studies described in the previous sections demonstrated that the changes in the absorption band are due to the interaction of the ligand with the binding site. These changes

Table 5
HILL COEFFICIENTS (n_H)
FOR NAD$^+$ BINDING TO THE
ENZYME

Competing ligand	$n_H(NAD^+)$
ADP-Rib	0.58 ± 0.03
ATP	0.88 ± 0.04

Note: The Hill coefficients were obtained from the slope of the Hill plots at midsaturation with NAD$^+$ (Figure 3). The values are mean $\pm$ SD of three separate determinations.

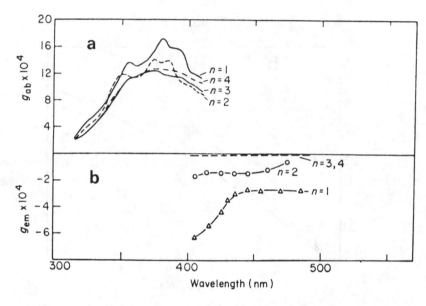

FIGURE 4.   CD and CPL spectra of ϵNAD$^+$ bound to the enzyme. The CD and CPL measurements were performed at 24°C (pH 7.5) as described.[13] The enzyme concentration was $(1.5 \text{ to } 2.0) \times 10^{-4}$ in all cases. (a) CD spectra of Enzyme(ϵNAD$^+$)$_n$ complexes. The fluorescence emission anisotropy factor ($g_{ab} = \dfrac{\Delta\epsilon}{\epsilon}$) is plotted against wavelength. (b) CPL spectra of Enzyme(ϵNAD$^+$)$_n$ complexes. The fluorescence emmission anisotropy factor, g_{em}, is plotted against wavelength. The excitation wavelength was 320 nm. Free ϵNAD$^+$ had no CPL signal ($g_{em} = 0$).

reflect conformational changes in the coenzyme binding sites; the dependence of these changes on the number of bound ϵNAD$^+$ molecules demonstrates unambiguously the sequential nature of these changes. This fact is manifested even more dramatically by the optical activity of the bound ϵNAD$^+$ in the excited state as measured by the CPL spectrum (Figure 4). Since the CPL signal measures the asymmetry of the binding site in the vicinity of the fluorophore (which resides in the adenine portion of the coenzyme analog), the results demonstrate marked conformational changes *in the adenine subsite* as a function of saturation by ϵNAD$^+$. The sequential nature of these alterations (g_{em} differs for n = 1, n = 2, and n = 3,4) is apparent. The observed changes do not fit the two-classes-of-sites model, which would predict identical spectroscopic properties for Enzyme(ϵNAD$^+$)$_1$ and Enzyme(ϵNAD$^+$)$_2$.

Table 6
INTRINSIC DISSOCIATION
CONSTANTS FOR AAD$^+$ BINDING TO
THE ENZYME

	Constants from difference spectra titration (M)	Constants from fluorescence quenching titration (M)
K$'_1$	$<10^{-6}$	$<10^{-6}$
K$'_2$	$<10^{-6}$	$<10^{-6}$
K$'_3$	1.5×10^{-5}	2.7×10^{-5}
K$'_4$	2.3×10^{-5}	4.1×10^{-5}

Note: Binding was measured by spectroscopic titrations following the difference spectrum band formed upon binding of AAD$^+$ at 350 nm ($\epsilon = 1350 \, M^{-1} \, cm^{-1}$) or the quenching of the enzyme fluorescence ($\lambda_{ex} = 307$ nm) at 368 nm.[21] Both signals were found to be proportional to the binding of AAD$^+$ to the enzyme.[21] The binding of the first 2 mol of AAD$^+$ was too tight to permit accurate determination; thus, the values of the intrinsic K$'_1$ and K$'_2$ according to the sequential model could only be estimated to be below $10^{-6} \, M$. The values obtained using both methods are in reasonable agreement.

In order to explore the molecular events at the nicotinamide subsite, analogous experiments were performed employing AAD$^+$, which contains a fluorescent pyridine derivative.[21] This coenzyme analog is a competitive inhibitor of NAD$^+$ in the GAPDH-catalyzed reaction,[21] and binds to the enzyme in a negatively cooperative manner (Table 6). Titration of the enzyme at a high concentration with AAD$^+$ enabled measurements of the CD and CPL spectra.[21] The CD spectra of the difference spectral band at different degrees of saturation with AAD$^+$ are shown in Figure 5. The CD spectrum appears to change between one, two, and three bound AAD$^+$ moles per mole enzyme, while the spectra of Enzyme(AAD$^+$)$_3$ and Enzyme(AAD$^+$)$_4$ are similar. Unlike the situation with the CD measurements, the CPL spectra of the enzyme-bound AAD$^+$ are similar within the experimental error for all degrees of saturation by AAD$^+$ (Figure 6). Since AAD$^+$ is fluorescent exclusively at the pyridine moiety, the CPL spectrum of bound AAD$^+$ reflects the asymmetry of the nicotinamide subsite (free AAD$^+$ has no CPL signal). The fact that the CPL spectrum of bound AAD$^+$ does not depend on the degree of saturation suggests that the conformation of the nicotinamide subsites is identical for the complexes Enzyme(AAD$^+$)$_1$ through Enzyme(AAD$^+$)$_4$, in spite of the negatively cooperative binding of AAD$^+$ to the enzyme. These findings indicate that if AAD$^+$ binding induces any conformational changes at the nicotinamide subsite, those changes are local and are not transmitted to the nicotinamide subsites of neighboring subunits.[21] Since the catalytic center in GAPDH is located at the nicotinamide subsite, this notion is supported by the report[22] that the catalytic constant per active site of the GAPDH-catalyzed reaction is independent of the degree of saturation by NAD$^+$.

The results of the spectroscopic studies (and especially the CPL experiments) demonstrate that the binding of NAD$^+$ analogs to the enzyme is accompanied by conformational changes, which are transmitted from one subunit to the other through the adenine subsites.[13] On the other hand, the induction of conformational changes between the nicotinamide subsites does not appear to be involved in the mechanism generating the negative cooperativity in coenzyme binding.[21] However, the different binding patterns of NAD$^+$ analogs that differ from NAD$^+$ only in the pyridine moiety demonstrate that the structure of the pyridine derivative has a significant effect on the cooperativity[11,12] (Figures 1 and 3; 2, 3, and 6).

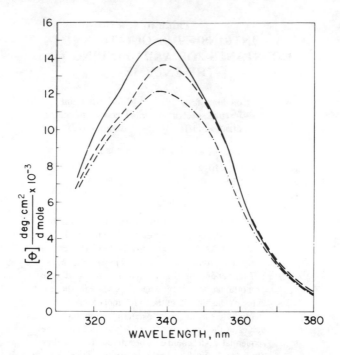

FIGURE 5. CD spectra of AAD$^+$ bound to the enzyme. The enzyme concentration was 7.9×10^{-5} M, and the measurements were performed at 25°C (pH 7.5) as described.[13] The molar elipticity, [θ], was computed according to bound AAD$^+$ concentration. The values of [θ] for n = 1 (——), n = 2 (---), and n = 3,4 (-··-) are plotted, where n is the number of AAD$^+$ molecules bound per tetramer.

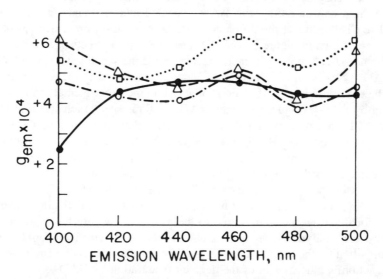

FIGURE 6. CPL spectra of AAD$^+$ bound to the enzyme. The enzyme concentration was 7.5×10^{-5} M, and the measurements were performed at 25°C (pH 7.5) as described.[21] Fluorescence was excited at 325 nm. The fluorescence anisotropy factor, g_{em}, is plotted against the emission wavelength. The contribution of free AAD$^+$ to the fluorescence (it does not contribute to the fluorescence anisotropy signal itself) was subtracted.[21] The spectra shown are those of Enzyme(AAD$^+$) complexes containing one (●—●), two (○-··-○), three (△---△), and 3.6 (□···□) bound AAD$^+$ molecules per tetramer. The experimental error is estimated as ±10% for n = 1,2, and ±15% for n = 3 and 3.6.

Kinetic and equilibrium data on the binding of a large number of NAD$^+$ analogs modified at the pyridine nucleus to rabbit muscle GAPDH as well as to other GAPDHs[23] support the view[13,21] that the pyridine containing moiety plays a key role in the mode of coenzyme binding. The authors, for example, conclude[23] that the methyl group of the acetyl group on the pyridine ring in APAD$^+$ cannot assume the same configuration of the carboxamide residue in the NAD$^+$ molecule. The oxygen of the carboxamide residue in NAD$^+$ makes a hydrogen bond contact with Asn-313 at a distance of 31Å. This hydrogen bond is believed to be important in NAD$^+$ bonding.[24] This configuration makes the carboxamide nitrogen within a hydrogen bonding distance to the phosphate group. When the carboxamide nitrogen is replaced by a methyl group (as in APAD$^+$), this contact is most probably avoided by a rotation which can still keep the pyridinium ring at the same distance from Cys-149, thus retaining full catalytic activity and the characteristic spectral band at 350 nm (the "Racker band").[11,25] This rotation requires a glycosidic bond readjustment and modifies, therefore, the detailed mode of interaction between APAD$^+$ and GAPDH as compared to the NAD$^+$-GAPDH interaction.[23] These subtle differences are reflected in the difference between the mode of binding of the two ligands, but not in their catalytic efficiency. Thus, as explained, the turnover number per coenzyme (k_{cat}) is identical for NAD$^+$ and APAD$^+$, whereas the former binds noncooperatively and the latter with extreme negative cooperativity and the latter binds noncooperatively. Experiments on a large number of NAD$^+$ analogs where the carboxamide residue is replaced by other substituents support this analysis. Details on the mechanism of negative cooperativity have also been learned from the use of spin-labeled NAD$^+$ molecules.[26-28] For improved sensitivity and resolution a ^{15}N,^{2}H-labeled NAD$^+$ derivative (SL-NAD$^+$) was examined for its interaction with rabbit muscle and rat blood cell GAPDH.[29] The use of this specific analog allowed the investigators to probe in detail intersubunit distances between NAD$^+$ binding domains. It was found that the properties examined were identical for the GAPDH-coenzyme complexes in its crystalline state and in solution. This finding legitimizes the comparison between properties of the enzyme in its crystalline state and in solution. Quantitative analysis of the data suggests that the first spin-labeled NAD$^+$ molecule induces a conformational change which diminishes the following binding steps. The dibound species GAPDH(SL-NAD$^+$)$_2$ possess three isomers, as expected from the tetrahedral symmetry of the molecule.[24] One can calculate from the EPR data the relative distribution of the species.[29]

In summary, the SL-NAD$^+$ used by Beth et al.[29] binds with strong negative cooperativity (Hill coefficient n_H = 0.42), and its mode of interaction with GAPDH can be analyzed quantitatively by the EPR technique. These data strongly support the conclusions drawn from experiments using other approaches. It has also been pointed out that spin-labeled AMP, ATP, and ADP-Rib bind noncooperatively and do not give rise to dipolar interactions between the bound ligands. These findings support other data[21] that the nicotinamide moiety is detrimental to the mode of interaction between the adenine subsite and the protein.

B. Evidence for Ligand-Induced Conformational Change Using Fluorescent Alkylating Agents

The studies described in the former section suggested the existence of ligand-induced conformational changes in the rabbit muscle enzyme. Moreover, they demonstrated a dependence of the effects of the competing ligands on their exact structure, suggesting differences in the interactions of the various NAD$^+$ analogs with the enzyme. To obtain a better understanding of the factors controlling the conformational changes, it was desired to compare the results of the competition experiments with measurements of conformational changes using an independent physical method. Such studies are enabled by the technique of circularly polarized luminescence (CPL) which is extremely sensitive to conformational changes in the vicinity of the fluorescent marker.[30]

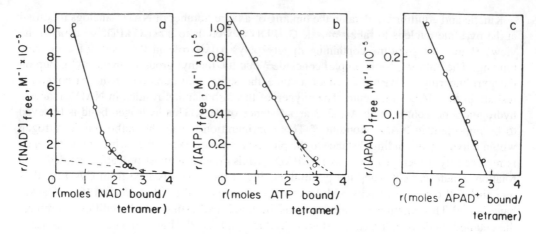

FIGURE 7.   Binding of NAD^+ analogs to Enzyme(alkyl)$_1$. The enzyme (0.1 mM) was reacted with equimolar concentration of I-AEDANS at pH 7.5 in the dark, and excess reagent was removed by passage on a Sephadex G-50 column.[31] The binding of NAD^+ and of ATP was measured by the dialysis rate method (described in Figure 1), which measures binding to both alkylated and nonalkylated sites. $APAD^+$ binding was measured by following the difference spectrum at 350 nm (described in Figure 1); since the alkylated sites do not give rise to this band, this method measures binding to nonalkylated sites only. (a) NAD^+ binding. Enzyme concentration was 11.4 μM. The [^{14}C]-NAD^+ concentration was 7.45 μM (302 mCi/mmol). (b) ATP binding. Enzyme concentration was 40 μM. The [α-^{32}P]-ATP concentration was 2.6 μM (7.7 Ci/mmol). All solutions contained 5 mM phosphate. (c) $APAD^+$ binding. Enzyme concentration was 40 μM.

In order to measure only those conformational changes which are transmitted from one subunit to another (eliminating those occurring in the coenzyme binding subunit itself), it was chosen to label a fraction of the active sites on the enzyme with a fluorescent alkylating agent whose reaction with an active site results in a marked decrease of the affinity of the reacted sites toward the coenzyme.[31] This phenomenon, which was observed with a number of alkylating and acylating agents,[32-38] ensures that coenzyme analogs bind preferentially to the unreacted sites of the tetramer. The preferential binding of the coenzyme to the nonalkylated sites allows the study of the spectral alterations in the CPL signal of the fluorescent alkylating agent attached to neighboring sites.

The fluorescent alkylating agent employed in the CPL studies[31] was *N*-iodoacetyl-*N'*-(5-sulfo-1-naphtyl)-ethylenediamine (I-AEDANS). Alkylation of the apoenzyme by this reagent was accompanied by the loss of enzymatic activity. The inactivation was not linear with the number of alkylated sites, and showed half-of-the-sites reactivity[3] — namely, the first two sites of the tetramer reacted much more readily with the alkylating agent, and alkylation of two sites resulted in an almost complete inactivation of the enzyme. This phenomenon was encountered earlier with the reaction of several other alkylating or acylating agents with this enzyme,[3,32,36-38] and was interpreted to reflect a conformational change induced by the reagent and transmitted to a neighboring vacant subunit, thus inactivating it.[35,37] The enzyme sites reacting with the I-AEDANS are most likely the four active sites (each possessing one reactive SH group), since reaction with an excess of the reagent yielded 4.0 to 4.3 mol alkyl bound per mol tetramer.[31]

The binding of NAD^+ analogs (NAD^+, $APAD^+$, and ATP) to GAPDH containing an average* of one AEDANS moiety per tetramer [Enzyme(alkyl)$_1$] is described in Figure 7 and Table 7. The nonalkylated sites bind NAD^+, $APAD^+$, and ATP with much higher

---

*   Reaction for equimolar concentration of the tetrameric enzyme with the alkylating reagent yields a mixture of enzyme species containing 0 to 4 alkyl groups. When alkylation occurs randomly, the main constituents are the mono- and dialkylated species (the relative distribution is 81:108:54:12:1 for 0,1,2,3, and 4 alkyl groups/tetramer).[31] In cases of site heterogeneity or negative cooperativity, the distribution will be even more in favor of the mono- and dialkylated species.

Table 7
INTRINSIC DISSOCIATION CONSTANTS FOR THE BINDING OF NAD$^+$ ANALOGS TO ENZYME(ALKYL)$_1$

Ligand	Intrinsic dissociation constants (*M*)			
	K'_1	K'_2	K'_3	K'_4
NAD$^+$	1.4×10^{-6}	1.4×10^{-6}	8.5×10^{-6}	5.0×10^{-5}
ATP	2.6×10^{-5}	2.6×10^{-5}	2.6×10^{-5}	1.5×10^{-4}
APAD$^+$	8.2×10^{-5}	8.2×10^{-5}	8.2×10^{-5}	N.D.

Note: The constants were derived from the data in Figure 7.[31] K'_i are the intrinsic dissociation constants for the binding of the i th ligand molecule to the enzyme. K'_4 is a lower limit for the intrinsic dissociation constant of the ligand to the alkylated site (K'_d), since when there is site heterogeneity (in this case, formed by the alkylation), K'_4 represents an average of the intrinsic dissociation constants characterizing the binding to the different sites.[31] Since an alkylated site has a lower affinity, K'_d is higher than K'_4. N.D. = not determined.

affinity than the alkylated sites. Thus, when Enzyme(alkyl)$_1$ is titrated with NAD$^+$ and its analogs, most of the ligand binds to the nonalkylated sites, and the effects of ligand binding to these sites on the CPL or fluorescence signals of the alkyl group in the neighboring site can be followed.[31]

The effect of the binding of NAD$^+$ analogs on the CPL spectrum of the alkyl group in Enzyme(alkyl)$_1$ is shown in Figure 8. The binding of NAD$^+$ and of ADP-Rib induces a significant change in the spectrum of the alkyl moiety. Interestingly, the effects of NAD$^+$ and ADP-Rib on the signal are similar, although ADP-Rib binds to the enzyme noncooperatively (Figure 1). ATP binding has a different effect on the spectrum, whereas changes induced by APAD$^+$ binding are insignificant (Figure 8). In order to demonstrate that most of the effect on the CPL signal is due to ligand binding to the nonalkylated sites, a control experiment was performed using Enzyme(alkyl)$_4$ (Figure 9). In this case, the only effects observed are those of ligand binding to the *alkylated* sites (no nonalkylated sites are present). No change in the spectrum was observed upon the addition of NAD$^+$ or ADP-Rib. In contrast, the addition of ATP caused a significant change, which is clearly due to binding to the alkylated NAD$^+$ sites, as the phosphate sites were blocked by 5 m*M* phosphate (Figure 9). Thus, while the effects of NAD$^+$ and ADP-Rib on the CPL spectrum of Enzyme(alkyl)$_1$ are exclusively due to binding to neighboring nonalkylated sites, the ATP effect may contain a significant contribution due to direct interaction with the alkylated site. This indicates that ATP binding to the NAD$^+$ sites has a character very different from that of the other NAD$^+$ analogs tested, probably due to the high negative charge on the ATP molecule.

The immediate conclusion from the data in Figures 8 and 9 is that the binding of NAD$^+$ analogs can induce conformational changes in the neighboring (alkylated) subunits, which result in alterations in the CPL spectrum of the alkyl group.[31] These findings clearly support the KNF model, and are in accord with ^{19}F-NMR studies performed on lobster muscle GAPDH reacted with 2 mol of 3,3,3-trifluorobromoacetone,[38] which demonstrated the propagation of conformational changes between the subunits upon NAD$^+$ binding. Moreover, the above results (Figures 8 and 9) demonstrate that the type of conformational change depends strongly on the exact structure of the NAD$^+$ analog. Thus, for example, NAD$^+$ and ADP-Rib affect the CPL spectrum of the fluorescent alkyl group in the neighboring subunit, while the binding of APAD$^+$ fails to do so (Figure 8).

The marked effect of ADP-Rib on the CPL spectrum suggests that although this ligand binds to the native enzyme noncooperatively,[11,12] it is capable of inducing conformational

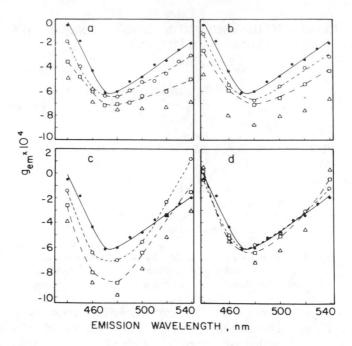

FIGURE 8.   Effect of NAD$^+$ analogs on the CPL of Enzyme(alkyl)$_1$. The experimental details of the CPL measurements are given elsewhere.[31] The excitation wavelength was 330 nm. The fluorescence emission anisotropy factor [$g_{em} = \Delta F/(F/2)$] is plotted agaist the emission wavelength. The occupancy of the nonalkylated sites by NAD$^+$, ATP, and APAD$^+$ was derived from the binding data given in Figure 4 and Table 7.[31] Occupancy by ADP-Rib was computed using the intrinsic dissociation constant characterizing its binding to the apoenzyme (Table 3). The spectra shown are those of 40 μM Enzyme(alkyl)$_1$ alone (●—●), with 1.0 mol (○--○), 2.0 mol (□-·-□), and 2.4 mol (△···△) ligand bound to nonalkylated sites per mole tetramer. (a) NAD$^+$ (b) ADP-Rib (c) ATP (+ 5 mM phosphate) (d) APAD$^+$; 5 mM phosphate by itself had no effect on the CPL spectrum, and the effect of ATP alone was similar to that of ATP + phosphate.

changes which are transmitted to neighboring subunits. As for APAD$^+$, the lack of change in the CPL spectrum upon its binding does not necessarily mean that it does not induce any conformational change, since its binding to the nonalkylated sites of Enzyme(alkyl)$_1$ resulted in significant quenching of the alkyl group fluorescence.[31] It therefore appears that not only NAD$^+$ binding, but also the binding of the noncooperative analogs ADP-Rib and APAD$^+$ (as well as ATP), incude conformational changes which are transmitted to neighboring subunits. Moreover, the different effects of ADP-Rib and APAD$^+$ on the CPL spectrum (Figure 8) and on the fluorescence intensity[31] of the fluorescent alkyl group demonstrate that the two ligands induce different conformational changes, whose nature depends strongly on the structure of the NAD$^+$ analog. The induction of different conformational changes by the two noncooperative ligands is in agreement with their different effects on the binding of εNAD$^+$ to the enzyme where competition by APAD$^+$ abolishes the negative cooperativity in εNAD$^+$ binding, while ADP-Rib has no effect[11,12] (Figure 2, Table 4, see Section III.A).

C. Evidence from Crystallographic Data

Comparison of the crystal structure of *Bacillus stearothermophilus* apo-GAPDH with GAPDH(NAD$^+$)$_1$ reveals that the binding of the NAD$^+$ molecule induces nonequivalence in the GAPDH tetramer. The subunit containing the bound NAD$^+$ adopts a conformation

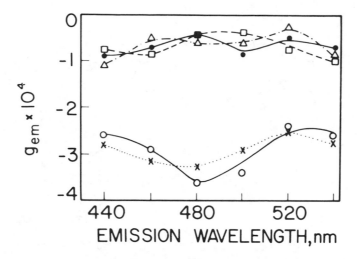

FIGURE 9. Effect of NAD$^+$ analogs on the CPL of Enzyme(alkyl)$_4$. Experimental conditions are as in Figure 8. g_{em} is the fluorescence emission anisotropy factor [$\Delta F/(F/2)$]. The spectra shown are those of 10 μM Enzyme(Alkyl)$_4$ in the absence of ligands (●—●), and in the presence of ligand concentrations corresponding to those accessible to the alkylated sites when 2.0 mol NAD$^+$ (□---□), 2.0 mol ADP-Rib (△-··-△), 2.0 mol ATP (○—○), and 2.0 mol ATP + 5 mM phosphate (x...x) are bound to the nonalkylated sites.[31] 5 mM phosphate alone had no effect on the CPL spectrum.

very similar to the subunit conformation of the holoenzyme; while the other three unliganded subunits possess a conformation more similar to that of the apoenzyme.[39,40] The apoenzyme is symmetric when all the subunits are identical. These results support the view that NAD$^+$ binding gives rise to sequential ligand-induced structural changes of the tetramer. These sequential conformational changes are the basis for the negative cooperativity in coenzyme binding.

Rossmann and his colleagues reported asymmetry within the lobster muscle holoenzyme.[19,42] Crystallographic data on the apoenzyme indicate that the major conformational changes, which typify the holoenzyme to apoenzyme transition, is mainly due to the removal of the *last* NAD$^+$ molecule.[24] Thus, as in the case of *B. stearothermophilus*, the binding of the first coenzyme *induces* the conformational change that breaks up the symmetry of the tetramer. In summary, crystallographic data, as well as a large variety of studies performed on enzyme in solution, support the negative cooperativity model of coenzyme binding to GAPDH (Figure 10).

IV. PHYSIOLOGICAL SIGNIFICANCE

Negative cooperativity in ligand binding is a rare phenomenon occurring in very few enzyme systems, and is occasionally observed in certain hormone-receptor systems.[41] Since the evidence for negative cooperativity in GAPDH is overwhelming, it is useful to discuss briefly its possible physiological significance. First, negative cooperativity is, in principle, a valid mechanism to boost the affinity toward a ligand. Hence, the high affinity of GAPDH toward the first two NAD$^+$ molecules is at the "expense" of the affinity toward the last two molecules. This mechanism insures a constant supply of NAD$^+$ to GAPDH, and is of a relative advantage of this enzyme over the other dehydrogenases. GAPDH is a key enzyme in glycolysis, with the flow of substrate through this pathway depending heavily on the GAPDH reaction. This is probably the reason as to why the enzyme concentration in the

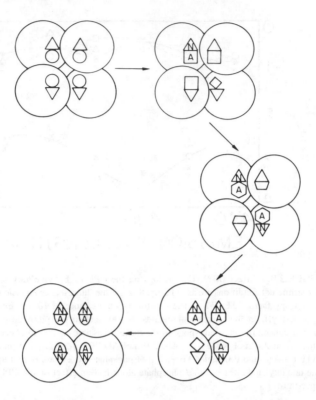

FIGURE 10. Schematic representation of conformational changes in the enzyme following coenzyme binding. The apoenzyme is presented as a symmetric tetrahedron, but the interactions between the two pairs of subunits located diagonally to each other were taken to be stronger, according to the crystallographic data on preferential interactions across the molecular symmetry axis R.[24] The adenine subsite in the apoenzyme is represented by a circle and the nicotinamide subsite by a triangle (upper left). The coenzyme is designated NA, where N is the nicotinamide portion and A is the adenine moiety. Coenzyme binding induces sequential conformational changes, which propagate through the adenine subsites. The nicotinamide subsites do not participate in those changes, but their interactions with the pyridine moiety determine the orientation of the adenine portion in the adenine subsite.

tissue is about 0.1 mM.[1] The high concentration of the enzyme and the very high affinity toward NAD$^+$ assure that the GAPDH reaction will be used effectively by the cell. Another aspect of the issue is the finding that AMP, ADP, and ATP have a very similar effect on the binding of coenzyme to GAPDH.[11] Since AMP, ADP, and ATP all seem to bind with equal affinity to GAPDH,[11] this suggests that they all bind by using identical chemical interactions. This observation immediately suggests that NAD$^+$ binding to GAPDH is independent of the intracellular fluctuations in the ATP to ADP plus AMP ratio. As in the presence of 0.1 M salt ATP binding is weak,[11] this further supports the view that ATP levels do not play a regulatory role in GAPDH function. Furthermore, most of the intracellular ATP is complexed to Mg^{2+} and it is known that MgATP binds much less tightly to the enzyme than ATP.[32,42] This makes ATP an even less likely candidate for GAPDH regulation. However, in view of the observation that acyl-GAPDH binds ATP tighter than ADP, while NAD$^+$ binds with equal affinity to the two species,[43] it is still possible that the ATP/ADP ratio can regulate the progress of the GAPDH reaction.

REFERENCES

1. **Levitzki, A.,** Quantitative aspects of allosteric mechanisms, in *Molecular Biology, Biochemistry and Biophysics,* Vol. 28, Springer-Verlag, Heidelberg, 1978.
2. **Levitzki, A. and Koshland, D. E., Jr.,** The role of negative cooperativity and half-of-the-sites reactivity in enzyme regulation, in *Current Topics in Cell Regulation,* Vol. 10, Stadman, E. R. and Horecker, B. L., Eds., 1976, 1.
3. **Seydoux, F., Malhotra, O. P., and Bernhard, S. A.,** Half site reactivity, *CRC Crit. Rev. Biochem.,* 2, 227, 1974.
4. **Monod, J., Wyman, J., and Changeux, J. P.,** On the nature of allosteric transitions: a plausible model, *J. Mol. Biol.,* 12, 88, 1965.
5. **Koshland, D. E., Jr., Nemethy, G., and Filmer, D.,** Comparison of experimental binding data and theoretical models in proteins containing subunits, *Biochemistry,* 5, 365, 1966.
6. **Koshland, D. E., Jr.,** The molecular basis for enzyme regulation, in *The Enzymes,* Vol. 1, 3rd ed., Boyer, P. D., Ed., Academic Press, New York, 1970, 341.
7. **Henis, Y. I. and Levitzki, A.,** Ligand competition curves as a diagnostic tool for delineating the nature of site-site interactions: theory, *Eur. J. Biochem.,* 102, 449, 1979.
8. **Henis, Y. I. and Sokolvsky, M.,** Muscarimic antagonists induce different receptor conformations in rat adenohypophysis, *Mol. Pharmacol.,* 24, 357, 1982.
9. **Dahlquist, F. W.,** Cooperativity in ligand binding to proteins showing association-dissociation phenomena: an application of the quantitative interpretation of the Hill coefficient, *Methods Enzymol.,* 48, 270, 1978.
10. **Greenfield, J. C., Leonard, N. J., and Gumport, R. I.,** Nicotinamide 3, N^4-ethenocytosine dinucleotide, an analog of nicotinamide adenine dinucleotide: synthesis and enzyme studies, *Biochemistry,* 14, 698, 1975.
11. **Henis, Y. I. and Levitzki, A.,** The sequential nature of the negative cooperativity in rabbit muscle glyceraldehyde-3-phosphate dehydrogenase, *Eur. J. Biochem.,* 112, 59, 1980.
12. **Henis, Y. I. and Levitzki, A.,** Mechanism of negative cooperativity in glyceraldehyde-3-phosphate dehydrogenase deduced from ligand competition experiments, *Proc. Natl. Acad. Sci. U.S.A.,* 77, 5055, 1980.
13. **Schelssinger, J. and Levitzki, A.,** Molecular basis of negative cooperativity in rabbit muscle glyceraldehyde-3-phosphate dehydrogenase, *J. Mol. Biol.,* 83, 547, 1974.
14. **Luisi, P. L., Baici, A., Bonner, F. J., and Aboderin, A. A.,** Relationship between fluorescence and conformation of ϵNAD^+ bound to dehydrogenases, *Biochemistry,* 14, 362, 1975.
15. **Conway, A. and Koshland, D. E., Jr.,** Negative cooperativity in enzyme action: the binding of diphosphopyridine nucleotide to glyceraldehyde-3-phosphate dehydrogenase, *Biochemistry,* 7, 4011, 1968.
16. **deVijlder, J. J. M. and Slater, E. C.,** The reaction between NAD^+ and rabbit muscle glyceraldehyde-3-phosphate dehydrogenase, *Biochim. Biophys. Acta,* 167, 23, 1968.
17. **Booth, R., Box, F. W., Muller, M. E., and Peterson, T. I.,** Forecasting by Generalized Regression Methods: Non-linear Estimation, Share Program No. 687, WL-NL-I, IBM, Mimes, Princeton, New Jersey, 1959.
18. **Adair, G. S.,** The hemoglobin system. VI. The oxygen dissociation curve of hemoglobin, *J. Biol. Chem.,* 63, 529, 1925.
19. **Olsen, K. W., Garavito, M. R., Sabesan, M. N., and Rossmann, M. G.,** Studies on coenzyme binding to glyceraldehyde-3-phosphate dehydrogenase, *J. Mol. Biol.,* 107, 577, 1976.
20. **Eby, D. and Kirtley, M. E.,** Cooperativity and non-cooperativity in the binding of NAD^+ analogues to rabbit muscle glyceraldehyde-3-phosphate dehydrogenase, *Biochemistry,* 15, 2168, 1976.
21. **Henis, Y. I. and Levitzki, A.,** The role of the nicotinamide and the adenine subsites in the negative cooperativity of coenzyme binding to glyceraldehyde-3-phosphate dehydrogenase, *J. Mol. Biol.,* 117, 699, 1977.
22. **Teipel, J. and Koshland, D. E., Jr.,** The effect of NAD^+ on the catalytic efficiency of glyceraldehyde-3-phosphate dehydrogenase from rabbit muscle, *Biochim. Biophys. Acta,* 198, 183, 1970.
23. **Wallen, L. and Branlant, G.,** NAD^+ analogue binding to glyceraldehyde-3-phosphate dehydrogenase, *Eur. J. Biochem.,* 137, 67, 1983.
24. **Moras, D., Olsen, K. W., Sabesan, M. N., Buehner, M., Ford, G. C., and Rossmann, M. G.,** Studies of asymmetry in the three-dimensional structure of lobster D-glyceraldehyde-3-phosphate dehydrogenase, *J. Biol. Chem.,* 250, 9137, 1975.
25. **Kaplan, N. O., Giotti, M. M., and Stolzenback, F. E.,** Studies on the interactions of dihydropyridine nucleotide analogs with dehydrogenases, *Arch. Biochem. Biophys.,* 69, 441, 1957.
26. **Gloggler, K. G., Balasubramanian, K., Beth, A. M., Park, J. H., and Trommer, W. E.,** The role of the nicotinamide moiety for negative cooperativity in glyceraldehyde-3-phosphate dehydrogenase studied by spin labelled cofactors, *Biochim. Biophys. Acta,* 706, 197, 1982.
27. **Deporade, M. P., Gloggler, K. G., and Trommer, W. E.,** Isolation and properties of glyceraldehyde-3-phosphate dehydrogenase from a sturgeon from the Caspian Sea and its interactions with spin labelled NAD^+ derivatives, *Biochim. Biophys. Acta,* 659, 442, 1981.

28. **Gloggler, K. G., Balasubramanian, K., Beth, A., Fritzsche, T. M., Park, H. J., Pearson, D. E., Trommer, W. E., and Venkataramu, S. D.,** The synthesis of deuterium-substituted spin-labelled analogues of AMP and their use in ESR studies of lactate dehydrogenase, *Biochim. Biophys. Acta,* 701, 224, 1982.

29. **Beth, A. M., Robinson, B. H., Cobb, C. E., Dalton, L. R., Trommer, W. E., Birktoft, J. J., and Park, J. H.,** Interactions and special arrangement of spin-labelled NAD^+ bound to glyceraldehyde-3-phosphate dehydrogenase, *J. Biol. Chem.,* 259, 9717, 1984.

30. **Steinberg, I. Z.,** Fluorescence polarization: some trends and problems, in *Biochemical Fluorescence: Concepts,* Chen, R. F. and Edelhoch, H., Eds., Marcel Dekker, New York, 1975, 79.

31. **Henis, Y. I., Levitzki, A., and Gafni, A.,** Evidence for ligand-induced conformational changes in rabbit muscle glyceraldehyde-3-phosphate dehydrogenase, *Eur. J. Biochem.,* 97, 519, 1979.

32. **Schwendimann, B., Ingbar, D. M., and Bernhard, S. A.,** On the function of half-site reactivity: intersubunit NAD^+-dependent activation of acyl-glyceraldehyde-3-phosphate dehydrogenase reduction by NADH, *J. Mol. Biol.,* 108, 123, 1976.

33. **Bode, J., Blumenstein, M., and Raftery, M. A.,** Non-identical alkylation sites in rabbit muscle glyceraldehyde-3-phosphate dehydrogenase, *Biochemistry,* 14, 1146, 1975.

34. **Seydoux, F. J., Keleman, N., Kellersohn, N., and Roucous, C.,** Specific interactions of 3-phosphoglycerol-glyceraldehyde 3-phosphate dehydrogenase with coenzymes, *Eur. J. Biochem.,* 64, 481, 1976.

35. **Levitzki, A.,** Half-of-the-sites and all-of-the-sites reactivity in rabbit muscle glyceraldehyde-3-phosphate dehydrogenase, *J. Mol. Biol.,* 90, 451, 1974.

36. **Malhotra, O. P. and Bernhard, B. A.,** Spectrometric identification of an active site-specific acyl glyceraldehyde-3-phosphate dehydrogenase, *J. Biol. Chem.,* 243, 1243, 1968.

37. **Levitzki, A.,** Ligand-induced half-of-the-sites reactivity in rabbit muscle glyceraldehyde-3-phosphate dehydrogenase, *Biochem. Biophys. Res. Commun.,* 54, 889, 1973.

38. **Long, J. W. and Dahlquist, F. W.,** Evidence for induced interactions in the anti-cooperative binding of nicotinamide adenine dinucleotide to sturgeon muscle glyceraldehyde-3-phosphate dehydrogenase, *Biochemistry,* 16, 3792, 1977.

39. **Leslie, A. G. and Wonacott, A. J.,** Coenzyme binding in crystals of glyceraldehyde-3-phosphate dehydrogenase, *J. Mol. Biol.,* 165, 375, 1983.

40. **Leslie, A. G. and Wanocott, A. J.,** Structural evidence for ligand-induced sequential conformational changes in glyceraldehyde-3-phosphate dehydrogenase, *J. Mol. Biol.,* 178, 743, 1984.

41. **Levitzki, A.,** Negative cooperativity at the insulin receptor, *Nature (London),* 289, 442, 1981.

42. **Oguchi, M., Meriwether, B. P., and Park, J.,** Interaction between adenosine triphosphate and glyceraldehyde-3-phosphate dehydrogenase, *J. Biol. Chem.,* 248, 5562, 1973.

43. **Malhotra, O. P. and Bernhard, S. A.,** Activation of a covalent enzyme substrate bond by non-covalent interaction with an effector, *Proc. Natl. Acad. Sci. U.S.A.,* 70, 2077, 1973.

Chapter 7

CITRATE SYNTHASE

P. D. J. Weitzman

TABLE OF CONTENTS

I. INTRODUCTION

Fifty years ago, Hans Krebs slotted into place the missing piece of the metabolic puzzle surrounding cell respiration and thereby discovered the citric acid cycle. He showed[1] that citrate could be synthesized by minced muscle and other animal tissues from oxaloacetate and some unknown product of carbohydrate metabolism and proposed a cyclic scheme in which the oxaloacetate formed as the end product of citrate oxidation could be recombined with an unknown metabolite to form further citrate. The scheme established a link between the metabolism of carbohydrate and of citrate and the plant acids. The detailed mechanism of citrate synthesis remained obscure for several years until coenzyme A was discovered and acetyl-CoA was shown to be the second substrate for the reaction producing citrate and catalyzed by citrate synthase (citrate oxaloacetate-lyase[CoA-acetylating], EC 4.1.3.7), originally termed "condensing enzyme":

$$\text{oxaloacetate} + \text{acetyl-CoA} + H_2O \rightarrow \text{citrate} + \text{CoA-SH} + H^+$$

It was subsequently shown that the citric acid cycle is the terminal oxidative pathway for all major foodstuffs and that it operates throughout nature in animals, plants, and microorganisms.

In addition to its role as an energy-yielding catabolic pathway, the cycle fulfills an anabolic function in providing the carbon skeletons of cell constituents and is clearly established at the very heart of intermediary metabolism. Although the formation of citrate was the final step in the citric acid cycle to be discovered, that step is, in a sense, the first one in the cyclic pathway. It is in the formation of citrate that carbon atoms, in the form of acetyl units, enter the citric acid cycle to be oxidized to CO_2 and produce reducing equivalents. Citrate synthase (CS) may be considered the "initiating" enzyme of the cycle and, not surprisingly, much attention has been given to its study. The different demands which diverse organisms make on the catabolic and anabolic functions of the cycle would be expected to be reflected in different regulatory mechanisms operating on CS. In turn, such regulatory differences would be expected to be associated with differences in the molecular structures of the enzyme from diverse sources. Studies on CS from a wide range of organisms have amply fulfilled such expectations and have revealed striking patterns of both structure and function.

Several reviews on CS[2-7] with complementary emphasis on its structure, catalytic function, and regulation are recommended to the reader for their comprehensive coverage, particularly as the present article cannot do justice to all facets of the enzyme nor to all the investigators who have contributed to our present knowledge. Rather, its purpose is to kindle the reader's interest in CS, to present a broad picture of the regulation of activity of the enzyme and of the molecular studies which have led to some understanding of function in structural terms, and to draw attention to some potentially fruitful areas for future progress.

It is a privilege for CS to be included in a short list of allosteric proteins chosen for this volume and to share company with such classic examples of the genre; I hope to show that such inclusion is justified. I also hope that, as with allosteric properties, the association of CS with the other examples in this book will result in cooperative interactions and synergistic benefits.

II. OBSERVATION

A. Assay of Enzyme Activity

Without doubt, progress in the study of CS has been greatly assisted by the availability of rather simple and convenient assay procedures. These have made possible not only the measurement of activity in partially or wholly purified enzyme preparations, as well as in crude cell-free extracts and even in permeabilized whole cells, but have also facilitated examination of the action of metabolic effectors and made it feasible to undertake investigations on CS from a wide range of organisms.

Although two ultraviolet spectrophotometric methods were first employed, a major advance was the introduction by Srere et al.[8] of a method based on the reaction of 5,5'-dithiobis-(2-nitrobenzoate) (DTNB) with the thiol group of CoA-SH; the stoichiometrically liberated thionitrobenzoate anion absorbs strongly at 412 nm and thus permits simple continuous colorimetric assay of the enzyme. However, the reactivity of DTNB with thiol groups can also result in reaction with the CS enzyme itself leading to inactivation or densensitization to effectors. To avoid this, Weitzman[9] introduced a polarographic assay based on the concentration-dependent electrochemical signal produced by CoA-SH at a dropping mercury electrode; this assay has several advantages including its suitability for the study of permeabilized whole cells as, unlike spectral methods, it is unaffected by turbidity.

B. Nonspecific Inhibition by Adenine Nucleotides

Hathaway and Atkinson[10] first demonstrated a direct inhibition of yeast CS by adenine nucleotides, effective in the order ATP > ADP > AMP. Similar inhibition of CS from a very wide range of organisms — microbial, plant, and animal — has been reported.[4,5] Since a major role of the citric acid cycle is in the production of energy as ATP (via NADH formation, electron transfer, and oxidative phosphorylation), it is tempting to identify the inhibition of CS by ATP as a physiological feedback control mechanism; several textbooks of biochemistry do indeed cite this inhibition as contributing to the overall control of the citric acid cycle and even accord it the gratuitous distinction of being an allosteric phenomenon. However, this is unjustified as no evidence has been presented for any allostery. The inhibition is competitive with acetyl-CoA and, in view of the structural similarity between acetyl-CoA and the adenine nucleotides, it is likely that they all bind to the same site on the enzyme. Supporting evidence comes from the linear relationship between the reported K_m values for acetyl-CoA (2 to 400 μM) for a large number of different CSs and their K_i values for ATP,[5] and from multiple-inhibition studies which indicate the isosteric nature of the ATP inhibition.[11] Reduction of ATP inhibition in the presence of Mg^{2+} ions and in permeabilized yeast cells or rat liver mitochondria, together with the fact that the inhibition has been observed even with CSs from organisms which do not use the citric acid cycle for energy generation (e.g., cyanobacteria), make it unlikely that the ATP effect is an *in vivo* regulatory mechanism.

Inhibition of CS by adenine nucleotides is accompanied by nonspecific inhibition by the nicotinamide nucleotides, in the order NADPH > NADH > $NADP^+$ > NAD^+. These compounds also compete for the acetyl-CoA site on the enzyme by virtue of their structural similarity to the substrate.[4,5]

C. Specific Inhibition by NADH

Consideration of the overall citric acid cycle highlights the importance of NADH as an "end product" of the cycle. Weitzman[12] found the *Escherichia coli* enzyme to be only weakly inhibited by ATP, but to be very sensitive to specific inhibition by NADH (virtually total inhibition at 0.1 mM NADH under assay conditions). The inhibition appeared to be competitive with acetyl-CoA, but no inhibition was observed with NADPH, NAD^+, or $NADP^+$. Furthermore, the *E. coli* enzyme was distinguished from the yeast and pig heart CSs inasmuch as the latter two showed no inhibition by NADH (except at very high concentrations when a low level of nonspecific inhibition may be observed — see previous section). It was therefore proposed that in bacteria it is the level of NADH which acts as a specific feedback regulator of CS and hence of the citric acid cycle. A few other bacterial CSs were studied and somewhat different behavior was observed with the enzyme from

Acinetobacter calcoaceticus (lwoffi). Although it, too, showed inhibition by NADH, low concentrations of AMP completely reversed the inhibition, whereas this was not the case with the *E. coli* enzyme. In view of these differences, Weitzman and Jones[13] carried out a large-scale survey of CS from a broad range of bacterial genera and demonstrated a striking pattern of regulatory behavior. Bacteria fall into two groups according to the susceptibility of their CSs to inhibition by NADH. The NADH-sensitive group contains only Gram-negative bacteria, while all the Gram-positive bacteria fall into the NADH-insensitive group. Thus, NADH inhibition of CS is not a universal bacterial feature, but is restricted to Gram-negative organisms. Moreover, the NADH-sensitive CSs could themselves be divided into two groups according to the effect of AMP on the inhibition. Those CSs that are reactivated (deinhibited) by AMP come from strictly aerobic bacteria (e.g., *Pseudomonas* species), whereas those which are unaffected by AMP occur in the facultatively anaerobic organisms (e.g., *E. coli* and other enterobacteria). The rationale proposed for this difference was based on the different energy metabolism in these two groups of bacteria. As the strict aerobes are dependent on the citric acid cycle for energy production, it is appropriate that AMP, an indicator of the energy state of the cell, should regulate the first enzyme of the cycle. By contrast, the facultative anaerobes are capable of generating energy by fermentation and of using the citric acid cycle only as a biosynthetic pathway; control of CS by AMP may thus be inappropriate in such organisms. Subsequent studies on other organisms have supported this scheme of regulatory patterns of CS (see Reference 5 for details and further references). Of course, in biology, we are accustomed to finding exceptions to general patterns and some bacterial CSs have proved anomalous. For example, *Acetobacter* species, *Thermus aquaticus,* and some *Thiobacillus* species, though all Gram-negative, have been found to produce CSs which are insensitive to NADH inhibition. However, no Gram-positive bacterial CS has been found to display NADH sensitivity. It therefore remains the case that only Gram-negative bacteria fall in the group showing NADH inhibition of CS and that all Gram-positive bacteria, together with all eukaryotes, belong to the NADH-insensitive CS group.

D. Biosynthetic Controls

In its biosynthetic role, the citric acid cycle produces oxoglutarate and succinyl-CoA used, respectively, for amino acid and porphyrin biosynthesis, and CS may show regulatory features which reflect its role in the generation of these substances. Strictly aerobic organisms which have an overriding dependence on the citric acid cycle for energy production may not so obviously exhibit biosynthetic control of CS, but organisms which derive energy from other metabolic pathways and use the citric acid cycle only for its biosynthetic function may well display more clearly such biosynthetic control.

As stated earlier, *E. coli* and related facultatively anaerobic organisms can derive energy from carbohydrate fermentation. Under such circumstances, the enzyme complex oxoglutarate dehydrogenase is absent and a branched noncyclic variant of the citric acid cycle operates to satisfy the biosynthetic demands for oxoglutarate and succinyl-CoA. In this scheme, oxoglutarate is the end product of a short biosynthetic pathway initiated by CS; succinyl-CoA, on the other hand, is produced by a reversed set of reactions from oxaloacetate. Following the report by Wright et al.[14] that oxoglutarate inhibits *E. coli* CS, Weitzman and Dunmore[15] surveyed a range of organisms and found oxoglutarate inhibition of CS only with the facultatively anaerobic Gram-negative bacteria; their aerobic counterparts, together with Gram-positive bacteria and eukaryotes, did not display this inhibitory effect. The likely physiological significance of the inhibition is supported by its occurrence only in organisms lacking oxoglutarate dehydrogenase and thus producing oxoglutarate as an end product.[5]

Another example of the use of the citric acid cycle reactions for biosynthesis only is provided by the cyanobacteria. These Gram-negative bacteria lack the enzyme oxoglutarate dehydrogenase, their incomplete cycle is not used for energy production, and cyanobacterial CSs are not inhibited by NADH.[5] Again, oxoglutarate is an end product of CS action, but, unlike the situation in *E. coli,* succinyl-CoA is also formed from citrate via isocitrate and cleavage by isocitrate lyase. Consistent with this, Lucas and Weitzman[16] found that, in addition to oxoglutarate inhibition of CS from cyanobacteria, succinyl-CoA is also an inhibitor of the enzyme from these organisms; the two inhibitors are competitive with oxaloacetate and acetyl-CoA, respectively. Moreover, the absence of succinyl-CoA inhibition of CS from other organisms again highlights the specificity of the effect on the cyanobacterial enzyme.

Overall, the conclusion is inescapable that, superimposed on the basic catalytic function of CS conserved throughout nature, there is a striking diversity of regulatory behavior which appears to reflect the particular metabolic individuality of different types of organisms.

E. Molecular Size

Prompted by the more complex regulatory properties displayed by Gram-negative bacterial CSs and by the knowledge that the molecular weight of at least one Gram-negative bacterial CS was around 250,000, whereas that of the pig heart enzyme was under 100,000, Weitzman and Dunmore[17] examined the molecular sizes of a range of CSs. The enzymes fall clearly into two groups — "large" and "small". The large enzymes ($M_r \sim 250,000$) are derived only from Gram-negative bacteria, while Gram-positive bacteria and eukaryotes produce only the small enzymes ($M_r \sim 100,000$). Subsequent studies on a wide range of both prokaryotic and eukaryotic CSs[4,5] have further confirmed this division of molecular sizes and have firmly established that the regulatory sensitivities to NADH, oxoglutarate, and succinyl-CoA are shown only by the large CSs with their more complex molecular structure.

Although there are no exceptions to the rule that eukaryotic and Gram-positive bacterial CSs are all of the small type, some reverse exceptions, i.e., Gram-negative bacterial CSs of small size, have been noted.[5,18] Without exception, these are all insensitive to NADH. Furthermore, it is noteworthy that although the Gram-negative *Acetobacter* produces NADH-insensitive CS, the enzyme is of the large type and some other similar examples have also been observed.[5,18] Thus, there are three major groups of CSs — large/NADH-sensitive, large/NADH-insensitive, and small/NADH-insensitive.

Weitzman and co-workers have isolated bacterial mutants producing altered CS.[19-21] Their strategy was first to produce CS-deficient mutants and then, by a second mutation, to generate revertants which had regained CS activity but not necessarily with the same regulatory properties as the original (wild-type) enzyme. The significant finding was that starting, for example, with *E. coli,* three distinct groups of mutant CSs were isolated — large/NADH sensitive, large/NADH-insensitive, and small/NADH-insensitive. The similarity between these three types of "artificial" enzymes and the three types of naturally occurring CSs is remarkable and extends beyond the gross properties of size and NADH sensitivity to kinetic characteristics and K_m and K_i values. The findings indicate that small genetic alterations of the amino acid sequence of the enzyme can lead to substantial changes in subunit arrangement and regulatory behavior. It may well be that nature has herself made use of such minor "tinkering" in order to generate the diversity of CSs outlined above. The differences observed between the CSs from Gram-negative bacteria, on the one hand, and Gram-positive bacteria and eukaryotes, on the other, may well stem from only minor, though crucial, amino acid changes and the different enzymes are likely to share many molecular features governing the fundamental catalytic process.

Despite the diversity of CS, the discussion here has implied the existence of only a single form of the enzyme within any one cell type. Although this is probably largely the case, some notable exceptions have been identified. Distinct forms of the enzyme have been found in the mitochondria and glyoxysomes of plant tissue, and both mitochondrial and cytoplasmic forms of the enzyme have been reported in yeast along with the presence of two distinct genes. However, all these forms appear to be of the small type. On the other hand, two forms of CS have been found in various *Pseudomonas* species — one is a large form inhibited by NADH and reactivated by AMP, and the other is a small form unaffected by these compounds.[18] It is not yet known whether these represent products of distinct genes. Interestingly, variation was observed in the relative proportions of the two forms with the stage of growth of the bacterial culture.[22] It is conceivable that both forms are derived from a single gene but that some form of differential modification occurs in response to physiological demands. If this were so, it would further emphasize the view that subtle structural changes are responsible for gross differences in subunit aggregation and regulatory responses.

The regulatory and structural diversity of CS revealed by the studies considered so far emphasize that there is no one single CS, and even the biochemist's favorite prokaryote, *E. coli,* has been exposed as far from representative of other bacteria. The investigation of CSs from a wide range of organisms has, however, proved a fruitful pursuit and, in the drive toward a molecular understanding of the regulatory processes described, comparative structure-function studies on diverse CSs are likely to make a significant contribution.

III. ANALYSIS

A. Reaction Mechanism

The work of Eggerer's group (see Reference 7) has shown that the overall reaction catalyzed by CS has three components: (1) proton abstraction from the methyl group of acetyl-CoA to form an enolate anion (enolase), (2) condensation with the carbonyl group of oxaloacetate to form citryl-CoA (ligase), and (3) hydrolysis to form citrate and CoA-SH (hydrolase). Oxaloacetate binds first to the enzyme and increases the binding constant for acetyl-CoA. With almost all CSs, the acetyl-CoA attacks the si-face of oxaloacetate, and configuration inversion at the methyl group occurs; in obligate anaerobic bacteria, the stereochemistry appears to be reversed and acetyl-CoA condenses with the re-face of oxaloacetate.[3]

B. Molecular Structure

The small CS was first shown to have a dimeric structure (subunit M_r ~50,000) in the case of the mammalian enzyme[2] and, more recently, in Gram-positive bacteria.[23] In contrast, some uncertainty has surrounded the subunit composition of the large CS.[4] The *E. coli* enzyme was originally reported to be a tetramer of identical subunits of M_r 60,000 to 65,000, which could equilibrate with inactive monomeric and octameric forms. Later studies by Tong and Duckworth[24] revised the subunit M_r to ~47,000. Cross-linking with dimethyl suberimidate followed by SDS-electrophoresis indicated six bands with M_r values integral multiples (1 to 6) of 47,000. These results indicate a hexameric structure for the *E. coli* enzyme. Examination of the native enzyme by sedimentation equilibrium ultracentrifugation indicated association-dissociation equilibria over the pH range 7 to 10. Thus, at pH 7.8, a mixture of oligomers was present with apparent M_r values ranging from under 90,000 to over 300,000; in the presence of 0.1 M KCl, however, all of the enzyme was converted homogeneously to the hexameric form (M_r ~280,000). At pH 9, the enzyme also appeared homogeneous, but was in the dimeric form (M_r ~98,000). Electrophoretic analysis of the cross-linked enzyme (see above) showed the even-numbered bands to be present in larger amounts than the odd-numbered ones, and the dimer was the major product of cross-linking. These results, together with the ultracentrifugation studies, suggest that the dimeric arrangement of subunits in the *E. coli* enzyme is particularly stable and may be equivalent to the eukaryotic dimeric enzyme.

Another large CS which has been examined is that from *Acinetobacter calcoaceticus.*[25] Unlike the *E. coli* enzyme, this does not show association-dissociation equilibria. Electrophoretic analysis of the cross-linked enzyme also indicated six bands consistent with a hexameric structure; the even-numbered bands were again more dominant and the dimer band was the major one. It is therefore likely that the large CSs all have a hexameric structure (of identical subunits) in which the dimeric arrangement is an important element. In the course of an electron microscopic study of large CS, Rowe and Weitzman[26] observed a small proportion of enzyme molecules showing definite threefold symmetry. As it was then believed that the enzyme was tetrameric, this threefold symmetry was accounted for in terms of a particular view of a tetrahedral structure. However, in the light of the now accepted hexameric structure, probably based on a trimer of dimers, the threefold symmetry observed is more readily explained.

The apparent importance of the dimeric arrangement as a constituent of the hexamer makes it likely that structural information gained from the study of dimeric small CS can be extrapolated to the large enzyme. Moreover, there may be substantial similarities between the tertiary structures of the monomer subunit of the small and large enzymes. Supporting evidence comes from the similarity in proteolytic fragmentation of various CSs[27,28] and from amino acid sequence analysis. Complete amino acid sequences have now been determined for the pig heart, yeast, and *E. coli* CSs,[29-32] the latter being facilitated by gene cloning.[33] Substantial homology exists between these three sequences, especially around residues believed to contribute to the active site (deduced from X-ray crystallographic studies[34]). The homology between the pig and yeast CSs is extensive (60%)[30] consistent with the structural and catalytic similarities between these two dimeric enzymes. The *E. coli* CS, on the other hand, displays less homology (27%), perhaps related to its quite different hexameric quaternary structure (and hence, intersubunit contacts) and its allosteric site for NADH. The amino acid sequences also permit definitive molecular weight assignments: the pig heart CS subunit contains 437 residues with $M_r = 48,969$,[29] while that of *E. coli* contains 426 residues[32] and has $M_r = 47,938$. By means of statistical estimation of likely sequence homology, Morse and Duckworth[35] have suggested that the *E. coli* and *Acinetobacter* CSs may be as much as 90% homologous and show 65 to 70% homology with the *Pseudomonas aeruginosa* enzyme. Thus, there is likely to be considerable conservation of structural features between the subunits of diverse CSs, associated with a conserved catalytic mechanism, but the differences may be related to the quaternary aggregation and the diversity of allosteric regulatory behavior. This would agree with the conclusions reached from the study of mutant CSs (Section II.E). Structural studies on the small CS may thus be of considerable relevance to an understanding of the molecular mechanism underlying the function of the large enzyme.

Kinetic studies on pig heart CS[36] indicated that the enzyme undergoes conformational changes during the catalytic cycle and led to a proposed mechanism in which the free enzyme represents the hydrolase activity (see Section III.A). Binding of oxaloacetate brings about a conformational change and converts the enzyme to a ligase which catalyzes the condensation reaction; conversion of the enzyme back to the hydrolase conformation accompanies expulsion of the products. It was speculated that the effect of oxaloacetate is to produce a movement of two domains in the enzyme toward each other, changing an "open" form to a "closed" form.

X-ray crystallographic studies on pig and chicken CSs have clearly revealed two domains in the enzyme and have indicated the nature of their relative movement.[34] The enzyme is predominantly α-helical; each subunit comprises 20 helices, the dimer being a tightly packed globular molecule. The two subunits in the dimer are interdependent, each contributing to the binding sites of the other subunit. This presumably underlies the significance of the dimer for enzymic activity and a similar situation in the large CSs may account for the apparent importance of the dimeric unit as a constituent of the hexamer. Each subunit contains a large and a small domain and two distinct crystal forms were observed. One form is "open" with a deep cleft between the two domains, whereas the other is "closed". It was proposed[34] that binding of oxaloacetate converts the open to the closed form and brings active site residues into the correct configuration for the catalytic mechanism and known stereochemistry. The mechanism of this domain closure has been analyzed further in terms of the cumulative effect of small conformational adjustments between close-packed helices.[37]

Only a very preliminary crystallographic study of the *E. coli* CS at low resolution has been reported,[38] but it is to be hoped that the impressive success achieved with the small CS will before long be matched by a detailed structural picture of the large enzyme. It may well transpire that a similar form of domain closure accompanies catalytic activity in the large CSs, and it is conceivable that the mechanisms of allosteric regulation of activity involve some interaction with such domain movements.

C. Allostery

In contrast to the nonspecific isosteric nucleotide inhibition, the specific inhibition of Gram-negative bacterial CSs by NADH has all the qualities of an allosteric phenomenon. It has been studied in a number of enzymes, but most extensively in the facultative anaerobe *E. coli* and the strictly aerobic genera *Acinetobacter* and *Pseudomonas;* much of the following discussion concerns the enzymes from these three sources. First, evidence came from studies on the *E. coli* enzyme[39] which was shown to be desensitized to NADH inhibition in the presence of 0.2 *M* KCl or at mildly alkaline pH (~9), though the catalytic activity was preserved; in fact, the presence of salt actually enhanced activity. Loss of sensitivity to NADH around pH 9 may reflect dissociation of the enzyme to the dimeric form as later indicated by ultracentrifugation.[24] Evidence that the desensitization of the enzyme by KCl is due to a salt-promoted conformational change was obtained from thermal inactivation studies, when the presence of KCl was shown to confer marked stability on the enzyme, and from treatment of the enzyme with thiol-blocking reagents (DTNB, *N*-ethylmaleimide, Hg^{2+}) which largely destroyed activity in the absence, but not the presence, of KCl. The form of dependence of inhibition of *E. coli* CS on NADH concentration is hyperbolic,[12] but this enzyme shows a sigmoidal dependence of activity on acetyl-CoA concentration (the substrate competitive with the inhibitor NADH).[40] Significantly, the presence of KCl abolishes this sigmoidicity and normal hyperbolic kinetics are observed. On the other hand, the *Acinetobacter* CS shows sigmoidal dependences of both inhibition and reactivation on NADH (which acts noncompetitively against acetyl-CoA) and AMP concentrations, respectively, but shows a hyperbolic dependence of activity on acetyl-CoA concentration. (Note that the original report[41] of such sigmoidicity was erroneously attributed to the *E. coli* enzyme but subsequently corrected[4] to the *Acinetobacter* enzyme.) In the nomenclature of allosteric R ↔ T equilibria, these properties suggest that the resting state of *E. coli* CS is the inactive T form, whereas that of the *Acinetobacter* enzyme is the active R conformation. Evidence for a conformational rearrangement of the latter enzyme on conversion to the inhibited condition is discussed later. The desensitization to NADH by KCl is not due to dissociation of the hexameric enzyme, but may reflect a modification of the structure or arrangement of subunits within the hexamer. Indeed, electron microscopic examination of *Acinetobacter* CS[26] has shown that 0.2 *M* KCl changes the mean diameter of the enzyme molecules from 9.8 to 10.9 nm. Spectroscopic examination of the *E. coli* enzyme also indicates that KCl promotes a conformational change and, as well as the change from sigmoidal to hyperbolic kinetics noted above, the values of K_m (or $S_{0.5}$) for both substrates are greatly reduced by the presence of KCl.[40]

The inhibition of CS from *E. coli* and related enterobacteria by oxoglutarate was also found to be abolished in the presence of KCl[14,15] and may therefore be an allosteric phenomenon. A similar conclusion may also apply to the succinyl-CoA inhibition of cyanobacterial CS which was reported to be considerably reduced by KCl and to vary with succinyl-CoA concentration in a sigmoidal fashion.[16]

The use of DTNB as an assay reagent led to the observation that some CSs may be desensitized to NADH by treatment with DTNB, thus implicating thiol groups, directly or indirectly, in the NADH effect. For example, CS from *Pseudomonas* may be desensitized without loss of activity and treatment of the modified enzyme with mercaptoethanol or dithiothreitol to reverse the modification restores the NADH inhibition.[4] Danson and Weitzman[42] observed differential effects of DTNB on the activity and NADH inhibition of *E. coli* CS; although the time-course of inactivation was exponential, that of desensitization to NADH showed a definite lag. Kinetic analysis indicated that it requires the modification of several thiol groups in the enzyme molecule to produce desensitization to NADH. Duckworth and co-workers[43] did not observe DTNB inactivation of the enzyme, and this may be due to differences between the CSs of different *E. coli* strains. They did, however, confirm the DTNB desensitization to NADH and the lag phenomenon. On the basis of their finding that the binding of one molecule of NADH per dimer inactivates that dimer, they suggested that the NADH site is located near the contact surface between the two subunits of the dimer so that only one of the potential sites may be occupied at any time. Reaction of one subunit per dimer with DTNB prevents NADH binding to that subunit, but not to the other subunit of the same dimer. Consequently, a dimer remains sensitive to NADH until both its subunits have been modified by DTNB; hence,the observed lag. More recently, they have suggested that the cysteine residue involved in the DTNB desensitization lies near the interface of the small and large domains in the enzyme subunit.[44] Response to DTNB is variable and the enzyme from some bacteria (e.g., *Acinetobacter*) is completely resistant to inactivation and desensitization by DTNB.[4,35]

Various other chemical treatments have been used to modify a range of CSs, and the residues implicated in activity and regulation have been summarized.[4] Photooxidative destruction of selected amino acid residues has also proved a useful experimental tool; application to the *E. coli* enzyme suggested the involvement of cysteine residues in the regulation by NADH and histidine residues in the inhibition by oxoglutarate. As mentioned above, *Acinetobacter* CS is not desensitized to NADH by DTNB, but desensitization has been achieved by photooxidation in the presence of oxaloacetate (to protect activity).[45]

Immuno-desensitization has also been demonstrated.[5] Rabbit antiserum prepared against purified *Acinetobacter* CS was tested at various concentrations for its effects on the enzyme; the NADH inhibition was abolished at a lower concentration of antiserum than was the activity, suggesting that the antiserum contained antibodies directed against the allosteric NADH site in addition to those specific for the active site.

A kinetic approach which has been used to identify the allosteric nature of effector action is multiple-inhibition analysis which involves examination of the inhibition of an enzyme by two inhibitors, each competitive with the same substrate. Graphical analysis then indicates whether the two inhibitors act at the same or different sites on the enzyme. Harford and Weitzman[11] examined the effects of bromoacetyl-CoA, ATP, and NADH (all competitive inhibitors with respect to acetyl-CoA) on the CSs of pig heart, *Bacillus megaterium* (Gram-positive), and *Pseudomonas aeruginosa* (Gram-negative). On the assumption that bromoacetyl-CoA binds to the acetyl-CoA site of the enzyme, it was concluded that ATP and NADH also interact with the acetyl-CoA site of the pig and *Bacillus* enzymes; the concentrations of nucleotides required are consistent with the isosteric nonspecific nucleotide inhibition discussed earlier. With the *Pseudomonas* enzyme, however, the analysis clearly indicated that although ATP acts at the acetyl-CoA site, the inhibition by NADH is effected through interaction at an allosteric site. On desensitization of this enzyme by treatment with DTNB, the allosteric interaction with NADH was completely abolished and only a very much less sensitive isosteric response to NADH was observed (200-fold increase of K_i). Multiple-inhibition analysis has also been used to study the oxoglutarate inhibition of *E. coli* CS.[51] Oxoglutarate and fluorooxaloacetate were both shown to be competitive inhibitors against oxaloacetate but the analysis indicated the nonidentity of their sites of interaction with the enzyme. On the assumption that fluorooxaloacetate binds to the oxaloacetate site, it may be concluded that oxoglutarate acts at an allosteric site. Finally, multiple inhibition of cyanobacterial CS, using ATP, succinyl-CoA, and bromoacetyl-CoA, has indicated that succinyl-CoA acts here as an allosteric inhibitor.[16]

Other evidence for allosteric behavior is derived from the isolation of mutant *E. coli* CSs with altered functional properties. The fact that a large CS inhibited by both NADH and oxoglutarate may be genetically desensitized to a large CS which is unaffected by NADH and oxoglutarate[19] is clearly consistent with the action of both these effectors at allosteric sites on the enzyme rather than at the active site. At the same time, it is noteworthy that this mutation was accompanied by a 14-fold reduction in the K_m for acetyl-CoA and an identical 14-fold reduction in the K_i for ATP. These results indicate an enhancement in the binding of nucleotides at the *active* site of the mutant enzyme and are in marked contrast to the effects of the mutation on the *allosteric* sites.

Although these various experimental approaches have been effective in demonstrating the allosteric nature of the NADH binding site in Gram-negative bacterial CSs, the question of whether or not the NADH and AMP binding sites on the aerobic CSs are identical has not been settled. Morse and Duckworth[35] have reported that AMP and related compounds also bind to the NADH site of *E. coli* CS and have attempted to account for the AMP reactivation of the aerobic CSs in terms of their lower affinity for NADH. They assumed that NADH and AMP would also interact with the same site on the *Acinetobacter* CS and that it is unnecessary to postulate a separate site for AMP, especially as AMP does not activate the *Acinetobacter* enzyme in the absence of NADH. Selective desensitization of CS to AMP without loss of NADH inhibition has not been reported and so there has been no reason to infer distinct sites for NADH and AMP, though the experimental evidence available does not rule this out. Our recent examination of an unusual CS from a mutant of *Pseudomonas aeruginosa* may shed some light on this problem. The enzyme is unusual in being activated by AMP in the absence of NADH;[22] NADH produces inhibition, but the percentage inhibition is unaffected by the presence of AMP.[49] Furthermore, reaction with DTNB and photooxidation both result in destruction of the NADH sensitivity with little loss of AMP activation. These findings suggest that AMP and NADH may act at distinct sites and, as it is unlikely that a *new* AMP site has been created in the mutant, it is probable that distinct sites exist also in the wild-type *Pseudomonas* enzyme and, by extrapolation, in other NADH/AMP-sensitive CSs. The effects of NADH and AMP on the active site are normally interactive, resulting in an apparent competition between the two effectors. In the mutant, however, the effects appear to have been uncoupled and the activating influence of AMP has become independent of the inhibition by NADH.

We may speculate that just as all CSs show a weak nonspecific nucleotide competition with acetyl-CoA but the Gram-negative bacterial CSs have evolved a specific allosteric NADH site, so there may be weak nonspecific nucleotide competition with NADH at this site, but the aerobic CSs may have evolved a distinct regulatory site for AMP.

Several approaches to the examination of conformational changes in response to effectors have been pursued. We have used thermal inactivation behavior as a simple indicator of such changes.[5] The CSs of both *Acinetobacter* and *Pseudomonas* were found to be substantially protected against thermal inactivation by either NADH or AMP; however, when these enzymes were first chemically desensitized to the regulatory effectors, no such protection was afforded. These results suggest that both NADH and AMP bring about rearrangements of enzyme conformation to produce a more thermostable form.

Rowe and Weitzman[26] used electron microscopic examination of CS in the absence and presence of effectors to obtain direct evidence for structural changes accompanying regulation of activity. The *Acinetobacter* enzyme was studied by negative staining and the mean diameter was found to increase significantly in the presence of NADH (from 9.8 to 12.2 nm). In the presence of both NADH and KCl (conditions which reverse the NADH inhibition) the mean diameter of enzyme molecules was reduced to 10.9 nm, a value identical with that measured in the presence of KCl alone. Measurement of the "triangular" structures exhibiting threefold symmetry (Section III.B) showed that the substructures (possibly dimers within the overall hexamer) move apart from each other in the presence of NADH. Platinum-shadowing was also used to examine the dimensions of enzyme molecules in various states. Again, a substantial "swelling" of the enzyme was observed in the presence of NADH which was completely reversed by the additional presence of AMP (consistent with the deinhibition by AMP observed kinetically). The specificity of NADH action as an inhibitor was mirrored by its specificity in producing these size changes; neither NAD^+ nor NADPH caused any change. Support for the electron microscopic results was obtained from ultracentrifugation experiments. NADH (but not NADPH) produced a 7% retardation of the sedimentation rate of the enzyme, consistent with molecular swelling to a less dense structure, and the additional presence of AMP abolished this retardation. Overall, these results indicate that an NADH-promoted swelling of the enzyme molecule accompanies inhibition of activity and reactivation by AMP is associated with contraction of the enzyme to its original state.

The association between these gross changes in molecular conformation and the allosteric regulation of enzyme activity have been further explored using the technique of cross-linking.[25] Cross-linking the native *Acinetobacter* enzyme with the cleavable bifunctional reagent dithiobis (succinimidyl propionate) was expected to "freeze" its conformation, prevent the subunits from moving part, and thus modify the regulatory behavior. Unexpectedly, the cross-linked enzyme was still inhibited by NADH and reactivated by AMP but, strikingly, the sigmoidal dependences of these two effects on NADH/AMP concentrations that are displayed by the native enzyme were converted to nonsigmoidal, hyperbolic relationships. Cleavage of the disulfide bonds in the cross-links with dithiothreitol resulted in restoration of the sigmoidal dependences on both effectors. These findings suggest that cross-linking imposes conformational constraints which prevent homotropic interactions between the effector sites while still allowing activity modulation to be expressed. The molecular allosteric processes associated with regulation are clearly complex.

More recent work in our laboratory[50] has extended these studies. It appears that partial cross-linking of *Acinetobacter* CS (perhaps only to dimers) is sufficient to abolish cooperativity. Whereas NADH protected the native enzyme against thermal inactivation, the cross-linked enzyme was not so protected, though cleavage of the cross-links restored NADH protection. The cross-linked enzyme was only slightly desensitized to NADH by KCl; cross-link cleavage again reversed the effect. Interestingly, cross-linking in the presence of KCl and subsequent removal of the salt by dialysis resulted in an enzyme form which was not inhibited by NADH, i.e., the enzyme was locked into the desensitized state; cleavage of the cross-links restored sensitivity to NADH. Preliminary electron microscopic examination of the cross-linked enzyme has indicated that, unlike the native enzyme, it does not "swell" in the presence of NADH. Thus, the swelling of the hexameric enzyme appears to be related to the cooperative interactions between subcomponents of the molecule which may be the means of relaying inhibitory responses across subunits.

IV. PROSPECT

This article has attempted to present a bird's eye view of the breadth of studies on CS and its allosteric regulation. Although the diversity of enzyme structure and function has been a major theme, recent progress in structure analysis has brought about a "reconciliation" of the different CSs and an appreciation of the fundamental unity which underlies their overt diversity. Investigations on CS are at a particularly exciting stage. We have well-established patterns of CS structures and allosteric properties, advances in gene cloning and sequencing offer the prospect of rapid determinations of more amino acid sequences, and we have a three-dimensional model of the dimeric unit. Fusion of these three elements is likely in the near future to result in real progress toward an understanding of the molecular mechanisms of conformational shifts and allosteric control of CS activity, and comparative examination of different CSs should clarify the basis of the regulatory variation between them. In many ways, CS is an excellent system for the analysis of allostery, and lessons learned from CS may also be of more general value in understanding other allosteric enzymes. The new area of site-directed mutagenesis and protein engineering is likely to play a significant part in advancing CS studies. From observations on both naturally occurring and artificially mutated CSs, we have concluded that minor changes in protein structure may have dramatic effects on allosteric behavior. The technology for altering amino acid sequences at will offers the prospect of identifying residues which play a crucial role in effector binding, conformational rearrangements, and cooperative interactions. A start has already been made in this direction[46,47] and progress is likely to accelerate in the future. Gene cloning and amplification also holds out the promise of large-scale enzyme purification, a prerequisite for extensive enzymological studies and crystallization for X-ray analysis.

This article has also attempted to show that the *chemical* study of the CS enzyme has been much assisted by *biological* considerations of its function in particular living cells. Future advances in our knowledge of CS may profitably continue to interweave the chemical and biological approaches. For example, there is growing awareness that conditions within the living cell are "crowded" and that interenzymic associations are probably important. The association of CS with other enzymes has been studied[6] and a multienzyme cluster of five citric acid cycle enzymes, including CS, has recently been identified in several bacterial species.[48] The possibility that such associations may modify the allosteric behavior of "free" CS or even introduce wholly new allosteric potential is a stimulating speculation.

Clearly, citrate synthase is not running out of questions to be answered.

NOTE ADDED IN PROOF

This article was completed in December 1986. In the following year, the 50th anniversary of the discovery of the citric acid cycle was celebrated with a symposium encompassing a range of aspects of the cycle.[52] Several contributions to that symposium concern citrate synthase and are recommended for additional information.

REFERENCES

1. **Krebs, H. A. and Johnson, W. A.,** The role of citric acid in intermediate metabolism in animal tissues, *Enzymologia,* 4, 148, 1937.
2. **Srere, P. A.,** The citrate enzymes: their structures, mechanisms, and biological functions, *Curr. Top. Cell. Regul.,* 5, 229, 1972.
3. **Srere, P. A.,** The enzymology of the formation and breakdown of citrate, *Adv. Enzymol.,* 43, 57, 1975.
4. **Weitzman, P. D. J. and Danson, M. J.,** Citrate synthase, *Curr. Top. Cell. Regul.,* 10, 161, 1976.
5. **Weitzman, P. D. J.,** Unity and diversity in some bacterial citric acid cycle enzymes, *Adv. Microb. Physiol.,* 22, 185, 1981.
6. **Beeckmans, S.,** Some structural and regulatory aspects of citrate synthase, *Int. J. Biochem.,* 16, 341, 1984.
7. **Wiegand, G. and Remington, S. J.,** Citrate synthase: structure, control, and mechanism, *Annu. Rev. Biophys. Biophys. Chem.,* 15, 97, 1986.
8. **Srere, P. A., Brazil, H., and Gonen, L.,** The citrate condensing enzyme of pigeon breast muscle and moth flight muscle, *Acta Chem. Scand.,* 17, S129, 1963.
9. **Weitzman, P. D. J.,** Polarographic assay for malate synthase and citrate synthase, *Methods Enzymol.,* 13, 365, 1969.

10. **Hathaway, J. A. and Atkinson, D. E.,** Kinetics of regulatory enzymes: effect of adenosine triphosphate on yeast citrate synthase, *Biochem. Biophys. Res. Commun.,* 20, 661, 1965.
11. **Harford, S. and Weitzman, P. D. J.,** Evidence for isosteric and allosteric nucleotide inhibition of citrate synthase from multiple-inhibition studies, *Biochem. J.,* 151, 455, 1975.
12. **Weitzman, P. D. J.,** Regulation of citrate synthase activity in *Escherichia coli, Biochim. Biophys. Acta,* 128, 213, 1966.
13. **Weitzman, P. D. J. and Jones, D.,** Regulation of citrate synthase and microbial taxonomy, *Nature (London),* 219, 270, 1968.
14. **Wright, J. A., Maeba, P., and Sanwal, B. D.,** Allosteric regulation of the activity of citrate synthetase of *Escherichia coli* by α-ketoglutarate, *Biochem. Biophys. Res. Commun.,* 29, 34, 1967.
15. **Weitzman, P. D. J. and Dunmore, P.,** Regulation of citrate synthase activity by α-ketoglutarate. Metabolic and taxonomic significance, *Fed. Eur. Biochem. Soc. Lett.,* 3, 265, 1969.
16. **Lucas, C. and Weitzman, P. D. J.,** Regulation of citrate synthase from blue-green bacteria by succinyl coenzyme A, *Arch. Microbiol.,* 114, 55, 1977.
17. **Weitzman, P. D. J. and Dunmore, P.,** Citrate synthases: allosteric regulation and molecular size, *Biochim. Biophys. Acta,* 171, 198, 1969.
18. **Mitchell, C. G. and Weitzman, P. D. J.,** Molecular size diversity of citrate synthases from *Pseudomonas* species, *J. Gen. Microbiol.,* 132, 737, 1986.
19. **Harford, S. and Weitzman, P. D. J.,** Mutant citrate synthases from *Escherichia coli, Biochem. Soc. Trans.,* 6, 433, 1978.
20. **Weitzman, P. D. J., Kinghorn, H. A., Beecroft, L. J., and Harford, S.,** Mutant citrate synthases from *Acinetobacter* generated by transformation, *Biochem. Soc. Trans.,* 6, 436, 1978.
21. **Danson, M. J., Harford, S., and Weitzman, P. D. J.,** Studies on a mutant form of *Escherichia coli* citrate synthase desensitised to allosteric effectors, *Eur. J. Biochem.,* 101, 515, 1979.
22. **Solomon, M. and Weitzman, P. D. J.,** Occurrence of two distinct citrate synthases in a mutant of *Pseudomonas aeruginosa* and their growth-dependent variation, *Fed. Eur. Biochem. Soc. Lett.,* 155, 157, 1983.
23. **Robinson, M. S., Danson, M. J., and Weitzman, P. D. J.,** Citrate synthase from a Gram-positive bacterium, *Biochem. J.,* 213, 53, 1983.
24. **Tong, E. K. and Duckworth, H. W.,** The quaternary structure of citrate synthase from *Escherichia coli* K12, *Biochemistry,* 14, 235, 1975.
25. **Mitchell, C. G. and Weitzman, P. D. J.,** Reversible effects of cross-linking on the regulatory cooperativity of *Acinetobacter* citrate synthase, *Fed. Eur. Biochem. Soc. Lett.,* 151, 260, 1983.
26. **Rowe, A. J. and Weitzman, P. D. J.,** Allosteric changes in citrate synthase observed by electron microscopy, *J. Mol. Biol.,* 43, 345, 1969.
27. **Bell, A. W., Bhayana, V., and Duckworth, H. W.,** Evidence for structural homology between the subunits from allosteric and non-allosteric citrate synthase, *Biochemistry,* 22, 3400, 1983.
28. **Mitchell, C. G. and Weitzman, P. D. J.,** Proteolysis of *Acinetobacter* citrate synthase by subtilisin, *Fed. Eur. Biochem. Soc. Lett.,* 151, 265, 1983.
29. **Bloxham, D. P., Parmelee, D. C., Kumar, S., Walsh, K. A., and Titani, K.,** Complete amino acid sequence of porcine heart citrate synthase, *Biochemistry,* 21, 2028, 1982.
30. **Suissa, M., Suda, K., and Schatz, G.,** Isolation of the nuclear yeast genes for citrate synthase and fifteen other mitochondrial proteins by a new screening method, *EMBO J.,* 3, 1773, 1984.
31. **Ner, S. S., Bhayana, V., Bell, A. W., Giles, I. G., Duckworth, H. W., and Bloxham, D. P.,** Complete sequence of the *glt* A gene encoding citrate synthase in *Escherichia coli, Biochemistry,* 22, 5243, 1983.
32. **Bhayana, V. and Duckworth, H. W.,** Amino acid sequence of *Escherichia coli* citrate synthase, *Biochemistry,* 23, 2900, 1984.
33. **Guest, J. R.,** Hybrid plasmids containing the citrate synthase gene (*glt* A) of *Escherichia coli* K12, *J. Gen. Microbiol.,* 124, 17, 1981.
34. **Remington, S., Wiegand, G., and Huber, R.,** Crystallographic refinement and atomic models of two different forms of citrate synthase at 2.7 and 1.7 Å resolution, *J. Mol. Biol.,* 158, 111, 1982.
35. **Morse, D. and Duckworth, H. W.,** A comparison of the citrate synthases of *Escherichia coli* and *Acinetobacter anitratum, Can. J. Biochem.,* 58, 696, 1980.
36. **Bayer, E., Bauer, B., and Eggerer, H.,** Evidence from inhibitor studies for conformational changes of citrate synthase, *Eur. J. Biochem.,* 120, 155, 1981.
37. **Lesk, A. M. and Chothia, C.,** Mechanisms of domain closure in proteins, *J. Mol. Biol.,* 174, 175, 1984.
38. **Rubin, B. H., Stallings, W. C., Glusker, J. P., Bayer, M. E., Janin, J., and Srere, P. A.,** Crystallographic studies of *Escherichia coli* citrate synthase, *J. Biol. Chem.,* 258, 1297, 1983.
39. **Weitzman, P. D. J.,** Reduced nicotinamide-adenine dinucleotide as an allosteric effector of citrate synthase activity in *Escherichia coli, Biochem. J.,* 101, 44c, 1966.
40. **Faloona, G. R. and Srere, P. A.,** *Escherichia coli* citrate synthase. Purification and the effect of potassium on some properties, *Biochemistry,* 8, 4497, 1969.

41. **Weitzman, P. D. J.,** Allosteric fine control of citrate synthase in *Escherichia coli, Biochim. Biophys. Acta,* 139, 526, 1967.

42. **Danson, M.J. and Weitzman, P. D. J.,** Thiol groups of *Escherichia coli* citrate synthase and their influence on activity and regulation, *Biochim. Biophys. Acta,* 485, 452, 1977.

43. **Talgoy, M. M., Bell, A. W., and Duckworth, H. W.,** The reactions of *Escherichia coli* citrate synthase with the sulfhydryl regents 5,5'-dithiobis-(2-nitrobenzoic acid) and 4,4'-diethiodipyridine, *Can. J. Biochem.,* 57, 822, 1979.

44. **Duckworth, H. W. and Bell, A. W.,** Identification of a cysteine residue which seems to mark the allosteric site of *Escherichia coli* citrate synthase, Poster (TH-252) presented at the 13th Int. Congr. Biochem., Amsterdam, August 1985.

45. **Weitzman, P. D. J., Ward, B. A., and Rann, D. L.,** Effects of photo-oxidation on the catalytic and regulatory properties of citrate synthase from *Acinetobacter lwoffi, Fed. Eur. Biochem. Soc. Lett.,* 43, 97, 1974.

46. **Duckworth, H. W. and Anderson, D.,** *In vitro* mutagenesis of *Escherichia coli* citrate synthase, Poster presented at the Harden Conf. Protein Engineering, Wye, U.K., September 1985.

47. **Handford, P. A. and Ner, S. S.,** Site-directed mutagenesis of *Escherichia coli* citrate synthase, *Biochem. Soc. Trans.,* 14, 1224, 1986.

48. **Barnes, S. J. and Weitzman, P. D. J.,** Organization of citric acid cycle enzymes into a multienzyme cluster, *Fed. Eur. Biochem. Soc. Lett.,* 201, 267, 1986.

49. **Solomon, M. and Weitzman, P. D. J.,** to be published.

50. **Lloyd, A. J. and Weitzman, P. D. J.,** *Biochem. Soc. Trans.,* 15, 840, 1987.

51. **Harford, S. and Weitzman, P. D. J.,** unpublished observations.

52. **Kay, J. and Weitzman, P. D. J., Eds.,** *Krebs' Citric Acid Cycle — Half a Century and Still Turning,* The Biochemical Society, London, 1987.

Chapter 8

RIBONUCLEOTIDE REDUCTASE

Staffan Eriksson and Britt-Marie Sjöberg

TABLE OF CONTENTS

I. INTRODUCTION

A. General Background

In nature there is only one pathway of deoxyribonucleotide synthesis: the direct reduction of the corresponding ribonucleotides. This essential reaction is catalyzed by the enzyme ribonucleotide reductase. Deoxyribonucleotides are highly specialized metabolites and serve only as building blocks for DNA. Their concentrations are very low in nongrowing cells and the activity of ribonucleotide reductase is thus a prerequisite for DNA synthesis and multiplication of all cells. A close correlation of the regulation of DNA synthesis and ribonucleotide reduction has been observed in many systems. However, there is no evidence that the supply of deoxyribonuclcotidcs by itsclf regulates DNA synthesis.

Ribonucleotide reductases are subject to complex allosteric regulation mediated by positive and negative effectors. These bind to regulatory sites that are distinct from the catalytic site. The regulatory properties of the enzyme are probably an evolutinary adaptation to the fact that a single enzyme catalyzes the first reaction in each of four parallel pathways. The end products of these pathways (deoxynucleoside triphosphates) serve as regulators of both the substrate specificity and the overall activity of ribonucleotide reductase. This regulation allows the enzyme to produce a balanced supply of all four DNA precursors. Consequently, an inadequate allosteric regulation of ribonucleotide reductase leads to decreased fidelity of DNA replication. The allosteric behavior of the mammalian enzyme has also provided an explanation for hereditary immunodeficiency diseases caused by a lack of purine metabolizing enzymes.

This article will focus on the allosteric control mechanism of ribonucleotide reductases. No attempt is made to cover all published results. For further information the reader is referred to reviews by Thelander and Reichard, Holmgren, Lammers and Follmann, Reichard, and Hogenkamp and McFarlan.[1-5]

B. Mechanism of the Reaction

The mechanism of ribonucleotide reduction involves a direct replacement of the hydroxyl-group at the $2'$-position of the ribose moiety by a hydrogen derived from solvent. NADPH is the ultimate reducing agent, but several proteins participate as electron carriers (Figure 1A). The first step in the reaction is an abstraction of the $3'$-hydrogen of the substrate by an organic free radical which is present as a stable entity of the holoenzyme.[6] At the next stage, release of the $2'$-hydroxyl gives rise to an intermediate substrate cation radical (Figure 1B). Product formation involves regeneration of the enzyme associated free radical through electron transfer to the substrate radical and stereospecific introduction of hydrogen in the $2'$-position of the substrate at the expense of reduced dithiols on the enzyme (Figure 1B). This mechanism is apparently general for all types of ribonucleotide reductases even though the structure and stability of the free radical(s) may be quite different in the different classes of reductases (see Section II).[6-8]

C. Hydrogen Donor Systems

Ribonucleotide reductase requires cyclic reduction of a disulfide present in the active site of the enzyme. Dithiols, such as reduced lipoate or dithiothreitol, can serve as direct hydrogen donors for the purified ribonucleotide reductases, but there exist two physiological hydrogen donor systems in all cells investigated so far (Figure 1A).[9] The first system to be identified was the thioredoxin system. Thioredoxin is a small protein (108 amino acid residues in *Escherichia coli*), which contains a redox-active disulfide. The reduced form of thioredoxin serves as an efficient hydrogen donor for ribonucleotide reductases. The oxidized thioredoxin is reduced in the cell at the expense of NADPH by a specific enzyme, thioredoxin reductase. This enzyme contains FAD (flavin adenine dinucleotide) and has a molecular mass (in *E.*

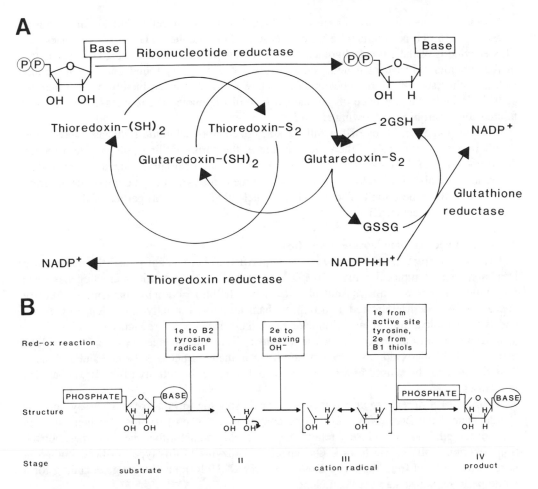

FIGURE 1. Enzymatic reduction of ribonucleotides. (A) Components involved in the electron transfer from NADPH. (B) Proposed reaction mechanism as typified for ribonucleotide reductase from *E. coli*.

coli) of 66 kDa. The second system is the glutaredoxin system, where glutaredoxin (85 amino acid residues in *E. coli*) also contains a redox-active disulfide and in reduced form can serve as hydrogen donor for ribonucleotide reductase. In this case, the oxidized glutaredoxin is reduced by glutathione provided by gluthathione reductase and NADPH (Figure 1A).[2,9] The true *in vivo* hydrogen donor system is presently unknown, but both systems may possibly substitute for each other under different growth conditions.

II. ENZYME AND GENE STRUCTURE

Two types of ribonucleotide reductases have been characterized in great detail. One class, as represented by the enzyme from *Lactobacillus leichmannii*, is monomeric and dependent on adenosylcobalamin (AdoCbl) as a dissociable cofactor. The substrates for this type of reductase are ribonucleoside triphosphates and several prokaryotic species, as well as a few eukaryotes, contain a ribonucleotide reductase of this type. The other class of reductase, as represented by the enzyme from *E. coli,* consists of two nonidentical homodimeric subunits. One of the subunits contains a binuclear iron center and a tyrosyl free radical necessary for enzyme activity. The other subunit contains binding sites for the substrates, which are ribonucleoside diphosphates, and separate binding sites for the allosteric nucleoside triphosphate effectors. This type of ribonucleotide reductase is found in some prokaryotes and bacteriophages, in most eukaryotes including mammalian cells, and in some animal viruses.[1-5] A third type of reductase has been described in a number of Gram-positive bacteria. This enzyme is of the second oligomeric type but contains manganese instead of iron in one of the subunits.[10] In recent genetic studies, it was unexpectedly found that *E. coli* strains lacking either subunit of ribonucleotide reductase can grow anaerobically but not aerobically.[11,11a] Thus, the presence of yet another type of ribonucleotide reductase may be postulated for anaerobically growing *E. coli*.

In Table 1 the structures of the fully characterized ribonucleotide reductases have been listed. In a number of cases, the gene structure of the corresponding polypeptide has been determined and in these cases a direct comparison of analytical and deduced sequence data is possible (see also Section IV). In this section, some features of the structure of the different reductases will be described. The models and nucleotide binding properties of the enzymes are presented in Section III.

A. Monomeric Ribonucleotide Reductases

The enzyme from *L. leichmannii* has been purified to homogeneity, and the active form is a monomer of molecular mass 76 kDa (Table 1).[12,13] No subunit interaction has been found, but the enzyme requires AdoCbl for activity.[14] In this reaction mechanism, the AdoCbl coenzyme has been implicated as a radical chain initiator, formally equivalent to the iron and tyrosyl radical containing subunit of the oligomeric type of reductases.[7] AdoCbl has also been shown to undergo a reductase dependent isotope exchange between 5′-methylene tritium of the 5′-deoxyadenosyl moiety and water in the presence of substrate and dithiols.[15] This reaction has been used as a simple assay to determine the activity of the B_{12}-dependent reductases.

A survey of different organisms was performed using the tritium exchange assay, and the B_{12}-dependent reductase was found to be common among prokaryotes. No general rules were observed, and even closely related species could contain either the monomeric or the iron-containing oligomeric form.[16] One eukaryotic enzyme of this type has been studied in pure form, isolated from the algae *Euglena gracilis*.[17] Unfortunately, no gene coding for a monomeric reductase has yet been cloned.

B. Oligomeric Ribonucleotide Reductases

The enzyme from *E. coli* is the best studied reductase of this group. It consists of two nonidentical subunits, proteins B1 and B2. The B1 subunit is a dimer of two identical polypeptide chains each containing 761 amino acid residues with a polypeptide molecular mass of 86 kDa.[1,18,19] The B2 subunit is also a dimer of two identical polypeptide chains each containing 375 residues with a polypeptide molecular mass of 43.3 kDa (Table 1).[18] A characteristic property of the enzyme is that the two subunits easily dissociate. Each subunit alone is inactive, but, in the presence of Mg^{2+}, they interact as a 1:1 complex to form the active enzyme.[20,21] The B1 and B2 polypeptides are encoded by two adjacent genes, *nrdA* and *nrdB*, located at 49 min on the *E. coli* map.[22] The genes are part of an operon, in which *nrdA* is the promoter proximal gene.[23] The two genes have been cloned and sequenced, and recently, part of the *nrdA* sequence has been redetermined (see Section IV).[19]

Infection of *E. coli* with certain T bacteriophages induces the synthesis of a phage specific ribonucleotide reductase.[24] The bacteriophage T4 enzyme has been purified to homogeneity and consists of two α-polypeptides (of 86 kDa molecular mass) and two β-polypeptides (of 45.4 kDa molecular mass), together forming a tight complex (Table 1).[24,25] The T4 *nrdA* and *nrdB* genes are adjacent on the chromosome. Both genes were recently cloned and sequenced, and they show considerable homology with the corresponding *E. coli* genes.[26,27]

The characterized mammalian ribonucleotide reductases consist of two nonidentical dimeric subunits called proteins M1 (polypeptide molecular mass 84 kDa) and M2 (polypeptide

Table 1
ENZYME AND GENE STRUCTURE OF RIBONUCLEOTIDE REDUCTASES

Source	General structure	Subunit trivial name	α-Polypeptide (molecular mass, Da)		Subunit trivial name	β-Polypeptide (molecular mass, Da)		Ref.
			Analytical	DNA sequence		Analytical	DNA sequence	
Lactobacillus leichmannii	α[a]		76,000[a]					12
Escherichia coli	α₂β₂	B1	85,000	85,700	B2	39,000	43,400	18, 19, 55
Bacteriophage T4	α₂β₂	α₂	84,000	86,000	β₂	43,500	45,400	24—26
Calf thymus	α₂β₂	M1	84,000		M2	44,000	45,100	29, 30
Mouse	α₂β₂	M1	84,000	90,200	M2	44,000	44,400	30, 33, 34
Spisula solidissima	α₂β₂	M1	86,000		M2, p42	41,000		36, 95
Herpes simplex virus type 1	α₂β₂	144K, Vmw136, RR1, H1	136,000	124,000	p39, Vmw38, RR2, H2	38,000	38,000	40—42, 45
Herpes simplex virus type 2		140K, 1PC10	138,000	125,100	38K, p41	38,000	37,700	40, 41

[a] A structural homology between the monomeric α-polypeptide and α₂ of the oligomeric enzymes has so far only been established for the active site sequence.[79]

molecular mass 44 kDa).[1] The enzyme has been highly purified from calf, mouse (Table 1), and rat.[28-32] The mouse genes for the two subunits were recently cloned, sequenced, and mapped to different chromosomes.[33-35] The coding region for protein M2 from the surface clam *Spisula solidissima* (Table 1) and from *Saccharomyces cerevisiae* were recently identified and sequenced.[36-38] All the deduced primary structures show considerable homology with the *E. coli* enzyme.

Several eukaryotic viruses induce a virus-specific ribonucleotide reductase. None of these enzymes has been isolated in pure form, nor has the subunit structure been definitely determined. On the other hand, the genes coding for the large and the small subunit of the herpes simplex viruses type 1 and 2 (HSV) (Table 1), the Epstein Barr virus (EBV), the Varicella-Zoster virus (VZV), and the vaccinia virus (VV) have been identified and sequenced.[39-45] Again, considerable sequence homology was found between these viral proteins and the *E. coli* and mammalian enzymes.

III. ALLOSTERIC CONTROL

Regulatory effects of nucleotides on ribonucleotide reductase enzyme activity were first described by Reichard et al. in extracts of chick embryos.[46] They found that CDP and GDP reductions were stimulated by dTTP addition. Later, a somewhat more purified enzyme from rat Novikoff ascites tumor was studied by Moore and Hurlbert, and a complex pattern of activation-inhibition of the reduction of all four substrates by nucleoside triphosphates was revealed.[47] However, the isolation of pure enzyme from these sources turned out to be very difficult, and therefore, the regulation of the enzymes from *E. coli* and *L. leichmannii* was the first to be characterized in great detail in 1966.[48-50] Since then, the allosteric control of a number of reductases from different sources has been studied. In this section, a general description of allosteric regulation and interspecies comparison will be presented (see Table 2).

A. The *Lactobacillus leichmannii* Enzyme

Figure 2a shows a model of the monomeric ribonucleotide reductase from *L. leichmannii*. The allosteric properties of this reductase have been reviewed previously.[1-3,5] Equilibrium dialysis experiments defined one common binding site for deoxyribonucleoside triphosphates with K_d values from 9 to 80 μM. This site is indicated as a regulatory site, but it also binds substrate ribonucleoside triphosphates with a 100- to 1000-fold lower affinity.[13] Binding of deoxyribonucleosides to the regulatory site greatly improves the enzyme's affinity for AdoCbl as well as stimulates the reduction of a certain substrate. The presence of dATP stimulates reduction of CTP, while dCTP stimulates UTP reduction, and dGTP stimulates ATP reduction (Table 2). The effector dTTP was initially shown to stimulate GTP reduction, but later it was shown to be a positive effector only for ITP reduction.[12-14] Various combinations of effectors may give weak inhibition but there is no strong negative effector known for the *Lactobacillus* enzyme.[5,50-53]

Very little AdoCbl is bound to the enzyme when the regulatory site is unoccupied. The presence of an effector, e.g., dGTP, reduces the apparent K_d value from 8 to 0.3 μM for the coenzyme.[53] There is no base specificity for this stimulation of coenzyme binding. Experiments using sugar-modified adenine nucleotides have indicated the existence of two functional parts of the regulatory site on the *Lactobacillus* reductase.[51] O-methyl-ATP stimulated CTP reduction like dATP, but did not promote the hydrogen exchange reaction of AdoCbl. Arabinonucleotides, on the other hand, inhibited reduction, but facilitated hydrogen exchange.[52] These data indicate that the monomeric reductases may posses two different regulatory functions similar to the *E. coli* and mammalian enzymes (see Section III.B and D). However, direct binding studies in the absence of AdoCbl identified only one binding site.[13]

Table 2
STIMULATION OF RIBONUCLEOTIDE REDUCTASE-CATALYZED REACTION RATES BY EFFECTOR NUCLEOTIDES[a]

Effector added	Lactobacillus leichmannii[3,51] Substrate reduction				Escherichia coli[49,51,59,82] Substrate reduction				Bacteriophage T4[24,61] Substrate reduction				Calf thymus[66] Substrate reduction			
	CTP	UTP	GTP	ATP	CDP	UDP	GDP	ADP	CDP	UDP	GDP	ADP	CDP	UDP	GDP	ADP
none	36	14	100*	20	25	7	21	12	17	10	17	10	<2	<2	2	4
ATP	100				110	120			100*	100			100*	65		
dATP					130	120			100	87						
dCTP		37							100	60						
dTTP					100*	110	120	100	63	43	83	87			46(84)	4(9)
dGTP				100			88	110				93			6(9)	38(89)

a   Activities are given in percent. An asterisk (*) denotes the substrate and effector combination used for correlation of the other activities measured with the respective enzyme. Only the action of the prime effectors is indicated. Figures in parenthesis measured in the presence of 2 mM ATP.

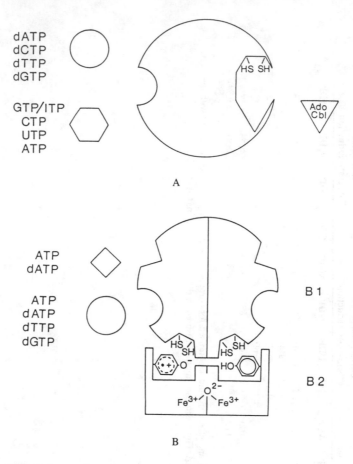

FIGURE 2.   Schematic models of (A) ribonucleoside triphosphate reductase from *L. leichmannii*, and ribonucleoside diphosphate reductases from (B) *E. coli*, (C) bacteriophage T4, and (D) calf thymus.[1,5,24,29]

The existence of a substrate binding site could not be demonstrated by direct binding studies, but kinetic experiments showed that the apparent K_m for GTP, the best substrate, was as high as 0.24 mM.[53] It is assumed that a single catalytic site binds all four substrates, but a relatively complex pattern was found since the substrates also bind to the regulatory site. For instance, reduction of GTP measured at low concentrations of AdoCbl and in the abscence of deoxynucleotide effector, showed substrate activation. No clear-cut product inhibition was found as these compounds bind only to the regulatory site.

While other B12-dependent ribonucleotide reductases generally are very similar to the *Lactobacillus* enzyme,[3,5] two microbial enzymes show properties distinct from either of the two main types of ribonucleotide reductases. The enzyme of the manganese type, e.g., from *Brevibacterium ammoniagenes,* consists of two subunits, and reduces ribonucleoside diphosphates. This enzyme shows a positive allosteric regulation similar to the mammalian enzyme, but without negative control.[5,10,10a] Ribonucleotide reductase from *Corynebacterium nephridii* has been purified and shown to differ from other reductases both with regard to subunit structure and substrate regulation. This subject has recently been elegantly reviewed by Hogenkamp and McFarlan.[5]

B. The *Escherichia coli* Enzyme

Figure 2b presents a model of the *E. coli* ribonucleotide reductase. Protein B1 isolated from overproducing cells carrying a recombinant plasmid is composed of two identical

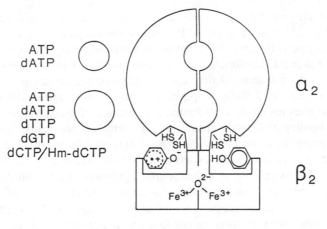

FIGURE 2C.

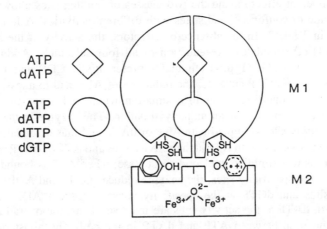

FIGURE 2D.

polypeptides originating from the *nrdA* gene.[54] When protein B1 is isolated from thymine-starved cells, its two chains differ considerably in their N-terminal sequences.[55] Because this heterogeneity does not seem to have any functional significance, it is unlikely that any nucleotide binding domain resides close to the N-terminus. Protein B1 contains two binding sites for the ribonucleoside diphosphate substrates as demonstrated by equilibrium dialysis experiments.[56] All four substrates are mutually competitive in binding studies, indicating that the substrates are bound to two common and identical sites.

The active site of the enzyme is formed by elements of both B1 and B2.[57] Protein B1 contains redox-active sulfhydryl groups, which reduce a stoichiometric amount of substrate.[21] Protein B2 contributes the unique tyrosyl radical, which most likely is directly involved in the catalytic mechanism (see Section I.B).[58] Protein B2 also contains two Fe(III) ions, which are antiferromagnetically coupled through a μ-oxo bridge.[1,6] The function of the iron center is both to generate and stabilize the essential tyrosyl radical.

Protein B1 contains two classes of allosteric effector binding sites (Figure 2b). From the genetic construction of B1, it is conceivable that each polypeptide contains one binding site of each type. It is the affinity for dATP that defines the two classes of sites; h sites have high ($K_d = 0.03$ μM) and l sites have low ($K_d = 0.1\text{-}0.5$ μM) affinity for dATP.[59] Competition experiments have revealed that the h sites also bind the allosteric effectors ATP, dTTP, and dGTP, while l sites bind only ATP and dATP. Binding of the two latter effectors to the l sites regulates the overall activity of the enzyme. These sites are therefore referred to as the activity sites: when ATP is bound, the enzyme is active, and when dATP is bound, the enzyme is inactive. In the latter case, the inhibitory effect is accompanied by aggregation of protein B1 into dimers and higher oligomers as demonstrated by centrifugation analysis.[20] These relatively unsophisticated allosteric properties are complemented by another unique type of regulation at the h or specificity sites of the enzyme. Binding of a certain effector to these sites results in a conformational change of the active site which leads to the preferential binding of a particular substrate. The changes are manifested as an increased affinity for a certain substrate in binding experiments with isolated B1, and as decreased K_m and increased K_{cat} values for the reduction of the same substrate in the complete enzyme system.[48,49,56,59] The magnitude of the allosteric regulation depends on the substrate concentration; at low substrate concentration, a 50- to 100-fold stimulation of the reaction can occur, but at high substrate concentrations the effect is an order of magnitude lower. The enzyme shows substrate saturation curves that are hyperbolic, both in the absence and presence of allosteric effectors. No homotropic cooperative effects have been found.[48,49]

The large number of effectors and the two classes of binding sites allow the enzyme to assume a multitude of conformational states with different activities. A few of these states are exemplified in Table 2. In the absence of effectors, the activity of the enzyme is low, but addition of dTTP stimulates the reduction of all four substrates. Addition of ATP or low concentration of dATP (1 μM) stimulates preferentially CDP and UDP reduction, whereas the presence of dGTP enhances the reduction of ADP and to a lesser extent that of GDP. The described conditions refer to a situation where only the specificity sites are occupied; however, *in vivo* this is an unlikely situation. The enzyme in the cell has, in all probability, its activity sites occupied with ATP or dATP and thus essentially three positive and one negative states exist; with ATP at the activity sites and ATP or dATP at the specificity sites, the enzyme is a pyrimidine nucleotide reductase; when ATP is bound to the activity sites and dTTP at the specificity sites, the enzyme reduces GDP (and ADP); and with ATP at the activity sites and dGTP at the specificity sites, it reduces ADP (and GDP). All combinations with dATP at the activity sites are inactive. Thus, the overall activity will be determined by the ratio between ATP and dATP in the cell. The allosteric properties of several ribonucleotide reductases may be integrated into a scheme linking the activity of the reductase to the requirements for all four precursors needed for DNA synthesis as discussed in Section VI.A.

Recently, it was shown that addition of a substrate (GDP) increased the binding of an effector (dTTP) to the specificity site of protein B1.[60] This result strengthens the hypothesis that specific states of the enzyme exist where both effector and substrate stabilize mutual binding.

C. Bacteriophage-Induced Enzymes

The general structure of T4 ribonucleotide reductase is depicted in Figure 2c. It shows the same general activity as the *E. coli* enzyme, but it differs in two distinct ways. It does not dissociate into its subunits as readily, and it has no major negative feedback effector.[61] Like the *E. coli* enzyme, the reductions of all four ribonucleoside diphosphates by T4 ribonucleotide reductase are stimulated by nucleoside triphosphate effectors, which act by changing the maximal velocity of the reaction and the apparent K_m for the substrates. The effectors dATP and ATP promote the reduction of CDP and UDP, but have minimal effects on the reduction of purine ribonucleotides. The other deoxynucleoside triphosphates dCTP, dTTP, and dGTP, are positive effectors for the reduction of all substrates. However, dCTP may have little physiological importance since its concentration in the T4-infected cell is very low, and the bacteriophage T4 chromosome contains 5-hydroxymethyl cytosine instead of cytosine. The effect of 5-hydroxymethyl-dCTP (Hm-dCTP) is similar to that of dCTP and stimulates the reduction of pyrimidine nucleotides.[24] The prime effector for the reduction of GDP is dTTP, and for the reduction of ADP it is dGTP.

Hm-dCTP and dATP are the only nucleoside triphosphates whose effects on T4 ribonucleotide reductase are qualitatively different compared to the *E. coli* enzyme. First, Hm-dCTP is essentially inert with the host enzyme. Second, even at high concentrations, dATP acts to stimulate pyrimidine reduction by the T4 reductase. Equilibrium dialysis experiments showed that dATP bound strongly to the T4 ribonucleotide reductase (K_d = 0.3 μM). Binding of dATP to another class of sites with considerably lower affinity was indicated from a Scatchard plot. Values of 0.9 high affinity sites and 1.5 to 2 total sites per B1 molecule were calculated.[24] It was also observed that T4 ribonucleotide reductase bound to dATP-Sepharose could only be desorbed by ATP or dATP, and not by dTTP or dCTP.[62] These results all indicate two types of effector binding sites on T4 ribonucleotide reductase, thereby stressing the functional similarity between the T4-induced and the *E. coli* B1 proteins.

Other T-even bacteriophages induce ribonucleotide reductases which are very similar to the T4 enzyme. Bacteriophage T5 induces another type of enzyme, which seems to prefer ribonucleoside triphosphates as substrates.[63]

D. Mammalian Enzymes

Figure 2d shows a model of the ribonucleotide reductase from calf thymus, and its general structure resembles that of the *E. coli* enzyme. Properties of the enzyme purified from several other mammalian sources agree with this model. Some investigators have reported the isolation of reductases with different subunit molecular weights and with the capacity to reduce only certain substrates.[64] However, recent studies with cDNA probes show no indication of more than one gene for the mouse protein M1.[65] Furthermore, the capacity of the purified calf thymus reductase to reduce all four substrates, as well as the results from several somatic cell mutants (see Section VI.B) clearly demonstrate that one enzyme and one genetic unit are responsible for the control and reduction of all four ribonucleoside diphosphates in mammalian cells.

Pure calf thymus protein M1 bound 1.8 mol dATP and 0.7 mol dTTP per 170 kDa protein when analyzed by equilibrium dialysis. Cooperativity of binding was observed with both effectors, and the final dissociation constants were 0.3 μM for dATP and 2 μM for dTTP.[29] These data as well as competition experiments indicate that M1 contains two classes of effector binding sites. One is specific for ATP and dATP, and the other binds ATP, dTTP, and dGTP. The latter is a specificity site and the former is an activity site as in the case of the *E. coli* enzyme. However, the stoichiometry differs in that protein M1 appears to contain only one type of site per dimer (as was also described for the T4 reductase in Section III.C). These results could, however, be due to the presence of a large fraction of inactive M1 in the pure preparation. Glycerol sedimentation analysis argues against this interpretation, since upon addition of dTTP all of the monomeric form of M1 (5.7S) is shifted to the dimer position (8.8S). Furthermore, addition of dATP induces the formation of an oligomeric form of M1 (15.2S).[29] These results demonstrate a homogeneous behavior of the M1, and illustrate a real difference in the number of binding sites compared to the *E. coli* enzyme; however, further studies are required to definitively resolve the question.

Kinetic studies with the mammalian reductases have also in general shown a similar pattern to the *E. coli* model. In the absence of positive effectors, the purified calf thymus enzyme exhibits very low activity with all substrates. Reduction of CDP and UDP is stimulated by ATP, reduction of GDP, by dTTP, and even more when both ATP and dTTP are present (Table 2). Addition of dGTP or even better dGTP plus ATP promotes reduction of ADP. The nucleotide dATP serves as general inhibitor, but the negative effects can be reversed by high concentrations of ATP.[66] Strong support for the validity of this model comes from studies of somatic cell mutants selected for their resistance to deoxyribonucleosides (see Section VI.B).

Photoaffinity studies using dTTP as ligand and partially purified mouse M1 fully support the model described above.[67] Similar photoaffinity studies with dATP as ligand gave a more complex result; first, the molar incorporation was about 100-fold lower than with dTTP, and second, binding to only a special part of the specificity site seemed to be obtained with dATP, since the incorporation could be blocked by the presence of dTTP and dGTP, but not ATP.[68] DeoxyATP photoincorporation into pure protein B1 from *E. coli* showed labeling of multiple peptides in apparent nonspecific fashion.[60,69] Therefore, further experiments with more reactive affinity binding ligands are needed to define the possible subsites of the specificity site as well as the structure of the activity site.

Partially purified human ribonucleotide reductase from lymphoid cells shows similar regulation as that described above for calf and mouse enzymes with the exception that GTP and dGTP are equally efficient as positive effectors in ADP reduction.[70-72] The possible role of GTP as an effector for the mouse or calf enzymes has not yet been studied in detail, but there was no effect of GTP on the photoaffinity incorporation of dTTP into partially purified mouse M1.[67]

It has been claimed that a partially purified ribonucleotide reductase isolated from Ehrlich tumor cells could utilize nucleoside diphosphates as effector. The various nucleoside diphosphates appeared to be effective in the same concentration range as the corresponding nucleoside triphosphates.[73] However, these results are difficult to evaluate since the stability of the added diphosphates was stated, but not shown to be sufficient to exclude phosphorylation of these compounds to the triphosphate level. In any case, the physiological role of diphosphates as effectors is most likely minimal since their intracellular concentration normally is an order of magnitude lower than the corresponding triphosphates.

E. Virus-Induced Enzymes

Ribonucleotide reductase from the mammalian herpes simplex and pseudorabies viruses all seem to possess the same general characteristics.[74,75] These viral reductases differ from other eukaryotic and prokaryotic reductases in being completely insensitive to allosteric inhibition by dATP and dTTP. They are also largely refractory to stimulation by ATP.

The viral enzymes, as in the case of all other ribonucleotide reductase, reduce both purine and pyrimidine ribonucleotides at one common site. They have remarkably low K_m values for CDP (0.65 μM) and GDP (1.2 μM), but retain relatively typical values for UDP (80 μM) and ADP (12 μM). For CDP, GDP, and ADP, these are the lowest K_m values hitherto observed for any ribonucleotide reductase. The dNPD products are competitive inhibitors *vs.* the substrates albeit at concentrations close to the millimolar range.[74]

Infection by the vaccinia virus induces a novel riboncleotide reductase activity, which is viral encoded. The vaccinia-induced activity is similar in several respects to other eukaryotic reductases, but it differs from the host enzyme in response to certain modulators of reductase activity and in affinity for at least one of its four substrates.[44a,76]

IV. STRUCTURE-FUNCTION RELATIONSHIPS OF IRON-DEPENDENT RIBONUCLEOTIDE REDUCTASES

A. The Large Subunit

The nucleotide sequence for the *nrdA* gene of E. *coli,* as originally published by Carlson et al.,[18] has recently been revised.[19] Even though modifications at the nucleotide level were minor (occasional insertions and deletions), the corrections involved several frameshifts, and thus had a major influence on the deduced amino acid sequence of the B1 subunit. An alignment of the corrected protein B1 sequence with corresponding large subunits from bacteriophage T4, mouse, HSV, EBV, and VZV reveals extensive homologies at the primary structure level.[27,33,39,42,43,45] There are approximately 6% conserved residues, which are mostly confined to highly homologous regions, as is evident from Figure 3. Since some of these organisms are evolutionarily very distant, it is likely that most conserved residues have a functional importance.

The similarities start around position 200 in protein B1 and extend almost to the C-terminus. The lack of sequence homology in the N-terminal region fits well with the unexpected finding that the herpes simplex subunits, as deduced from their gene sequences, contain an N-terminal domain of 350 to 400 residues which is not found in the other reductases.[42,45] This part of the herpes enzyme appears structurally distinct from the rest of the protein, and a sequence alignment of the large subunits of HSV type 1 and 2 shows that the C-terminal ends are more highly conserved than the N-terminal ends. It has also been observed that the herpes large subunits retain considerable catalytic activity even after extensive proteolytic degradation.[77] Additional functions have been suggested for the N-terminal segment of the herpes large subunits, e.g., it may have an important physiological role in a multienzyme complex (see also Section VI.D).[42] In the bacteriophage T4-infected system, Cook and Greenberg observed that a C-terminally truncated form of T4 B1 was still able to associate with a multienzyme complex in the presence of T4 B2, but had lost its capactity to bind to dATP-Sepharose.[25]

With the knowledge that the large subunit of ribonucleotide reductase contributes redox-active sulfhydryls to the composite active site, conserved cysteine residues are of special interest. There are three conserved cysteines in the alignment; Cys-225, Cys-439, and Cys-462 (in *E. coli* numbering), and all three are part of longer conserved stretches in the central portion of B1 (Figure 3). In addition, all deduced sequences known so far have the C-terminal consensus sequence -Cys-X-X(-X-X)-Cys-X(-X)-COOH.[78] Recently, Lin et al. showed that the two latter cysteines in the *E. coli* enzyme were specifically oxidized by incubation with substrates.[79] Some oxidation also occurred at Cys-225, Cys-230, and probably Cys-462. Concomitantly, it was found that substrate specific cysteine oxidation of the *L. leichmannii* enzyme occurred at the sequence -Cys-Glu-Gly-Gly-Ala-Cys-Pro-Ile-Lys-,[79] which has considerable homolgy to the C-terminal part of the *E. coli* enzyme (Figure 3). This observation fortifies earlier suggestions about the functional similarity between these two classes of ribonucleotide reductases.

Photoaffinity labeling of protein B1 by the allosteric effector dTTP, revealed specific cross-linking to Cys-292.[69,78] Similar experiments with mouse protein M1 gave much less efficient labeling.[67] These results may be rationalized from the alignment in Figure 3. Mouse and *E. coli* subunits, which are known to conform to a similar allosteric mechanism, align well in the area around Cys-292, but there is no corresponding cysteine in the mouse sequence. The HSV subunits, which deviate from the general allosteric scheme in being insensitive to regulation by deoxyribonucleoside triphosphates (see Section III.E), does not resemble either the mouse or the *E. coli* sequence in the area around Cys-292. They lack a sequence of eight amino acids upstream of this residue and there is no predicted β-turn (centered around Gly-295 in *E. coli* numbering) in the HSV coded enzymes.[80]

The importance of Cys-292 in B1 was further substantiated by protein engineering studies.[78] It was changed into an alanine residue and the resulting protein was purified and characterized *in vitro*. B1(Cys-292 → Ala) was not photoaffinity labeled in the presence of dTTP, but the mutated protein still had 60% subunit activity in the presence of the general allosteric effector ATP and 40% subunit activity in the presence of dTTP. This shows that the mutation did not confer any substantial conformational change of the substrate specificity site of B1. Further studies are clearly needed to assess whether Cys-292 actively stabilizes effector binding or merely provides a reactive sidechain in the vicinity of the specificity site of protein B1. Common denominators in the primary structure of several ATP-requiring enzymes were recently proposed.[81] It is interesting to note that some conserved areas in the alignment of the large subunits of ribonucleotide reductase (residues 253-255, 514-520, and 726-729 in Figure 3) show a certain similarity to these proposed nucleotide binding sequences.

```
                 1                                                                                          98
E.coli     MNQNLLVTKR DGSTERINLD KIHRVLDWAA EGL..HNVSI SQVELRSHIQ FYDGIKTSDI HETIKAAAD  LISRDAPDYQ YLAARLAIFH LRKKAYGQFE
Phage T4   .MQLINVIKS SGVSQSFDPQ KIIKVLSWAA EG....TSV  DPYELYENIK SYLRDGMTTD DIQTIVIKAA ANSISVEEPD YQYVAARCLM FALRKHVYGQ
Mouse      ...MHVIKR  DGRQERVMFD KITSRIQKLC YGLNMDFVDP AQITMKV.IQ GLYSGVFTVE LDTLAAETAA TLTTKHFDYA ILAARIAVSN LHKETKKVFS
VZV        .......MEF KRIFNTVHDI INRLCQHGYK EYIIPPESTT PVELMEYIST IVSKLKAVTR QDERVYRCCG ELIHCRINLR
HSV1  ....AIMLEYFCRC AREETKRVPP RTFGSPPRLT EDDFGLLNYA LVEMQRLCLD VPPVPPNAYM PYYLREYVTR LVNGFKPLVS RSARLYRILG VLVHLRIRTR
EBV        .......... .......... .......... .......... MATTSHVEHE LLSKLLIDELK VKANSDPEAD VLAGRLLHRL KAESVTHTVA
Consensus  ---------- ---------- ---------- ---------- ---------- ---------- ----YL---- ---------- ---------- ----------

                99                                                                                        181
E.coli     PPALYDHVVK MVEMGKYDNH LLEDYTEEEF KQMDTFI... ......DHDR DMTFSYAAVK QLEGKYLVQN RVTGEIYESA QFLYILVAAC LF.......
Phage T4   YEPRSFIDHI SYCVNAGKYD PELLSKYSAE EITFLESKI. ......KHER DMEFTYSGAM QLKEKYLVKD KTTGQIYETP QFAFMTIGMA IH.......
Mouse      DVMEDLYNYI NPHNGRHSPM VASSTLDIVM ANKDRLNSAI ......IYDR DFSYNYTGFK TLERSYILK. .INGKVAERP QHMLMRVSVG IH.......
VZV        SVSMETWLTS PILCLTPRVR QAIEGRRDEI RRAILEPFLK D...QYPA  LATLGLQSAL .LEEGKLESL COFFLRLAAT VT.TEIVNLP
HSV1       EASFEWLRS KEVALDFGLT ERLREHEAQL VILAQALDHY DCLIHSTPHT LVERGLQSAL KYEDFYLTK. .FGGMYMESV FQMYTRIAGF LA....CRA
EBV        EYLEVFSDKF YDEEFFQMHR DELETRVSAF AQSP...... ......AYER IVSSGYLSAL RYDTTYLYV. .GRSGKQESV QHFYMRLAGF CASTCLYAG
Consensus  ---------- ---------- ---------- ---------- ---------- ---------- ---YL----- ---------- ----E----- ----------

                182                                                                                       272
E.coli     ......SNY PRETRLQYVK RFYDAVSTFK ISLPTPIMSG VRPTRQFSS  CVLIEC..GD SLDSINATSS AIVKVVSQRA GIGINAGRIR ALGSPIRGE
Phage T4   ......QD  EPVDRIKHVI RFYEAVSTRQ ISLPTPIMAG CRTPTRQFSS CVVIEA..GD SLKSINKASA SIVEYISKRA GIGINVGMIR AEGSKIGMGE
Mouse      .KEDIDAAI ETYNLLSEKW FTHASPTLFN AGTNRPQISS CFLLSM.KDD SIEGIYDTLK QCALISKSAG GIGVAVSCIR ATGSYIAGTN
VZV        KIATLIPGIN DGYTWTDVCR VFFTALACOK IVPATPVMF  LGRETGATAS CYLMDPESIT VGRAVRAITG DVGTVLQSRG GVGISLQSLN LIPTENQTKG
HSV1       TRGMRHIALG REGSWWEMEK FFFHRLYDHQ IVPSTPAMLN CVLVNPQATT NKATLRAITS NVSAILARNG GIGLCVQAFN DSGPGTASVM
EBV        LRAALQRARP EIESDMEVFD YYFEHLTSQT VCCSTPFMRF AGVENSTLAS CILTTPDLSS EWDVTQALYR HLGRYLFQRA GVGVGVTGAG QDGKHISLLM
Consensus  ---------- ---------- ---------- ----P----- -------S-C- ---------- ---------- ---------- --G-G----- ----------

                273        (A)                                                                            367
E.coli     AFHTGCIPFY KHFQTAVKSC SQGV.RGGA ATLFYPMWHL EVESLLVLK. NNRGVEGNRV RHMDYGVQIN KIMYTRLLKG EDITLFSPSD VPGLYD...A
Phage T4   VRHTGVIPFW KHFQTAVKSC SQGI.RGGA ATAYYPIWHL EVENLLVLK. NNKGVEENRI RHMDYGVQLN DIMVERFGKN DYITLFSPHE MGGELYY..S
Mouse      GNSNGLVPML RVYNNTARYV DQGGNKRPGA FAIYLEPWHL DIFEFLDLK. KNTGKEEQRA RDLFFALWIP DLFMRKVETN QDWSLMCPNE CPG.....L
VZV        .....LLAV  LKLLDCMVMA INSDCERPTG VCVYIEPWHV DLQTVLATRG MLVRDEIFRC DLFFERYLSY LKGASNVQWT LFDNRAD.IL
HSV1       ......PA  KVLDSLVAA HNKESARPTG ACVYLEPWHT DVAVLRMKG  VLAGEEAQRC DNIFSALWMP DLFFKRLIRH LDGEKNVTWT LFDRDTSMSL
EBV        .RMINSHVEY HNYGCKRPVS VAAYMEPWHS QIFKFLETK. LPENHERCPG IFTGLFVPEL FFKLFRDTPW SDWYIFDPKD AGD.....L
Consensus  ---------- ---------- ---------- ----WH---- -----L---- ---------- ---------- ----E----- ---------- ----------

                368                                               (S)                                     467
E.coli     FFADQEEFER LYTKYEKDDS IRKQRVKAVE LFSLMMQERA STGRIYIQNV DHCNTHSPFD PAIAPVRQSN LCLETALPTK PLNDVNDENG E.LCTLSAF
Phage T4   YFKQDRFRE  LYEAAEKDPN IRKRKIKARE LFELLMTERS GTARIYVQFI DNTNNYTPFI REKAPIROSN LCCEIAIPT. .NDVNSPDA  EIGLCTLSAF
Mouse      DEVWGEEFEK LYESYEKQGR VRK.VKAQQ  LWYAIIVSQT ETGTPYMLYK DSCNRKS.NQ QNLGTIKGSN LCTEIVEYT. ......SKD  EVAVCNIASL
VZV        RTLHGEAFTS TYLRLEREG. LGVSSVPIQD IAFTIIRSAA VTGSPFLMFK TQGNAITGSN LCTEIVHPA. .......DAH DHQVCNIASI
HSV1       ADFHGEEFEK LYQHLEVMG. .FGEQIPIQE LAYGIVRSAA TTGSPFVMFK DAVNRHIYID LCTEIVHPA. ........SKR SSQCVCNIGSV
EBV        ERLYGEEFER EYRLVTAG.  KFCGRVSIKS LMFSIVNCAV KAGSPFILLK LQGEAMNAFN LCAEVLQPS. .........RK SVATCNIANI
Consensus  ----Y----- ---------- ---------- ---------- ---------- ---------- ---------- -N-LC-EI-- ---------- ---C-L----
```

```
          468                                                                               532
E.coli    NLGAINNLDE  LEELAILA..  .....VRALDA  LLDYQDYPIP  AAK.......  .RGAMGRRT  LGIGVTNFAY  YLANDGKRYS
Phage T4  VLDNFDWDQQ  DKINELAEVQ  .....VRALDN  LLDYGGYPVP  EA........  .EKAKKRRN  LGVGVTNYAA  WLASNFASYE
Mouse     ALNMYVTPEH  TYDFEKLAEV  TKVTVRNLNK  IIDINYYPIP  EAH.......  .LSNKRHRP  IGIGVQGLAD  AFILMRYPFE
VZV       NLTTCLSKGP  VSFNLNDIQL  TARITVIFLN  GVLAAGNFPC  KKS.......  CKGVKNNRS  LGIGTQGLHT  TCLRLGFDLI
HSV1      NLARCVSRQ.  TDFDFGRLRDA VQACVLMVNI  MIDSTLQPTP  QCT.......  .RGNDNLRS  MGIGVMGLHT  ACLKLGIDLE
EBV       CLPRCLVNAP  LAVRAGRADT  QGDELILALP  RLSVTLPGEG  AVGDGFSLAR  LRDATQCATF VVACSILQGS  PTYDSRDMAS MLGVQGLAD VFADLGWQYT
Consensus -L--------                                                                          -G-

          533                                                                               630
E.coli    DGSANNLTHK  TFEAIQYLL   KASNELA..K  EQGACPWFNE  TTYAKGILPI  DTYKKDLDTI ANEPLHYDWE  ALRESIKTHG  LRNSTLSALM  PSETSSQISN
Phage T4  D.ANDLTHE   LFERLQYGLI  KASIKLA..K  EKGPSEYYSD  TRWSRGELPI  DWYNKKIDGI AAPKYVCDWS  ALREDIKLFG  IRNSTLSALM  PCESSSQVSN
Mouse     SPEAQLLNKQ  IFETIYYGAL  EASCELA..K  EYGFYETYEG  SPVSKGILQY  DMMNVAPTDL ....WDWK    PLKEKIAKYG  IRNSLLIAPM  PTASTAQILG
VZV       SQPARRLNVQ  IAEIMLYETM  KTSMEMCKIG  GLAPFKGFTE  SKYAKGWLHQ  TLRDDICAYG ....DLPWC   LYNSQFLALM  PTVSSAQVTE
HSV1      SAEFQDLNKH  IAEVMLLSAM  KTSNALCV.R  GARPFWHFKR  SMYRAGRFHW  ERFPDARPRY ....EGEWE   MLRQSMVKHG  LRNSQFVALM  PTAASAQISD
EBV       DPPSRSINKE  IFEHMYFTAL  CTSSLIGL.H  TRKIFPGFKQ  SKYAGGWFHW  HDWAGTDLSI P....REIWS  RLSERIVRDG  LFNSQFIALM  PTSGCAQVTG
Consensus ---------E  ---S                                         ---W----  --NS--A-M  -P----Q--

          631  (K)                                                                          695
E.coli    ATNGIEPPRG  YVSIKASKDG  ILRQVVPDYE  HLHDAYEL..  .........   ...LWEMPG  NDGYLQLVGI  MQKFIDQSIS
Phage T4  STNGTEPPRG  PVSVKESKEG  SFNDVVPNIE  HNIDLYDY..  .........   ..TWKLAKKG NKPYLTQVAI  MLKWVCQSAS
Mouse     NNESIEPYTS  NIYTRRVLSG  EFQIVNPHLL  KDLITERGL.. ..WNEZMKN   QIIACNGSIQ SIPEIPDDLK  OKTVLKMAAE  RGAFIDQSQS
VZV       CSEGFSPIYN  NMFSKVTTSG  ELLRPNLDLM  DELRDMYSCE  EKRLEVINIL  EKNQWSVIRS QLYKTWEIS   QEDLVDMCAE  RAPFIDQSQS
HSV1      VSEGFAPLFT  NLFSKVTRDG  ETLRPNTILL  KELERTFS.G  KRLLEVMDSL  DAKQWSVAQA LPCLEPTHPL  RRFKTAFDYD  QKLLIDLCAD  RAPVDHSQS
EBV       CSDAFYPFYA  NASTKVTNKE  EALRPNRSFW  RH.......V  RLDDREALNL  VGGRVSCL.. ..PEALRQRY  LRFQTAFDYN  QEDLIQMSRD  RAPFVDHSQS
Consensus -------p--                                                                          -S-S

          696                                                                    761
E.coli    ANTNYDPSRF  PSGKVPMQOL  LKDLLTAYKF  GVKTLYYQN   TRDGAEDAQD  DLVPSIQDDG  CESGACKI
Phage T4  ANTYDPQIF   PKGKVPMSIM  IDDMLYGWYY  GIN.FYYHN   TRDGSGTDDY  EIETPKADE   AACKL
Mouse     LNIHIAE...  ...PNYGKL   TSMHFYGWKQ  GLKTGMYYLR  TRPAANPIQF  TLNKEKDK    EKALKEEEK
VZV       MTLFIEE..R  PDGTIPASKI  MALLIRAYKA  GLKTGMYYCK  IRKATNSGLF  AGEE.LICTS  CAL
HSV1      MTLVTE..K   ADGTLPASTL  VRLLVHAYKR  GLKTGMYYCK  VRKATNSGVF  GDDNIVCMS   CAL
EBV       .EDAARASTL  ANLLVRSYEL  GLKTMYYCR   IEKAADLGVM  ECKASAALSV  PREEQNRSP   AEQMPPRPME  PAQVAGPVDI  MSKGPGEGPG
Consensus                        --G--K--   --YY

          901                                              938
EBV       GMCVPGGLEV  CYKYRQLFSE  DDLLETDGFT  ERACESCQ
```

FIGURE 3. Amino acid sequence alignment of the large subunit of ribonucleotide reductase from *E. coli*, bacteriophage T4, mouse, Epstein Barr virus, herpes simplex virus, and Varicella-Zoster virus.[19,27,33,39,43,45] For simplicity only, the HSV type 1 sequence is shown. The N-terminal domain of the HSV sequence, which is unrelated to all other sequences shown here, has been left out. The numbering refers to the *E. coli* sequence. Mutant constructs mentioned in the text are shown above their respective location. Some pertinent areas of high homology have been boxed.

Two mutations which affect the B1 activity have been deduced as single amino acid substitutions.[78] These were mutants isolated according to a divergent phenotype and later extensively characterized. Their main lesions were drastically reduced subunit activities (6 to 11% as compared to wild-type enzyme) for all four substrates *in vitro,* and drastically lowered pyrimidine deoxynucleotide pools *in vivo.*[82] These lesions corresponded in one case to a substitution of the highly conserved Gly-410 by a serine, and in the other case to a substitution of Glu-636 by a lysine.[78] Glu-636 is conserved in the phage and mouse sequences, and it is adjacent to the invariant Pro-637 at the C-terminal end of an extensively conserved area starting around Trp-599 (Fiugre 3). These mutations will undoubtedly contribute to a future understanding of the molecular details of the three-dimensional structure of B1.

B. The Small Subunit

To date, ten different primary structures are known for homologous small subunits of the iron-containing ribonucleotide reductases. Initially, an alignment of four of these sequences, *E. coli,* clam, HSV, and EBV, were used to make certain predictions about important residues in the B2 protein.[83] The observation of highly homologous areas also in the small subunit have so far been corroborated by every single addition to the sequence alignment.[26,34,37,38,43,44] All together there are to date 18 invariant residues among the known small subunit sequences.

The tyrosyl radical residue was recently localized to Tyr-122 (in *E. coli* numbering) by protein engineering.[84] Since Tyr-122 is an invariant residue, these experiments clearly define the position of the active site tyrosyl radical in all the other small subunits sequenced so far.

The C-terminal end of the small subunit of ribonucleotide reductase seems to carry the main side chains responsible for the subunit interaction, as can be rationalized from three independent observations. Heat-induced *E. coli* cells produced, in addition to normal B2, a truncated form of B2, lacking 30 C-terminal residues.[85] The truncated form of B2 was shown to be similar to normal B2, but it did not combine with B1. It is thus likely that the two subunits of *E. coli* ribonucleotide reductase interact via the C-terminal part of B2. In addition, two independent studies on the herpes simplex reductase demonstrated that a nonapeptide corresponding to the C-terminal end of the B2 homolog behaved as a specific inhibitor of the enzyme by inhibiting the subunit interaction *in vitro.*[86,87] There is a potentially important therapeutic value in this observation.

V. GENE REGULATION

The main issue of this article is the allosteric regulation of ribonucleotide reductase. As extensively discussed in Section III, the allosteric effects are localized to the large subunits. However, other types of regulation of the small subunits occur at the transcriptional, translational, and posttranslational levels, and must be considered in order to complete the general picture of this complex enzyme.

A. Prokaryotic Systems

In *E. coli,* the synthesis of the two subunits of ribonucleotide reductase from the adjacent *nrdA* and *nrdB* genes is coordinately regulated and involves positive as well as negative elements. The enzyme appears to be synthesized only once during the cell cycle at a time that coincides with the onset of DNA replication. The start of transcription was mapped 110 base pairs upstream of *nrdA.*[23] According to the nucleotide sequence, the promoter is expected to be potent, but some form of repression must contribute to its low *in vivo* strength. In thymine-starved cells, increased reductase activity is due to increased rate of transcription of ribonucleotide reductase mRNA. A positive regulatory protein probably accumulates during inhibition of DNA replication and acts to induce reductase synthesis. These types of regulatory event should affect both subunits similarly. A few observations in the *E. coli* system imply specific regulatory effects at the B2 level. First, the *nrd* operon contains a long intergenic region comprising 2 to 3 REP (repetitive extragenic palindrome) sequences, which has the potential to form a stable hairpin structure.[18] Second, the specific proteolysis of B2 observed in heat-induced bacteria involves an endogenous serine protease which acts specifically to inactivate the B2 subunit of the complex.[85] Third, a tyrosyl radical regenerating enzyme system was recently characterized in *E. coli* cells.[88,89] All these data suggest different forms of regulation of the activity of the *E. coli* ribonucleotide reductase, all exerted at the posttranscriptional level via the B2 subunit.

The *nrdA* and *nrdB* genes of bacteriophage T4 are adjacent on the chromosome. Unexpectedly, nucleotide sequencing of the T4 *nrdB* gene revealed the presence of a 0.6 kb intron, which divides its coding sequence into two exons.[26,90] Only two other prokaryotic introns have been reported; one is in the nearby T4 thymidylate synthase gene.[91] The presence of the intron implies that the T4 *nrdA* and *nrdB* genes are differently expressed, similar to what has also been reported in eukaryotic organisms such as clams, sea urchins, and mouse cells (see Section V.B).

B. Cell Cycle Control

The concentration of intracellular ribonucleotides is relatively constant throughout the mammalian cell cycle, increasing in proportion to the cellular volume. However, the deoxyribonucleotide concentrations increase several fold as cells enter S-phase. The activity of ribonucleotide reductase must consequently increase during S-phase and this has been confirmed by several studies.[1-3]

The separation and purification of the two subunits of the mammalian reductase have facilitated detailed studies on the cell cycle control of this enzyme. Initially, it was found that the activity of protein M2 in mouse T lymphoma cells changed three- to fourfold as cells passed from G1 into S-phase, while the activity of protein M1 was practically constant.[92] This observation was later confirmed and considerably extended by studies using electron spin resonance spectroscopy to quantitate the M2 specific radical content in whole cells. A three- to sevenfold increase in M2 radical was observed as cells passed from G1 to S-phase of the cell cycle. With pulse-chase experiments, it could be demonstrated that this increase was due to *de novo* synthesis of the protein and that the half-life of M2 in these cells was 3 h.[93]

Protein M1 was determined by immunochemical methods and the level of M1 per milligram of protein was constant in extracts from cells in different phases of cell cycle.[94] Thus, the levels of protein M1 and M2 are regulated differently during the cell cycle. The variation in reductase activity is controlled by *de novo* synthesis and break-down of protein M2. The regulation of protein M2 during the cell cycle occurs most likely at the level of transcription. However, it has been shown that one of the most prominent proteins synthesized immediately after fertilization of clam or sea urchin oocytes is equivalent to protein M2.[36,95] In this case, the protein is made from preformed mRNA stored in the oocytes and therefore, the control is at the level of translation. Recent studies with hydroxyurea resistant cell lines with amplified protein M2 genes have shown that protein M1 and M2 expression is regulated both transcriptionally and posttranscriptionally in mammalian cells.[96,97]

The existence of high levels of protein M1 in cells not engaged in DNA synthesis suggest a possible function for this protein in G1 cells, separate from ribonucleotide reduction. However, protein M1 is not present in all cells since immunocytochemical studies of rat tissue show that M1 is only present in proliferating cells.[94,98] With the cloned genes for protein M1 and M2 it should be possible to investigate the molecular mechanisms underlying the cell cycle control of mammalian ribonucleotide reductase.

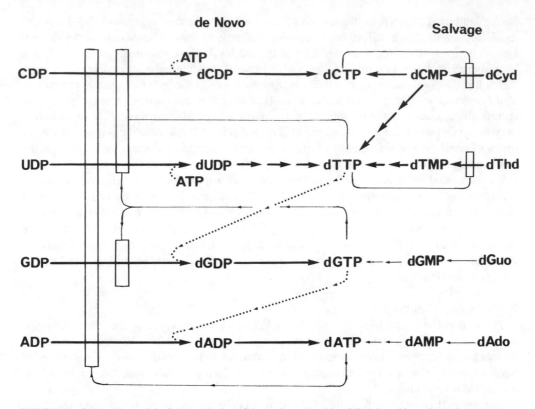

FIGURE 4. Schematic model of the regulation of deoxyribonucleoside triphosphate biosynthesis in mammalian cells. Feedback inhibition is denoted by solid lines ending in open rectangular boxes and allosteric activation is denoted by dashed lines.

VI. DEOXYRIBONUCLEOSIDE TRIPHOSPHATE BIOSYNTHESIS AND DNA SYNTHESIS

A. *In Vitro* and *In Vivo* Correlation

The biochemical characterization of the purified ribonucleotide reductases from *E. coli* and calf thymus has been used to propose a scheme that links ribonucleotide reduction to DNA synthesis (Figure 4). *De novo* synthesis of deoxyribonucleotides begins with the reduction of CDP and UDP by an ATP-activated enzyme. It proceeds to GDP reduction when sufficient levels of dTTP have been formed, and finally, ADP reduction is stimulated by a dGTP-activated enzyme. Accumulation of dATP as a result of decreased DNA synthesis leads to a general inhibition of the reductase. The stimulation of reduction of a certain substrate leads to the concordant inhibition of the reduction of the other substrates, e.g., accumulation of dTTP and dGTP inhibits CDP and UDP reduction.

How relevant is this "test tube" hypothesis to the *in vivo* situation? This issue has been extensively discussed in a recent review by Reichard.[99] The basic scheme in Figure 4 is clearly an oversimplification of the complex balance between various nucleotide metabolizing enzyme activities, which together determine the absolute intracellular DNA precursor concentration at a given time. However, a large number of observations, some of which will be described in this section, have shown that the principles outlined above can be used to explain the features of the deoxyribonucleotide metabolism in most cells.[1-3,99,100] To attain this goal the ribonucleotide reductase regulation in Figure 4 is complemented by some of the major routes of the salvage pathway: an alternative route, which in certain cells is important as a source of DNA precursors. Deoxyribonucleosides are taken up by a facilitated transport system and phosphorylated primarily by the thymidine and deoxycytidine kinases, the latter also being the principal phosphorylating enzyme for purine deoxyribonucleosides, but with about a 20- to 50-fold lower efficiency than for deoxycytidine. Through the action of the ubiquitous nucleotide kinases, deoxyribonucleoside triphosphates are formed.

Thymidine kinase is feedback inhibited by dTTP and deoxycytidine kinase by dCTP, but high intracellular concentrations of deoxyribonucleoside triphosphates can still accumulate by these routes. As can be predicted from the scheme, addition of thymidine to cells leads to depletion of the dCTP pool and increased dGTP and dATP pools. Addition of deoxyguanosine, which leads to an increased dGTP pool, results in a decreased dCTP pool and an increased dATP pool. Finally, addition of deoxyadenosine leads to decreased levels of all other DNA precursors in most cells investigated.[99,100] These deoxyribonucleotide pool changes are associated with growth inhibition, and selection of cells resistant to deoxynucleosides have provided evidence for the validity of this scheme as will be discussed in Section VI.B.

One of the end products, dCTP, is not active as effector for the *E. coli* enzyme or the mammalian reductase. However, dCTP serves as a major source for the synthesis of dTTP via the dCMP deaminase activity. This enzyme is allosterically controlled by dCTP and dTTP and is responsible for the regulation of the ratio between these two compounds in most mammalian cells. In this metabolic scheme, only the biosynthetic pathways have been indicated, but obviously several catabolic enzymes, e.g., nucleotide phosphatases are directly involved in maintaining certain final levels of the DNA precursors. "Substrate cycles", i.e., phosphorylation and dephosphorylation of deoxycytidine, deoxyuridine or thymidine, have been implicated as a major factor in maintaining a balance in the flux of nucleosides through these pathways.[101,102]

In *E. coli*, it was observed that overproduction of wild-type ribonucleotide reductase, cloned on a multicopy plasmid, resulted in three- to fourfold increased levels of all four deoxyribonucleoside triphosphates. Interestingly, four different point mutations in the B1 subunit all resulted in drastically changed DNA precursor pools *in vivo*.[78,82] These results confirm the *in vivo* significance of the allosteric regulation of ribonucleotide reductase also in prokaryotes.

B. Immunodeficiency Diseases and Somatic Cell Mutants

Inherited deficiencies of two enzymes of purine metabolism, adenosine deaminase, and purine nucleoside phosphorylase have been causally related to severe immunodeficiency diseases in humans.[100] The high levels of deoxyadenosine and deoxyguanosine found in patients with these enzyme defects have been shown to be the cytotoxic compounds. Intracellular accumulations of dATP and dGTP up to 50-fold higher than in normal cells have been observed. The reason for the selective immunotoxicity has been attributed to the fact that the activities of deoxyribonucleoside phosphorylating enzymes are high and nucleotide degrading enzymes are low in immature T cells of the thymus. Therefore, in affected patients lymphocytes may selectively accumulate dATP and dGTP and these compounds would inhibit ribonucleotide reductase and lead to depletion of the other DNA precursors needed for cell growth.

In order to gain insight into the mechanism underlying deoxynucleoside lymphotoxicity and substantiate the above hypothesis, a cell culture model was developed using S49 T lymphoma cells. This cell line was five- to tenfold more sensitive to the cytotoxic effect of deoxyadenosine and deoxyguanosine than most other cell lines in culture. A series of mutants resistant to deoxyribonucleosides were selected, and three principal types of resistance were found. One type was nucleoside transport deficient, the other type was nucleoside kinase deficient, and the third type showed normal capacity to accumulate deoxyribonucleoside triphosphates, but altered response in the other DNA precursor levels. Within this latter group, with a presumably altered ribonucleotide reductase, two different phenotypes were seen. One cell type was highly resistant to deoxyguanosine and thymidine, but sensitive to deoxyadenosine (dGuo-L and HAT-1.5-A), and the other cell type was resistant to deoxyadenosine and slightly resistant to the two other nucleosides (dGuo-200-1).[100,103] All three mutations gave rise to dominant or codominant changes in cell hybrids and it was possible to separate and identify both a normal and a mutated form of protein M1 in these cells.[31,104]

Cell types dGuo-L and HAT-1.5-A contain an altered M1 with a CDP reduction which was insensitive to inhibition by dGTP or dTTP.[103,104] This was dependent on the inability of mutated M1 to bind dTTP, as revealed by photoaffinity experiments.[67] The dGuo-L protein M1 also showed a decreased capacity to use ADP as substrate. Apparently, the substrate specificity site is altered so that efficient stimulation of purine reduction is abolished, which in turn leads to an inability of dTTP to inhibit CDP reduction.[31] The dGuo-L and HAT-1.5-A cells had altered intracellular deoxyribonucleotide pools with elevated pyrimidine and decreased purine deoxyribonucleoside triphosphates, as predicted from the properties of the altered protein M1.

The dGuo-200-1 cells contained a reductase abnormally resistant to inhibition by dATP.[31] Mutant protein M1 had a CDP reductase activity which was stimulated by dATP, unlike the wild-type enzyme.[67] It appears that this enzyme has lost the capacity to distinguish between dATP and ATP, but it is still sensitive to regulation by dGTP and dTTP. Thus, the activity site is altered in the dGuo-200-1 mutant, while the specificity site appears normal.[105] Recently, it was shown that the mutant phenotype is the result of a single point mutation converting Asp-57 to Asn.[105a] In dGuo-200-1 cells, all four DNA precursor pools were elevated three- to fourfold. Obviously, the relaxed feedback control in this situation leads to higher pools and resistance not only to deoxyadenosine, but also to other deoxyribonucleosides.[100] There are also other deoxyadenosine resistant cell lines selected from CHO (Chinese hamster ovary) cells with a reductase less sensitive to dATP inhibition.[106] Taken together these results provide strong genetic and biochemical evidence for a single genetic locus for protein M1, two independent allosteric regulatory domains of M1, and ribonucleotide reductase as the enzyme responsible for the *in vivo* regulation of DNA precursor synthesis as outlined in Figure 4.

Studies of *in vitro* DNA synthesis have shown that fidelity of replication is influenced by the relative concentrations of deoxyribonucleoside triphosphates, and in reconstituted prokaryotic replication systems, specific incorporation errors can be induced by an appropriate bias of the precursor pools. It is therefore not surprising that DNA precursor pool imbalances have been found to produce diverse genetic effects in mammalian cells in culture. This subject has recently been reviewed by Meuth.[107] A study of the ribonucleotide reductase mutants dGuo-L and dGuo-200-1 showed increased mutation rates at two loci, further demonstrating the central role of the allosteric control of ribonucleotide reductase in maintaining a balanced supply of the four deoxyribonucleotides.[108]

Addition of deoxycytidine resulted in protection of cells against the cytotoxicity of deoxynucleosides in most tissue culture systems tested. The mechanism(s) of protection involves phosphorylation of deoxycytidine by deoxycytidine kinase and thereby replenishing missing DNA precursors or alternatively, competition for phosphorylation with deoxyadenosine and deoxyguanosine, leading to inefficient triphosphate accumulation.[100] Addition of cytidine, but not uridine to human T lymphoma cells also led to reversal of deoxyguanosine and thymidine growth inhibition, comparable to that obtained with deoxycytidine.[109] Analysis of intracellular nucleotide pools showed that increased levels of cytidine ribonucleotides were sufficient to overcome the inhibitory effects of dGTP and dTTP on CDP reduction, i.e., allosteric inhibition *in vivo* was less pronounced at higher substrate concentration as was found with the purified enzyme.[66] These results may encourage the use of cytidine in combination with deoxycytidine (and preferably also a cytidine deaminase inhibitor to increase the half-life of these compounds) as a pharmacological regime in treatment of immunodeficiency diseases associated with increased deoxyribonucleotide levels.

C. Animal Virus Systems

Infection of cells with HSV changes the deoxyribonucleotide pool balance in these cells and induces a drastic increase in pyrimidines, particularly dTTP, by efficient salvage of the pyrimidines.[110] It is known that in HSV infected cells this high dTTP level does not lead to a decreased DNA synthesis as is the case in uninfected cells. These results suggested the existence of a virus induced ribonucleotide reductase insensitive to feedback control by dTTP (see Section III.E). Later it was shown that a temperature sensitive mutation in the herpes reductase gene results in a plaque reduction of virus at the nonpermissive temperature, which implies that deoxyribonucleotide synthesis via the *de novo* pathway is important for the virus.[111] In this context, the apparent lack of allosteric regulation of the viral enzyme may have a physological function, permitting rapid production of deoxyribonucleotides needed for viral DNA synthesis. The importance of this route in providing DNA precursors has been questioned since herpes also induces a DNase that breaks down host DNA and thus supplies an additional source of deoxyribonucleotides.[112] However, the induction of a viral reductase may be directly essential for the infection of differentiated cells such as nerve cells.

D. Compartments and Multienzyme Complexes

Deoxyribonucleotide pools are generally low compared to the demands of the replicative machinery. Likewise, the enzymatic activity of *E. coli* ribonucleotide reductase, as measured *in vitro*, was earlier estimated to be approximately an order of magnitude too low to sustain replication of the bacterial chromosome. This prompted investigations on ribonucleotide reductase activity in systems mimicking the intact cell, and in fact substantially more effective reduction of ribonucleotides was observed[113,114] A simplified purification scheme resulted in a considerably increased enzymatic activity, and levels of ribonucleotide reductase in wild-type bacteria determined immunologically were in fact higher than earlier predicted.[115] It was thus concluded that the estimated activity of *E. coli* ribonucleotide reductase would satisfy more than 50% of replicative needs of the bacterium, and there is consequently no need to invoke a highly active *in vivo* form of ribonucleotide reductase in *E. coli*.

Most studies on DNA synthesis *in vitro* indicate that much higher dNTP concentrations are needed to sustain a maximal incorporation rate than are actually available. For instance, T4-infected *E. coli* cells synthesize DNA at nearly ten times the rate of uninfected cells, but the total dNTP pools remain constant after T4 infection (except for the replacement of dCTP by Hm-dCTP). Thus, there must be a considerable increase in the rate of dNTP production and turnover. In fact, deoxyribonucleotide synthesis is the rate-limiting factor in the replication of bacteriophage T4 DNA, and T4 ribonucleotide reductase, in spite of its lack of feedback control, appears to be the limiting enzyme in the synthesis of dNTPs.[116] One way to maintain high dNTP concentrations is compartmentalization of pools. Evidence for this has been obtained in T4-infected *E. coli*.[116] A multienzyme complex of dNTP synthesizing enzymes channels DNA precursors to the replication forks in the infected cells. The complex is probably juxtaposed with the replication apparatus, such that dNTPs are efficiently generated at their sites of utilization. The activities of this aggregate are kinetically coupled to one another, so that distal DNA precursors (rNDPs) are incorporated into DNA more readily than proximal precursors (dNTPs). The T4 dNTP synthesizing complex includes eight phage-coded and two host-coded activities.

In eukaryotic DNA replication, chains replicate at $1/10$ the prokaryotic rate and *in vivo* systems saturate at dNTP concentrations 1 to 2 orders of magnitude lower than in prokaryotic systems. Therefore, there is no discrepancy between the rate of ribonucleotide reduction measured in eukaryotes and the amounts of product needed to support DNA synthesis. Still, a large number of investigations have described channeling of deoxyribonucleotides in eukaryotic cells, and the direct isolation of multienzyme aggregates of DNA precursor enzymes. These studies were reviewed by Mathews and Slabaugh,[117] and by Reichard.[99] The enzyme complex, named replitase, is formed only during S-phase and is composed of ribonucleotide reductase, thymidylate synthease, nucleotide kinases, and DNA polymerase.[118] The existence of this complex would explain a number of experiments in which incorporation of ribonucleotides into DNA was more efficient than incorporation of deoxyribonucleotides.[119] However, these results contradict evidence for the cytoplasmic localization of ribonucleotide reductase as demonstrated both by immunological techniques and rapid cell fractionation methods.[98,120] Therefore, at present, there is no direct data to support the replitase being a functional regulation point in the DNA precursor synthesis in eukaryotic cells. Instead, deoxyribonucleoside triphosphates are apparently synthesized in the cytoplasm and transported into the nucleus.

VII. GENERAL SUMMARY

Ribonucleotide reductase catalyzes the first unique step in DNA synthesis and provides the cell with a balanced supply of the four deoxyribonucleotides. The underlying regulatory mechanism resides in the molecular architecture of a single or at most two identical polypeptide chains. The allosteric mechanism consists of spatially separated catalytic and regulatory sites; the former being able to bind and distinguish the four substrates, and the latter interacting with at least four different effector nucleotides. Binding of effectors induce a conformation of the catalytic site favoring reduction of a particular substrate. The central role of this allosteric mechanism in the *in vivo* regulation of DNA precursor levels has been convincingly documented. With the gene structures defined, protein engineering experiments and crystallographic studies[121] will provide a detailed insight into the structural requirements for this unique class of allosteric enzymes. As evident from this article, except for the radical generating system, detailed results obtained with the prokaryotic enzymes will serve as a relevant model for other ribonucleotide reductases, since both the basic catalytic mechanism and allosteric control of this enzyme seem to have been preserved through otherwise substantial evolutionary changes in primary structure.

ACKNOWLEDGMENTS

We are grateful to Gunnel Johansson, Cecilia Carlvik-Bodén, and Lena Lindgvist for their patience and skill in preparing the manuscript and the illustrations. Work by the authors of this review was supported by grants from the Swedish Medical and Natural Science Research Councils, the Swedish Cancer Society, and Magn. Bergvall Foundation.

REFERENCES

1. **Thelander, L. and Reichard, P.,** Reduction of ribonucleotides, *Annu. Rev. Biochem.,* 48, 133, 1979.
2. **Holmgren, A.,** Regulation of ribonucleotide reductase, *Curr. Top. Cell. Reg.,* 19, 47, 1981.
3. **Lammers, M. and Follmann, H.,** The ribonucleotide reductases — a unique group of metalloenzymes essential for cell proliferation, *Stuct. Bonding (Berlin).,* 54, 29, 1983.
4. **Reichard, P.,** Regulation of deoxyribonucleotide synthesis, *Biochemistry,* 21, 3245, 1987.
5. **Hogenkamp, H. P. C. and McFarlan, S. C.,** Nature and properties of the bacterial ribonucleotide reductases, in *International Encyclopedia of Pharmacology and Therapeutics,* Cory, J. G., Ed., in press, 1988.

6. **Gräslund, A., Sahlin, M., and Sjöberg, B.-M.,** The tyrosyl free radical in ribonucleotide reductase, *Environ. Health Perspec.,* 64, 139, 1985.

7. **Ashley, G. W., Harris, G., and Stubbe, J.,** Current ideas on the chemical mechanism of ribonucleotide reductase, *Pharmacol. Ther.,* 30, 310, 1985.

8. **Ashley, G. W., Harris, G., and Stubbe, J.,** The mechanism of *Lactobacillus leichmannii* ribonucleotide reductase. Evidence for 3′ carbon-hydrogen bond cleavage and a unique role for coenzyme B_{12}, *J. Biol. Chem.,* 261, 3958, 1986.

9. **Holmgren, A.,** Thioredoxin, *Annu. Rev. Biochem.,* 54, 237, 1985.

10. **Willing, A., Follmann, H., and Auling, G.,** Ribonucleotide reductase of *Brevibacterium ammoniagenes* is a manganese enzyme, *Eur. J. Biochem.,* 170, 603, 1988.

10a. **Willing, A., Follmann, H., and Auling, G.,** Nucleotide and thioredoxin specificity of the manganese ribonucleotide reductase from *Brevibacterium ammoniagenes*, *Eur. J. Biochem.,* 175, 167, 1988.

11. **Hantke, K.,** Characterization of an iron sensitive Mud1 mutant in *E. coli* lacking the ribonucleotide reductase subunit B2, *Arch. Microbiol.,* 149, 344, 1988.

11a. **Lindahl, G., Sjöberg, B.-M., and Reichard, P.,** Mutants of *Escherichia coli* deficient in either subunit of ribonucleotide reductase are oxygen-sensitive, submitted to *Mol. Gen. Genet.,* 1988.

12. **Panagou, D., Orr, M. D., Dunstone, J. R., and Blakley, R. L.,** A monomeric allosteric enzyme with a single polypeptide chain. Ribonucleotide reductase of *Lactobacillus leichmannii, Biochemistry,* 11, 2378, 1972.

13. **Chen, A. K., Bahn, A., Hopper, S., Abrams, R., and Franzen, J. S.,** Substrate and effector binding to ribonucleoside triphosphate reductase of *Lactobacillus leichmannii, Biochemistry,* 13, 654, 1974.

14. **Singh, D., Tamao, Y., and Blakley, R. L.,** Allosterism, regulation and cooperativity: the case of ribonucleotide reductase of *Lactobacillus leichmannii, Adv. Enzymol. Reg.,* 15, 81, 1977.

15. **Hogenkamp, H. P. C., Ghamber, R. K., Brownson, C., Blakley, R. L., and Vitols, E.,** Cobamides and ribonucleotide reduction. VI. Enzyme-catalyzed hydrogen exchange between water and deoxycobalamine, *J. Biol. Chem.,* 243, 799, 1968.

16. **Gleason, F. K. and Hogenkamp, H. P. C.,** 5-deoxyadenosyl-cobalamin-dependent ribonucleotide reductase: a survey of its distribution *Biochem. Biophys. Acta,* 277, 466, 1972.

17. **Hamilton, F. D.,** Ribonucleotide reductase from *Euglena gracilis*. A 5-deoxyadenosylcobalamin-dependent enzyme, *J. Biol. Chem.,* 249, 4428, 1974.

18. **Carlson, J., Fuchs, J. A., and Messing, J.,** Primary structure of the *Escherichia coli* ribonucleoside diphosphate reductase, *Proc. Natl. Acad. Sci. U.S.A.,* 81, 4294, 1984.

19. **Fuchs, J. A.,** personal communciation.

19a. **Nilsson, O., Åberg, A., Lundqvist, T., and Sjöberg, B.-M.,** Nucleotide sequence of the gene for the large subunit of ribonucleotide reductase of *Escherichia coli,* (Correction), *Nucleic Acids Res.,* 16, 4174, 1988.

20. **Brown, N. C. and Reichard, P.,** Ribonucleoside diphosphate reductase. Formation of active and inactive complexes of proteins B1 and B2, *J. Mol. Biol.,* 46, 39, 1969.

21. **Thelander, L.,** Reaction mechanism of ribonucleoside diphosphate reductase from *Escherichia coli*. Oxidation-reduction-active disulfides in the B1 subunit, *J. Biol. Chem.,* 249, 4858, 1974.

22. **Fuchs, J. A., Karlström, H. O., Warner, H. R., and Reichard, P.,** Defective gene product in *dnaF* mutant of *Escherichia coli, Nature (London) New Biol.,* 238, 69, 1972.

23. **Tuggle, C. K. and Fuchs, J. A.,** Regulation of the operon encoding ribonucleotide reductase in *Escherichia coli:* evidence for both positive and negative control, *EMBO J.,* 5, 1077, 1986.

24. **Berglund, O.,** Ribonucleoside diphosphate reductase induced by bacteriophage T4: isolation and characterization of proteins B1 and B2, *J. Biol. Chem.,* 250, 7450, 1975.

25. **Cook, K. and Greenberg, G. R.,** Properties of bacteriophage T4 ribonucleoside diphosphate reductase subunits coded by *nrdA* and *nrdB* mutants, *J. Biol. Chem.,* 258, 6064, 1983.

26. **Sjöberg, B.-M., Hahne, S., Mathews, C. Z., Mathews, C. K., Rand, K. N., and Gait, M. J.,** The bactriophage T4 gene for the small subunit of ribonucleotide reductase contains an intron, *EMBO J.,* 5, 2031, 1986.

27. **Tseng, M.-J., Hillfinger, J. M., Walsh, A., and Greenberg, G. R.,** Total sequence, flanking regions and transcripts of bacteriophage T4 *nrdA* gene, coding for α chain of ribonucleoside diphosphate reductase, *J. Biol. Chem.,* 263, 16242, 1988.

28. **Engström, Y., Eriksson, S., Thelander, L., and Åkerman, M.,** Ribonucleotide reductase from calf thymus. Purification and properties, *Biochemistry,* 18, 2948, 1979.

29. **Thelander, L., Eriksson, S., and Åkerman, M.,** Ribonucleotide reductase from calf thymus. Separation of the enzyme into two nonidentical subunits, proteins M1 and M2, *J. Biol. Chem.,* 255, 7426, 1980.

30. **Thelander, M., Gräslund, A., and Thelander, L.,** Subunit M2 of mammalian ribonucleotide reductase. Characterization of a homogeneous protein isolated from M2-overproducing mouse cells, *J. Biol. Chem.,* 260, 2737, 1985.

31. **Eriksson, S., Gudas, L. J., Ullman, B., Clift, S. M., and Martin, D. W., Jr.,** DeoxyATP-resistant ribonucleotide reductase of mutant mouse lymphoma cells. Evidence for heterozygosity for the protein M1 subunits, *J. Biol. Chem.,* 256, 10184, 1981.

32. **Moore, E. C.,** Components and control of ribonucleotide reductase system of the rat, *Adv. Enzymol. Reg.,* 15, 101, 1977.

33. **Caras, I., Lewinson, B. B., Fabry, W., Williams, S. R., and Martin, D. W., Jr.,** Cloned mouse ribonucleotide reductase subunit M1 cDNA reveals amino acid sequence homology with *Escherichia coli* and herpes ribonucleotide reductase, *J. Biol. Chem.,* 260, 7015, 1985.

34. **Thelander, L. and Berg, P.,** Isolation and characterization of expressible cDNA clones encoding the M1 and M2 subunits of mouse ribonucleotide reductase, *Mol. Cell. Biol.,* 6, 3433, 1986.

35. **Yang-Feng., T. L., Barton, D. E., Thelander, L., Lewis, W. H., Srinivasan, P. R., and Francke, U.,** Ribonucleotide reductase M2 subunit sequences mapped to four different chromosomal sites in human and mice: functional locus identified by its amplification in hydroxyurea-resistant cell lines, *Genomics,* 1, 77, 1987.

36. **Standart, N. M., Bray, S. J., George, E. L., Hunt, T., and Ruderman, J. V.,** The small subunit of ribonucleotide reductase is encoded by one of the most abundant translationally regulated maternal RNAs in clam and sea urchin eggs, *J. Cell. Biol.,* 100, 1968, 1985.

36a. **Standart, N. M.,** personal communication.

37. **Elledge, S. J. and Davis, R. M.,** Identification and isolation of the gene encoding the small subunit of ribonucleotide reductase from *Saccharomyces cerevisiae:* DNA damage-inducible gene required for mitotic viability, *Mol. Cell. Biol.,* 7, 2783, 1987.

38. **Hurd, H. K., Roberts, C. W., and Roberts, J. W.,** Identification of the gene for the yeast ribonucleotide reductase small subunit and its inducibility by methyl methanesulfonate, *Mol. Cell. Biol.,* 7, 3673, 1987.

39. **Gibson, T., Stockwell, P., Ginsburg, M., and Barrell, B.,** Homology between two EBV early genes and HSV ribonucleotide reductase and 38K genes, *Nucleic Acids Res.,* 12, 5087, 1984.

40. **Dutia, B. M.,** Ribonucleotide reductase induced by herpes simplex virus has a virus specified constituent, *J. Gen. Virol.,* 64, 513, 1983.

41. **McLaughlan, J. and Clements, J. B.,** DNA sequence homology between two co-linear loci on the HSV genome which have different transforming abilities, *EMBO J.,* 2, 1953, 1983.

42. **Swain, M. and Galloway, D.,** Herpes simplex virus specifies two subunits of ribonucleotide reductase encoded by 3'-coterminal transcripts, *J. Virol.,* 57, 802, 1986.

43. **Davidson, J. A. and Scott, J. E.,** The complete DNA sequence of Varicella-Zoster virus, *J. Gen. Virol.,* 67, 1759, 1986.

44. **Slabaugh, M., Roseman, N., Davis, R., and Mathews, C.,** Vaccinia virus-encoded ribonucleotide reductase: sequence conservation of the gene for the small subunit and its amplification in hydroxyurea resistant mutants, *J. Virol.,* 62, 519, 1988.

44a. **Tengelsen, L. A., Slabaugh, M. B., Bibler, J. K., and Hruby, D. E.,** Nucleotide sequence and molecular genetic analysis of the large subunit of ribonucleotide reductase encoded by vaccinia virus, *Virology,* 164, 121, 1988.

44b. **Schmitt, J. F. C. and Stunnenberg, H. G.,** Sequence and transcriptional analysis of the vaccinia virus *Hind*III I fragment, *J. Virol.,* 62, 1889, 1988.

45. **Nikas, I., McLauchlan, J., Davidson, A. J., Taylor, W. R., and Clements, J. B.,** Structural features of ribonucleotide reductase, *Proteins,* 1, 376, 1986.

46. **Reichard, P., Canellakis, Z. N., and Canellakis, E. S.,** Studies on possible regulatory mechanism for the biosynthesis of deoxyribonucleic acid, *J. Biol. Chem.,* 236, 2514, 1961.

47. **Moore, E. C. and Hurlbert, R. B.,** Regulation of mammalian deoxyribonucleotide biosynthesis by nucleotides as activators and inhibitors, *J. Biol. Chem.,* 241, 4802, 1966.

48. **Larsson, A. and Reichard, P.,** Enzymatic synthesis of deoxyribonucleotides IX. Allosteric effects in the reduction of pyrimidine ribonucleotides by the ribonucleoside diphosphate reductase system of *Escherichia coli, J. Biol. Chem.,* 241, 2533, 1966.

49. **Larsson, A. and Reichard, P.,** Enzymatic synthesis of deoxyribonucleotides X. Reduction of purine ribonucleotides: allosteric behavior and substrate specificity of the enzyme system from *Escherichia coli, J. Biol. Chem.,* 241, 2540, 1966.

50. **Goulian, M. and Beck, W. S.,** Purification and properties of cobamide-dependent ribonucleotide reductase from *Lactobacillus leichmannii, J. Biol. Chem.,* 241, 4233, 1966.

51. **Follmann, H. and Hogenkamp, H. P. C.,** Interaction of ribonucleotide reductase with ribonucleotide analogs, *Biochemistry,* 10, 186, 1971.

52. **Ludwig, W. and Follmann, H.,** Inhibition of bacterial ribonucleotide reductase by arabinonucleotides, *Eur. J. Biochem.,* 91, 493, 1978.

53. **Vitols, E., Brownson, C., Gardiner, W., and Blakely, R. L.,** Cobamides and ribonucleotide reduction. V. Kinetic study of the ribonucleoside triphosphate reductase of *Lactobacillus leichmannii, J. Biol. Chem.,* 242, 3035, 1976.

54. **Sjöberg, B.-M., Eriksson, S., Jörnvall, H., Carlquist, M., and Eklund, H.,** Protein B1 of ribonucleotide reductase. Direct analytical data and comparisons with data indirectly deduced from the nucleotide sequence of the *Escherichia coli nrdA* gene, *Eur. J. Biochem.,* 150, 423, 1985.

55. **Thelander, L.,** Physicochemical characterization of ribonucleoside diphosphate reductase from *Escherichia coli, J. Biol. Chem.,* 248, 4591, 1973.

56. **von Döbeln, U. and Reichard, P.,** Binding of substrates to *Escherichia coli* ribonucleotide reductase, *J. Biol. Chem.,* 251, 3616, 1976.

57. **Thelander, L., Larsson, B., Hobbs, J., and Eckstein, F.,** Active site of ribonucleoside diphosphate reductase from *Escherichia coli.* Inactivation of the enzyme by 2'-substituted ribonucleoside diphosphates, *J. Biol. Chem.,* 251, 1398, 1976.

58. **Ehrenberg, A. and Reichard, P.,** Electron spin resonance of the iron-containing protein B2 from ribonucleotide reductase, *J. Biol. Chem.,* 247, 3485, 1972.

59. **Brown, N. C. and Reichard, P.,** Role of effector binding in allosteric control of ribonucleoside dihosphate reductase, *J. Mol. Biol.,* 46, 39, 1969.

60. **Eriksson, S.,** Direct photoaffinity labeling of ribonucleotide reductase from *Escherichia coli.* Evidence for enhanced binding of the allosteric effector dTTP by the presence of substrates, *J. Biol. Chem.,* 258, 5674, 1983.

61. **Berglund, O.,** Ribonucleoside diphosphate reductase induced by bacteriophage T4. II. Allosteric regulation of substrate specificity and catalytic activity, *J. Biol. Chem.,* 247, 7276, 1972.

62. **Berglund, O.,** Ribonucleoside diphosphate reductase induced by bacteriophage T4. III. Isolation and characterization of proteins B1 and B2, *J. Biol. Chem.,* 250, 7450, 1975.

63. **Eriksson, S. and Berglund, O.,** Bacteriophage-induced ribonucleotide reductase systems. T5- and T6-specific ribonucleotide reductase and thioredoxin, *Eur. J. Biochem.,* 46, 271, 1974.

64. **Youdale, T., MacManus, J. P., and Whitfield, J. F.,** Rat liver ribonucleotide reductase; separation, purification and properties of two nonidentical subunits, *Can. J. Biochem.,* 60, 463, 1982.

65. **Brissenden, J. E., Caras, I., Thelander, L., and Francke, U.,** The structural gene for the M1 subunit of ribonucleotide reductase maps to chromosome 11, band p15, in human and to chromosome 7 in mouse, *Exp. Cell Res.,* 174, 302, 1988.

66. **Eriksson, S., Thelander, L., and Åkerman, M.,** Allosteric regulation of calf thymus ribonucleoside diphosphate reductase, *Biochemistry,* 18, 2948, 1979.

67. **Eriksson, S., Caras, I. W., and Martin, D. W., Jr.,** Direct photoaffinity labeling of an allosteric site on subunit protein M1 of mouse ribonucleotide reductase by dTTP, *Proc. Natl. Acad. Sci. U.S.A.,* 79, 81, 1982.

68. **Caras, I. W. and Martin, D. W., Jr.,** Direct photoaffinity labeling of an allosteric site of subunit protein M1 of mouse ribonucleotide reductase by dATP, *J. Biol. Chem.,* 257, 9508, 1982.

69. **Eriksson, S., Sjöberg, B.-M., Jörnvall, H., and Carlquist, M.,** A photoaffinity-labeled allosteric site in *Escherichia coli* ribonucleotide reductase, *J. Biol. Chem.,* 261, 1878, 1986.

70. **Chang, C.-H. and Cheng, Y.-C.,** Substrate specificity of human ribonucleotide reductase from Molt-4F cells, *Cancer Res.,* 39, 5081, 1979.

71. **Chang, C.-H. and Cheng, Y.-C.,** Effects of nucleoside triphosphate on human ribonucleotide reductase from Molt-4F cells, *Cancer Res.,* 39, 5087, 1979.

72. **Harrington, J. A., Miller, W. H., and Spector, T.,** Effector studies of 3'-azidothymidine nucleotides with human ribonucleotide reductase, *Biochem. Pharmacol.,* 36, 3757, 1987.

73. **Corey, J. G., Rey, D. A., Carter, G. L., and Bacon, P. E.,** Nucleoside 5'-diphosphates as effectors of mammalian ribonucleotide reductase, *J. Biol. Chem.,* 260, 12001, 1985.

74. **Averett, D. R., Lubbers, C., Elion, G. B., and Spector, T.,** Ribonucleotide reductase induced by herpes simplex type 1 virus. Characterization of a distinct enzyme, *J. Biol. Chem.,* 258, 9831, 1983.

75. **Lankinen, H., Gräslund, A., and Thelander, L.,** Induction of a new ribonucleotide reductase after infection of mouse L cells with pseudorabies virus, *J. Virol.,* 14, 893, 1982.

76. **Slabaugh, M. B. and Mathews, C. K.,** Vaccina virus-induced ribonucleotide reductase can be distinguished from host cell activity, *J. Virol.,* 52, 501, 1984.

77. **Ingemarson, R. and Lankinen, H.,** The herpes simplex virus type 1 ribonucleotide reductase is a tight complex of the type composed of 40K and 140K proteins, of which the latter show multiple forms due to proteolysis, *Virology,* 156, 417, 1987.

78. **Nilsson, O., Lundqvist, T., Hahne, S., and Sjöberg, B.-M.,** Structure-function studies of the large subunit of ribonucleotide reductase from *Escherichia coli, Biochem. Soc. Trans.,* 16, 91, 1988.

79. **Lin, A.-N.I., Ashley, G. W., and Stubbe, J.,** Location of the redox-active thiols of ribonucleotide reductase: sequence similarity between the *Escherichia coli* and *Lactobacillus leichmannii* enzymes, *Biochemistry,* 26, 6905, 1987.

80. **Kierdaszuk, B. and Eriksson, S.,** Direct photoaffinity labeling of ribonucleotide reductase from *Escherichia coli* using dTTP: characterization of the photoproducts, *Biochemistry,* 27, 4952, 1988.

81. **Walker, J. E., Saraste, M., Runswick, M. J., and Gay, N. J.,** Distantly related sequences in the α- and β-subunits of ATP synthetase, myosine, kinases and other ATP-requiring enzymes and a common nucleotide binding fold, *EMBO J.,* 1, 945, 1982.

82. **Platz, A., Karlsson, M., Hahne, S., Eriksson, S., and Sjöberg, B.-M.,** Alterations in intracellular deoxyribonucleotide levels of mutationally altered ribonucleotide reductase in *Escherichia coli, J. Bacteriol.,* 164, 1194, 1985.

83. **Sjöberg, B.-M., Eklund, H., Fuchs, J. A., Carlsson, J., Standart, N. M., Ruderman, J. V., Bray, S. J., and Hunt, T.,** Identification of the stable free radical tyrosine residue in ribonucleotide reductase, *FEBS Lett.,* 183, 99, 1985.

84. **Larsson, Å. and Sjöberg, B.-M.,** Identification of the stable free radical tyrosine residue in ribonucleotide reductase, *EMBO J.,* 5, 2037, 1986.

85. **Sjöberg, B.-M., Karlsson, M., and Jörnvall, H.,** Half-site reactivity of the tyrosyl radical of ribonucleotide reductase from *Escherichia coli, J. Biol. Chem.,* 262, 9736, 1987.

86. **Cohen, E. A., Gaudreau, P., Brazeau, P., and Langlier, Y.,** Specific inhibition of herpes virus ribonucleotide reductase by a nonapeptide derived from the carboxy terminus of subunit 2, *Nature (London),* 321, 441, 1986.

87. **Dutia, B. M., Frame, M. C., Subak-Sharpe, J. H., Clark, W. N., and Marsden, H. S.,** Specific inhibition of herpesvirus ribonucleotide reductase by synthetic peptides, *Nature (London),* 321, 439, 1986.

88. **Eliasson, R., Jörnvall, H., and Reichard, P.,** Superoxide dismutase participate in enzymatic formation of the tyrosine radical of ribonucleotide reductase from *Escherichia coli, Proc. Natl. Acad. Sci. U.S.A.,* 83, 2373, 1986.

89. **Fontencave, M., Eliasson, R., and Reichard, P.,** NAD(P)H:Flavin oxidoreductase of *Escherichia coli:* a ferric iron reductase participating in the generation of the free radical of ribonucleotide reductase, *J. Biol. Chem.,* 262, 12325, 1987.

90. **Gott, J. M., Shub, D., and Belfort, M.,** Multiple self-splicing introns in bacteriophage T4: evidence from autocatalytic GTP labeling of RNA in vitro, *Cell,* 47, 81, 1986.

91. **Chu, F. K., Maley, G. F., West, D. K., Belfort, M., and Maley, F.,** Characterization of the intron in the phage T4 thymidylate synthase gene and evidence for its self-excision from the primary transcript, *Cell,* 45, 157, 1986.

92. **Eriksson, S. and Martin, D. W., Jr.,** Ribonucleotide reductase in cultured mouse lymphoma cells. Cell cycle dependent variation in the activity of subunit protein M2, *J. Biol. Chem.,* 256, 9436, 1981.

93. **Eriksson, S., Gräslund, A., Skog, S., Thelander, L., and Tibukait, B. C.,** Cell cycle dependent regulation of mammalian ribonucleotide reductase. The S phase-correlated increase in subunit M2 is regulated by de novo protein synthesis, *J. Biol. Chem.,* 259, 11695, 1984.

94. **Engström, Y., Eriksson, S., Jildevik, I., Skog, S., Thelander, L., and Tribukait, B.,** Cell cycle dependent expression of mammalian ribonucleotide reductase. Differential regulation of the two subunits, *J. Biol. Chem.,* 260, 9114, 1985.

95. **Standart, N.M., Hunt, T., and Ruderman, J. V.,** Differential accumulation of ribonucleotide reductase subunits in clam oocytes: the large subunit is stored as a polypeptide, the small subunit as untranslated mRNA, *J. Cell. Biol.,* 103, 2129, 1986.

96. **Wright, J. A., Alam, T. G., McClarlty, G. A., Tagger, A. Y., and Thelander, L.,** Altered expression of ribonucleotide reductase and role of M2 gene amplification in hydroxyurea-resistant hamster, mouse, rat and human cell lines, *Somatic Cell Mol. Genet.,* 13, 155, 1987.

97. **McClarlty, G. A., Chan, A. K., Engström, Y., Wright, J. A., and Thelander, L.,** Elevated expression of M1 and M2 components and drug-induced posttranscriptional modulation of ribonucleotide reductase in a hydroxyurea-resistant mouse cell line, *Biochemistry,* 26, 8004, 1987.

98. **Engström, Y., Rozell, B., Hansson, H. A., Stemme, S., and Thelander, L.,** Localization of ribonucleotide reductase in mammalian cells, *EMBO J.,* 3, 863, 1984.

99. **Reichard, P.,** Interactions between deoxyribonucleotide and DNA synthesis, *Annu. Rev. Biochem.,* 57, 349, 1988.

100. **Martin, D. W., Jr. and Gelfand, E.,** Biochemistry of diseases of immunodevelopment, *Annu. Rev. Biochem.,* 50, 845, 1981.

101. **Nicander, B. and Reichard, P.,** Dynamics of pyrimidine deoxynucleoside triphosphate pools in relationship to DNA synthesis in 3T6 mouse fibroblasts, *Proc. Natl. Acad. Sci. U.S.A.,* 80, 1347, 1983.

102. **Nicander, B. and Reichard, P.,** Evidence for the involvement of substrate cycles in the regulation of deoxyribonucleoside triphosphate pools in 3T6 cells, *J. Biol. Chem.,* 260, 9216, 1985.

103. **Ullman, B., Gudas, L. J., Clift, S. M., and Martin, D. W., Jr.,** Isolation and characterization of purine-nucleoside phosphorylase-deficient T-lymphoma cells and secondary mutants with altered ribonucleotide reductase: genetic model for immunodeficiency disease, *Proc. Natl. Acad. Sci. U.S.A.,* 76, 1074, 1979.

104. **Ullman, B., Gudas, L. J., Caras, I. W., Eriksson, S., Weinberg, G. L., Wormsted, M. A., and Martin, D. W., Jr.,** Demonstration of normal and mutant protein M1 subunit of deoxy-GTP resistant ribonucleotide reductase from mutant mouse lymphoma cells, *J. Biol. Chem.,* 256, 10189, 1981.

105. **Eriksson, S., Gudas, L. J., Clift, S. M., Caras, I. W., Ullman, B., and Martin, D. W., Jr.**, Evidence for genetically independent allosteric regulatory domains of the protein M1 subunit of mouse ribonucleotide reductase, *J. Biol. Chem.*, 256, 10193, 1981.

105a. **Caras, I. W. and Martin, D. W., Jr.**, Molecular cloning of the cDNA for a mutant mouse ribonucleotide reductase M1 that produces a dominant mutator phenotype in mammalian cells, *Mol. Cell. Biol.*, 8, 2698, 1988.

106. **Meuth, M. and Green, H.**, Alterations leading to increased ribonucleotide reductase in cells selected for resistance to deoxyadenosine, *Cell*, 3, 367, 1974.

107. **Meuth, M.**, The molecular basis of mutations induced by deoxyribonucleoside triphosphate pool imbalances in mammalian cells, *Exp. Cell Res.*, in press, 1988.

108. **Weinberg, G., Ullman, B., and Martin, D. W., Jr.**, Mutator phenotypes in mammalian cell mutants with distinct biochemical defects and abnormal deoxyribonucleoside triphosphate pools, *Proc. Natl. Acad. Sci. U.S.A.*, 78, 2447, 1981.

109. **Dahbo, Y. and Eriksson, S.**, On the mechanism of deoxyribonucleoside toxicity in human T-lymphoblastoid cells. Reversal of growth inhibition by addition of cytidine, *Eur. J. Biochem.*, 150, 429, 1985.

110. **Jamieson, A. T. and Bjursell, G.**, Deoxyribonucleoside triphosphate pools in herpes simplex type 1 infected cells, *J. Gen. Virol.*, 31, 101, 1976.

111. **Preston, V. G., Palfreyman, J. W., and Dutia, B.**, Identification of a herpes simplex type 1 polypeptide which is a component of the virus-induced ribonucleotide reductase, *J. Gen. Virol.*, 65, 1457, 1984.

112. **Nutter, L. M., Grill, S. P., and Cheng, Y.-C.**, The source of thymidine nucleotides for virus DNA synthesis in herpes simplex virus type 2-infected cells, *J. Biol. Chem.*, 260, 13272, 1985.

113. **Warner, H. R.**, Properties of ribonucleoside diphosphate reductase in nucleotide-permeable cells, *J. Bacteriol.*, 115, 18, 1973.

114. **Eriksson, S.**, Ribonucleotide reductase from *Escherichia coli:* demonstration of a highly active form of the enzyme, *Eur. J. Biochem.*, 56, 289, 1975.

115. **Eriksson, S., Sjöberg, B.-M., Hahne, S., and Karlström, O.**, Ribonucleoside diphosphate reductase from *Escherichia coli.* An immunological assay and a novel purification from an overproducing strain lysogenic for phage λdnrd, *J. Biol. Chem.*, 252, 6132, 1977.

116. **Mathews, C. K.**, Enzymatic channeling of DNA precursors, in *Genetic Consequences of Nucleotide Pool Imbalances,* de Serres, F. J., Ed., Plenum Press, New York, 1985, 47.

117. **Mathews, C. K. and Salbaugh, M. B.**, Eukaryotic DNA metabolism. Are deoxyribonucleotides channeled to replication sites?, *Exptl. Cell Res.*, 162, 285, 1986.

118. **Reddy, G. V. P. and Pardee, A. B.**, Multienzyme complex for metabolic channeling in mammalian DNA replication, *Proc. Natl. Acad. Sci. U.S.A.*, 77, 3312, 1980.

119. **Reddy, G. V. P. and Pardee, A. B.**, Coupled ribonucleoside diphosphate reduction channeling and incorporation into DNA of mammalian cells, *J. Biol. Chem.*, 257, 12526, 1982.

120. **Leeds, J. M., Slabaugh, M. B., and Mathews, C. K.**, DNA precursor pool and ribonucleotide reductase activity: distribution between nucleus and cytoplasm of mammalian cells, *Mol. Cell. Biol.*, 5, 3443, 1985.

121. **Joelson, T., Uhlin, U., Eklund, H., Sjöberg, B.-M., Hahne, S., and Karlsson, M.**, Crystallization and preliminary crystallographic data of ribonucleotide reductase protein B2 from *Escherichia coli, J. Biol. Chem.*, 259, 9076, 1984.

Chapter 9

ALLOSTERIC INTERACTIONS IN DEOXYCYTIDYLATE AMINOHYDROLASE (EC 3.5.4.12)

Edward P. Whitehead and Mosè Rossi

TABLE OF CONTENTS

I. SUMMARY

Inhibition of deoxycytidylate aminohydrolase (dCMPase) by dTTP, and its activation by dCTP, are mechanisms controlling the balance of cellular concentrations of the two pyrimidine deoxynucleoside 5'-triphosphate DNA precursors. The specific binding sites for the triphosphates on dCMPase are distinct from the substrate binding sites. The enzyme is a hexamer with six binding sites for substrate and six for each allosteric modifier. Binding and kinetic studies show that the interactions are typical of a simple "K system" in which the allosteric inhibitor can completely displace substrate or activator. The interactions are consistent with a scheme of "exclusive binding" of dTTP to one conformation, and dCMP, dCTP, and also the substrate analog dAMP, to another. Deoxynucleoside-5'-monophosphates (except dAMP) can apparently bind to both these conformations. Conformational studies make certain the existence of conformations of the enzyme besides these two, but their role in the allosteric interactions is not clear. The cooperativity is consistent only with a "sequential" conformational transition model; models based on symmetrical transitions or association-dissociation equilibria can be excluded. By treatment of the enzyme with glutaraldehyde in the presence of either dCTP or dTTP, conformationally frozen enzyme, with the properties expected of an enzyme permanently fixed in the conformations induced by these modifiers, can be obtained. The behavior of nonnatural substrates such as 5-aza-dCMP, arabinosyl cytidine 5'-monophosphate, dCMP-Hg-SCH$_2$CH$_2$OH, and CMP is more complex than that of dCMP, and allosteric modifiers exert "V effects" as well as "K effects" on them.

II. INTRODUCTION

Deoxycytidylate aminohydrolase (dCMPase), catalyst of the reaction:

$$dCMP \rightarrow dUMP + NH_3 \qquad (1)$$

is widely distributed in the animal kingdom (Reference 1 and references therein). Its physiological role does not seem to have been definitively settled. It is not indispensible to life since *Eschericia coli*[2] lacks it, as do some mutant human and hamster cells,[3,4] but it is found in larger amounts in all rapidly growing tissues.[5] The enzyme enables dTTP to be formed from dCMP precursor according to the metabolic scheme:

$$
\begin{array}{c}
dCMP \longrightarrow dCDP \longrightarrow dCTP \\
\end{array}
\qquad (2)
$$

dCMPase

dCMP ⟶ dCDP ⟶ dCTP

activation

inhibition

dUMP ⟶ dTMP ⟶ dTDP ⟶ dTTP

However, dTTP can also be formed without the intervention of dCMPase from UDP precursor via UDP reductase. Of key significance is probably the fact of the activation of dCMPase by dCTP and its inhibition by dTTP at micromolar concentrations, effects with whose physicochemical mechanism this review will be largely concerned. The most likely guess is that, thanks to these effects, dCMPase permits optimal regulation of the balance of dCTP and dTTP concentrations. In support of this supposition, mutant cells deficient in dCMPase have been shown to have a reduced dTTP pool,[3,4] become thymidine-auxotrophic when cytidine nucleotide pools increase[4] (due apparently to CTP inhibition of UDP reductase), and have an increased spontaneous mutation rate due undoubtedly to a known mechanism,[6-8] whereby the error rate of DNA polymerase increases when the ratios of deoxytri-phosphate concentrations are unbalanced. This regulatory role of an enzyme is probably important for chemotherapy.[9]

It was found[1] that the best source for preparation of the homogeneous enzyme was donkey spleen, and all the work reviewed here refers to enzyme from this source. However, the human,[10] chick, and T2 phage[11] dCMPases have controlling modifier interactions and molecular weights very similar to the donkey enzyme. This similarity is known to extend to the binding site numbers in the case of the T2 enzyme and to subunit composition in the T2 and chick enzyme.[11]

This review will emphasize the information obtained from steady-state kinetic and binding studies. These studies are admittedly very far from being complete at the present moment. Where sufficient appropriate data are available, we have used Linkage analysis, i.e., one based on the Linkage Equations of Wyman[12,13] for their interpretation. The advantages of this approach for allosteric enzyme kinetics have been discussed elsewhere.[14,15] Implicit in its use is the assumption that the catalytic steady-state can be treated as an equilibrium; the justification of this assumption is that it seems to work, allowing quite simple schemes to account for the kinetics. A few of these schemes have been subjected to quite exacting tests. For many of the phenomena, however, particularly those found with nonnatural substrates, critical kinetic tests of our hypotheses have not yet been applied. It is nevertheless useful to propose simple provisional schemes that explain at least the available data.

III. SALIENT PROPERTIES OF DEOXYCYTIDYLATE AMINOHYDROLASE

Catalysis of Reaction 1 by dCMPase does not require addition of metal or other cofactors to the reaction mixture, and no evidence has been found for the involvement of pyridoxal phosphate or other cofactor or prosthetic group. Metal ions are, however, required for the action of the main allosteric effectors with which we shall be concerned here: these are activation by dCTP and inhibition by dTTP.[16] Magnesium ions were used in all the work reviewed here; we shall, for convenience, talk of the effects of "dTTP" where their magnesium complexes should be understood. We recall that although dCMPase is embedded in a somewhat complicated set of reactions leading to deoxytriphosphate precursors of DNA, the biological sense of these two effects is broadly evident from Equation 2.

The only substrate known to be hydrolyzed at an efficiency comparable to that of dCMP is 5-methyl-dCMP, a slightly better substrate than dCMP. The various effectors of dCMPase produce effects very similar with this substrate to those with dCMP.

The molecular weight of the enzyme is about 110 kda. It is composed of six, apparently identical polypeptide chains of molecular weight 18 kda. This conclusion comes from cross-linking the subunits with glutaraldehyde followed by dissociation in sodium dodecyl sulfate (SDS), whereupon electrophoretic bands corresponding to monomer, dimer, trimer, tetramer, pentamer, and hexamer are observed.[17]

Although dCMPase was, to our knowledge, the second enzyme (after aspartate transcarbamylase) in which direct physical studies showed the allosteric and substrate sites to be distinct (see Section IV), the tertiary structure of dCMPase is still quite unknown. Crystals suitable for X-ray diffraction have not been obtained. About a third of the sequence has been determined.[18]

IV. BINDING SITES AND BINDING STUDIES

The number of binding sites per enzyme molecule for the allosteric effectors and for substrate analogs was originally estimated, using the Hummel and Dreyer gel filtration method, as four.[19] This figure does not tally with a hexameric structure, and the question has been reinvestigated using equilibrium dialysis for binding measurements.[20] Experiments with one effector present alone have shown that, at saturation, a molecule of dCMPase in fact binds six molecules of an allosteric effector, dCTP or dTTP, or six of the substrate analog, dAMP. The error in the earlier estimates was caused simply by a colorimetric method giving a misestimation of protein concentration.

Two points are important about the earlier[19] binding site work. The first is that the figure of four is quoted in almost all our references, including those on -SH groups, and this should be adjusted in reading any of these. But, secondly, other than the absolute figures, almost none of the conclusions of these papers, which mostly depend on ratios, are altered by the new values.

Thus, measurements of binding both of deoxytriphosphate modifiers and of substrate analogs (deoxynucleoside-5'-monophosphates) when present together established that the modifier binding sites are distinct from the substrate sites. At $6\mu M$ dGMP, a rather low amount of this nucleotide was bound. However, when dTTP and dGMP were present together, roughly equivalent amounts of each were bound, almost as much of each as of dTTP bound when alone. Similarly, when dCTP and dAMP were together, they were bound in amounts equivalent to each other, and to the amount of dCTP bound when alone. It is expected that dGMP and dAMP be bound at the substrate sites, and this supposition is consistent with the kinetic and other evidence.

The above experiments then very strongly indicate that (independently of the absolute number of sites) the binding site for substrate is distinct and separate from the site(s) for the triphosphate modifiers, and that the ratio of modifier (dTTP or dCTP) sites to substrate sites is 1:1.

Other experiments in this series showed antagonisms between pairs of effectors which are quite consistent with kinetic data (see Section V.B). Thus, dTTP antagonizes the binding of dAMP, and dCTP that of dGMP.[19] As expected, the allosteric activator dCTP totally displaces the allosteric inhibitor dTTP. Whether the dCTP site is identical to or distinct from the dCTP site cannot be concluded from this data and is not yet known.

V. KINETICS

The pattern of interactions between the natural substrate dCMP, inhibitor, dTTP, and activator dCTP is rather simple,[16] while the kinetics involving nonnatural substrates reveals a wider variety of phenomena.

A. Kinetics with the Natural Substrate and Modifiers

At any concentration of dCMP, a high enough concentration of dTTP is able to inhibit completely. The inhibition can always in turn be overcome by increasing the dCMP concentration. The action of dCTP is to increase activity at nonsaturating levels of substrate (of course these levels correpsond to higher dCMP concentrations when dTTP is present). If substrate is already saturating, dCTP has scarcely any effect on activity. Likewise, increasing either dCTP or dTTP concentrations overcomes the effect of the other. The velocity, V, obtained at sufficiently high dCMP concentration, then, is independent of dCTP or dTTP concentration.

All of this is the typical behavior of a classical K system, i.e., one that behaves as if the only effects of the allosteric modifiers were on substrate binding and not on catalytic efficiency.

Moreover, it is a particularly simple example of such a system, so far as concerns the three main biological effectors, in that the inhibitor can completely displace both substrate and activator, while these last two can only assist each others binding and never displace each other. The binding studies[19] have shown the complete mutual displacement of activator and inhibitor. Substrate-inhibitor interactions cannot be studied in this way, but the general similarity of inhibitor-activator interaction with substrate-inhibitor interaction, and various other strong arguments developed below point convincingly to the ability of dCMP and dTTP to displace each other completely. The effects of nucleoside monophosphates are less simple and absolute except, apparently for dAMP (see Section V.B). With some nonnatural substrates, behavior is also more complex and there are V effects as well as K effects (see Section V.C).

Given the above facts and that the activator and inhibitor site(s) is (are) distinct from the substrate site it is natural to formulate an exclusive binding model for the enzyme, i.e., a model in which dCTP and dCMP are bound exclusively to one conformation and dTTP exclusively to another. The completeness of inhibition by dTTP would not itself exclude that dCMP can be bound to the dTTP binding conformation if this is supposed catalytically inactive. However, in that case, V would be dependent on modifier concentration, contrary to experimental observation. In the postulated exclusive binding mechanism, the kinetics and binding interactions would be exactly the same whether the dCTP and dTTP sites were identical or not. In fact, the kinetic and binding data give us no clue on this point and one can only guess that identity of these sites is more likely, since three specific sites are rather a lot to accommodate on an 18-kda polypeptide.

The cooperative phenomena are also typical of what is expected for a simple exclusive binding K system — in fact, dCMPase was the first enzyme for which these, now classical, effects were observed. The substrate concentration-velocity curve in the normal assay conditions in the absence of allosteric effectors is approximately described as a curve of Hill slope (h) in the range 2 to 2.6 — moderate cooperativity. dCTP decreases this cooperativity and at full activation abolishes it. This is interpreted as all the enzyme being stabilized in the substrate binding conformation by the activator so that there are no conformational transitions to cause cooperativity. Thus, it is suggested, though not proved, that the dCTP and dCMP binding conformations are identical; this supposition is consistent with all evidence to date, in particular with kinetic observations with substrate analogs (see Section V.B). On the other hand, dTTP increases substrate cooperativity. It shifts the curves indefinitely to the right, but at about 2 μM dTTP a limiting value of about 4 for h was observed.[16] Further increase of dTTP concentration shifted the curve futher to the right without further increase in cooperativity, i.e., the curves beyond this point (in Hill plots or plots of v (see Table 1) against logarithm of substrate concentration) are parallel.

The inhibitor-induced increase of substrate cooperativity is understandable in terms of a minimal exclusive binding model

$$RX \rightleftharpoons R \rightleftharpoons T \rightleftharpoons TY \qquad (3)$$

As inhibitor concentration is increased, favoring the T conformation, the binding of a given amount of x involves an increasing amount of cooperative $T \rightarrow R$ transformation, and thus of free energy of interaction. (It is not necessary to this explanation that the $R \rightleftharpoons T$ transformation be the symmetric transformation proposed by Monod, Wyman, and Changeux;[21] it suffices that it be positively cooperative. Other details, such as whether or not R or T are really one single conformation, or whether there are other nonbinding conformations, are also inessential to the qualitative phenomenon.)

When substrate and inhibitor concentrations are both quite high, so that there are practically no subunits without either a substrate or an inhibitor molecule bound, the exclusive binding model of Equation 3 simplifies to

$$RX \rightleftharpoons TY \qquad (4)$$

Thus, each molecule of substrate or inhibitor bound in these conditions should displace one molecule of its antagonist, i.e., $dX/dY = -1$. The experimental fact of the invariance of cooperativity in these conditions indeed shows that dX/dY is *constant*, independent of X.

For the fact of the curves being parallel is equivalent to saying that $(\partial \ln x / \partial \ln y)_x$ is independent of X. Hence, by the Linkage Equation 5, $(\partial Y / \partial X)_y$ is independent of X. To show that one molecule that is bound really does displace one molecule of its antagonist in our system, it is very convenient to note that $(\partial X / \partial Y)_y$ is equal simply to the ratio of Hill coefficients in the substrate saturation curve to that in the inhibition curve. For the thermodynamic Linkage equation is[12,13]

$$\left(\frac{\partial X}{\partial Y}\right)_y = -\left(\frac{\partial \ln y}{\partial \ln x}\right)_x \tag{5}$$

If we assume that v depends only on X, then the right band side of Equation 5 is $-(\partial \ln y / \partial \ln x)_v$, which equals

$$\frac{\partial \log[v/(V - v)] / \partial \log x}{\partial \log[v/(V - v)] / \partial \log y} \tag{6}$$

by a general mathematical relationship. The numerator of Equation 6 is by definition the Hill coefficient of substrate; the denominator in the present circumstances where v at y = 0 is V, and at high y v = 0 corresponds to the usual definition of the Hill coefficient for inhibition (except that we find it convenient to let this be negative for an inhibitor). Thus,

$$(\partial X / \partial Y)_y = h_x / h_y$$

When, as is true of Equation 4 the degree of inhibition depends uniquely on the amount of Y bound, then we can similarly obtain

$$(\partial X / \partial Y)_x = h_x / h_y$$

so we shall write these results simply as

$$dX/dY = h_x / h_y \tag{7}$$

Now experimentally $h_x = 4$, and $h_y = -4$[14] in these conditions, so by Equation 7, dX/dY = 4/-4 = -1, as the exclusive binding Equation 4 predicts.

Equation 7 is useful in understanding a new phenomenon found[22] when dTTP concentration is extended beyond those of Reference 16. h_x further increases up to a new level of about eight (very strong cooperativity) at around 60 μM dTTP. However, h_y remains -4 even at very high substrate concentrations. Thus, dX/dY = 8/-4 = -2, i.e., the equilibrium becomes

$$\text{E-dCMP}_2 + \text{dTTP} \rightleftharpoons \text{E-dTTP} + 2 \text{ dCMP}$$

or

$$\text{R-dCMP}_2 + \text{dTTP} \rightleftharpoons \text{T-dTTP} + 2 \text{ dCMP}$$

at these high concentrations. There must be a second site on the polypeptide chain capable of binding substrate. This second site is noncatalytic, because V is no different at these high substrate concentrations than at the lower concentrations. It can be surmised, given the resemblance between dCMP and dCTP, that the inactive dCMP binding site is identical to the dCTP binding site. This leads to a prediction. In the presence of dCTP, which would occupy the second site and prevent substrate binding, there the equilibrium should become

$$\text{E-dCMP-dCTP} + \text{dTTP} \rightleftharpoons \text{E-dTTP} + \text{dCTP} + \text{dCMP}$$

and by Equation 7, h_x/h_y should again become -1. Experimentally, it has been found that dCTP does reduce h_x to 4, while h_y remains -4, so this prediction of the model is verified.

The model also makes some obvious predictions about the cooperativity of the activator, e.g., that at high dTTP concentrations, $(h_{dCMP} + h_{dCTP}) = -2h_{dTTP} = 8$. In experiments done since the above section was written, we found[22] this relation does not hold. This shows both the need for a more complicated model of substrate-activator interaction than that needed to explain the data reviewed in this chapter, and at the same time, the usefulness of Equation 7 for tests of allosteric models with minimal assumptions.

The structural similarity of the substrate and activator, and the binding of substrate at the activator site lead to the speculation that the activator site has been derived from the substrate site of dCMPase in the course of evolution through gene duplication. Evolution of an allosteric site in this way in another enzyme, phosphofructokinase, has been proposed on the basis of its structure.[23] Evolution of allosteric control sites from substrate sites, in many cases from the substrate sites, in many cases from the substrate sites that they control, is a very plausible hypothesis for their origin.

Another conclusion that follows from the results reviewed is that the Monod-Wyman-Changeux (MWC) model of symmetric transitions can be excluded. Such a model predicts[24] that at high concentrations the substrate and inhibitor cooperativity should be the maximum possible, i.e., $h_x = -h_y = n$, the number of binding sites.

This is easily understood. The Hill model assumes the existence of only two forms of the enzyme, EX_o and EX_n. The MWC model assumes the existence of only two conformational forms: R_oT_n and R_nT_o. R_oT_n cannot bind substrate, while at high substrate concentrations, all the n substrate sites of R_nT_o will be occupied by substrate, so under these conditions the Hill and MWC assumptions become practically identical.

To reinforce this with some formulae, at high substrate concentration ($Kx \gg 1$) the binding equation of the MWC model for exclusive binding:

$$\frac{X}{n} = \frac{Kx(1 + Kx)^{n-1}}{(1 + Kx)^n + L(1 + K'y)^n}$$

assumes the form of Hill's equation

$$\frac{X}{n} = \frac{x^n}{A + x^n}$$

where $A = L(1 + K'y)^n/K^n$. Also, at constant x, when $Kx \gg 1$ and $K'y \gg 1$,

$$\frac{X}{n - X} = By^{-n} + 1$$

where $B = L(Kx/K')^n$, so

$$h_y = \frac{d \log[X/(n - X)]}{d \log y} = \frac{d \log[v/(V - v)]}{d \log y} = -n$$

When only one site per subunit is involved, it has been found that the limiting h is 4 for substrate (or -4 for inhibitor), whereas $n = 6$ as discussed previously. Thus, the MWC model is rejected as inconsistent with the kinetic cooperativity observations. This rejection would be unaffected by the existence of a third conformation suggested by conformational probes (see Section VI) binding neither substrate nor inhibitor. Such a conformation would not be significantly present at the concentrations of substrate and inhibitor where cooperativity becomes invariant.

Models depending on association-dissociation equilibria of the oligomeric protein can also be excluded. In the conditions of the kinetic experiments described, dependence of catalytic velocity on enzyme concentration is linear, and no other effects of enzyme concentration on kinetics have been observed.

We are left then with a model of sequential conformational transitions, allowing structures of the stable hexameric enzyme with some subunits in the inhibitor binding and others in the substrate and activator binding conformations. It is not possible with data of the resolution presently available to say anything more than this about the subunit interactions.

B. Interactions of Substrate Analogs and Allosteric Effectors

Deoxynucleoside-5'-monophosphates would be expected to be competitors for substrate sites, and indeed all the natural dNMPs are inhibitors of dCMPase. The implication of binding at substrate sites is strongest of all for dTMP and dUMP, products of the enzymatic reaction according as 5-CH$_3$-dCMP or dCMP is substrate. The detailed kinetics of their interactions with substrate and allosteric effectors is different for different analogs.[25]

Inhibition by dGMP, as well as by 2'-3' GMP, is potentiated by dTTP. This can be explained by supposing that these analogs bind preferentially to the dTTP binding conformation (the ability of dGMP and dTTP to bind together to dCMPase was also noted in Section IV). dAMP on the other hand behaves as if it had predominantly affinity for the activator and substrate binding conformation (again this has been seen in the binding as well as the kinetic studies). One of the properties of a substrate analog that does this is its ability at low substrate concentrations to activate an enzyme with cooperative substrate kinetics, although at higher concentrations inhibition is manifest. This phenomenon has been known, and in outline explained, since the studies of Douglas, Haldane, and Haldane[26] in 1912. The activation results from positive interactions of bound competitor molecules with substrate molecules bound at other substrate sites. At higher concentrations, the effects of competition between substrate and analog for substrate sites predominate.

dAMP is the analog that displays the greatest degree of activation, though dTMP and dUMP also activate to some extent. dAMP is also the analog that has been studied in most detail with a view to examining a detailed allosteric mechanism.[15] The quantitative kinetics, analyzed using a linkage approach, accorded with the hypothesis that dCMP and dAMP are bound to the same site in one and the same conformation.

The kinetic approach was the following: it can be shown using linkage equations or otherwise[14,27,28] that the above hypothesis predicts the relation:

$$v/V = \frac{x\, f(x + ay)}{(x + ay)} \tag{8}$$

where a is the ratio between the association constants of analog and substrate for binding to the substrate binding conformation.

Equation 8 can be experimentally tested. In the absence of the competitor it reduces to

$$v/V = f(x) \tag{9}$$

and thus the function f can be *known empirically*; it is the substrate concentration-velocity curve determined in the absence of the competitor. If then a is known, v/V at any combination of x and y can be predicted from Equation 8. For this purpose, it is not necessary to have any model or formula giving f. The ratio a can be obtained by ordinary inhibition experiments at high concentrations where by Equation 9 f = 1; hence,

$$v/V = x/(x + ay) \tag{10}$$

ordinary hyperbolic inhibition. This was observed for the dCMP-dAMP pair. The value of a usually obtained was about 5, i.e., substrate sites have greater affinity for dAMP than for dCMP. An important point about f is that one can change it profoundly by introducing a new effector into the system. One needs no knowledge of the mechanism of action of the new effector — it does not even matter whether or not this binds to substrate sites. One needs merely to redetermine f, i.e., to simply measure the curve of v against x in the new conditions. It was found[15] that Equation 8 adequately predicted the effects of dAMP at various dCMP concentrations, not only in the absence of other effectors, but also in the presence of dUMP, dTTP, and (a special case) dCTP, all of which radically change f. The constant a used for all the predictions was obtained in the absence of the third effector. Thus, the kinetics of dAMP is consistent with the hypothesis that this analog and dCMP bind to one and the same conformation which maintains its identity in the presence of other effectors.

In the presence of dCTP where, as already discussed, dCMP kinetics is Michaelian, the dAMP-dCMP interaction simplifies to classical competitive inhibition. Moreover the ratio K_m/K_i is found equal to the constant a determined, as described above, in the absence of dCTP. This is further evidence that the dCMP and dAMP binding conformation now identified is identical to the dCTP binding conformation.

The predictions of Equation 8 are certainly falsifiable, as its failure to account for the interaction of dUMP and dCMP shows;[15] for example, Equation 10 is not obeyed at high concentrations of dCMP and dUMP — the curves are somewhat cooperative. However, at lowish [dUMP]/[dCMP] ratios (corresponding roughly to v > V/2) there is little cooperativity. This is interpreted as dUMP binding predominantly to substrate sites in the substrate binding conformation at low [dUMP]/[dCMP] ratios. The cooperativity observed at higher [dUMP]/[dCMP] ratios is explained by supposing that dUMP is also able to bind to some other conformation(s) (maybe identical to the dTTP binding conformation) and thereby induce cooperative conformational transitions. (A substrate analog able to bind only to the dTTP binding conformation would give inhibition kinetics of a type indistinguishable from those of dTTP itself.)

In the presence of dCTP, the dUMP-dCMP interaction simplifies to classical competitive inhibition. The ratio K_m/K_i then found (0.77 in the experimental conditions used) is quite close to the ratio of affinities of the two monophosphates in the absence of dCTP, at high dCMP concentrations and low [dUMP]/[dCMP] ratios. Again, this suggests that the dCMP and dCTP binding conformations are identical.

C. Nonnatural Substrates

Nonnatural substrates display a greater variety of interactions than does dCMP. Their explanation generally requires more than two conformations. As we shall compare the catalytic activity of dCMPase toward these substrates with that toward dCMP, we mention here that the turnover number of the enzyme for dCMP under conditions of maximal activity is about 2000 s^{-1}.

CMP is a poor substrate of dCMPase: when present alone its affinity and its maximum aminohydrolysis rate are both about a hundredth of that of dCMP.[28] However, dCTP greatly activates the hydrolysis of CMP. This activation affects V as well as the affinity. In the presence of dCTP, excess substrate inhibition is found with CMP, unlike dCMP. dTTP is inhibitor, and dCTP overcomes the dTTP inhibition, with CMP as well as with dCMP substrate.

A minimal scheme that explains these observations is

$$T \rightleftharpoons R \rightleftharpoons R'$$

where R' is a conformation with the greater affinity for CMP, but less catalytic efficiency than R which has affinity for dCTP. T is the usual dTTP binding conformation. Thus, CMP by itself would induce R' (inefficient catalysis). dCTP would induce R (efficient catalysis of CMP hydrolysis), but increase of CMP concentration would again pull the equilibrium toward R', decreasing the catalytic efficiency as observed. Because cooperativity of CMP alone is quite similar to that of dCMP alone, it is reasonable to assume the R' conformation is rather similar to R, and that the R $\longleftrightarrow$ R' transformation is noncooperative. These explanations have not yet however been critically tested and others are possible. It cannot, for instance, be excluded that CMP binding to dCTP sites plays a role, as we shall suggest below for aza-dCMP.

The behavior of arabinosyl-cytidine 5'-monophosphate as substrate resembles that of CMP: dCTP greatly increases both the affinity and V. Excess substrate inhibition has not been observed. The explanation can be similar to that for CMP except that the difference in affinity between the two substrate binding conformations would have to be less extreme.

5-aza-dCMP is a dCMPase substrate,[30] a fact that may have pharmacotherapeutic and cellular biological implications. Its maximum deamination rate is again about a hundredth of that of dCMP; however, its affinity for the enzyme is high. The kinetics in the absence of modifiers is apparently noncooperative with K_m approximately 0.1 mM. The hydrolysis of this substrate is inhibited by dTTP, and the inhibition is reversed by dCTP, effects similar to those with dCMP substrate. However, in the absence of dTTP, dCTP only weakly activates aza-dCMP hydrolysis. 5-aza-dCTP has effects very similar to those of dCTP with both dCMP and 5-aza-dCMP substrates. The effect of 5-aza-dCMP on dCMP hydrolysis (which can be assayed in the presence of the aza-dCMP) was activatory at low concentrations and inhibitory at high concentrations, reminiscent of the effects of dAMP, except that the aza-dCMP activation is much greater than that by dAMP.

It is not easy to rationalize this set of results in a simple scheme, particularly one that explains the lack of substrate cooperativity and the limited activation by activators. One rationalization would be to suppose that aza-dCMP can bind to the activator site and mimic activator, just as dCMP can (see Section V.A), but this time with higher affinity for the activator than for the substrate site. This and many other questions arising from this work could be and deserve to be, particularly in view of the applications, critically tested by kinetic methods, which has not yet been done as the substrate is not readily available.

Aza-deoxycytidine (aza-CdR) is widely used as an agent for producing cells in which cytosine bases of DNA are undermethylated.[31] This depends on the conversion of the aza-CdR to aza-dCTP which is incorporated into DNA in competition with dCTP. The incorporated 5-aza-cytosine of the DNA causes irreversible inactivation of DNA methyltransferase. The hypomethylation of the DNA changes the pattern of gene expression in the cells. An antileukemic action of aza-CdR also appears to depend on incorporation into DNA.[32] It was pointed out[9,30] that the kinetic effects described above could be a basis for a self-potentiating effect of aza-CdR. Aza-dCMP and aza-dCTP, by stimulating dCMPase which is a 100-fold more active on dCMP than on aza-dCMP, could drain the pool of the competitor (dCTP) for aza-dCTP incorporation into DNA.

Another very interesting substrate is dCMP-5-Hg-SCH$_2$CH$_2$OH. The normal effects of the allosteric modifiers are reversed with this substrate: dTTP activates and dCTP inhibits.[33] Both activation and inhibition are mainly effects on V, and there does not seem to be much effect on substrate affinity. So the behavior is as if there were a relatively small difference in the affinity of this substrate between the dCTP binding conformation and the dTTP binding conformation, the catalytic efficiency of the former being larger than that of the latter.

VI. STUDIES OF CONFORMATIONS

A. Optical and Chemical Studies

Optical and chemical properties have been exploited to investigate conformations of dCMPase.[34,35] The parameters investigated were the optical absorbance the fluorescence, the reactivity of the -SH groups, and susceptibility to proteases.

The idea of the experiments is that the simplest models of allosteric interaction postulate two preexisting conformations of an enzyme. Homotropic and heterotropic effects result from shifting the equilibrium between these conformations. Such a simple model predicts that the properties of unliganded enzyme be intermediate between those of allosterically activated and inhibited enzyme (in our case, operationally, the enzyme in the presence of dCTP and dTTP, respectively).

We will only detail the results with -SH group reactivity, because of its intrinsic interest and because the figures need revising in the light of the fresh knowledge of the enzyme's subunit composition (see Section III). Reinterpreting the original data then, there are per subunit three -SH groups that react fast with Ellman's reagent, 5-5′dithiobis(2-nitrobenzoate) or NbS_2 (out of six groups that react with this reagent when the enzyme is denatured in 4 M urea). dTTP protects one of these groups against reaction, dCTP protects two. Both of these modifiers protect almost completely against inactivation by NbS_2. Reaction of all three groups in the absence of the modifiers causes 80% inactivation of the enzyme.

The results of the investigations with dCMPase with all of the conformation probes mentioned above[34,35] can be summarized quite simply; for none of the parameters investigated was the unliganded enzyme's properties intermediate between those of activated and inhibited enzyme. Thus, the state of the enzyme in the absence of ligands cannot be an equilibrium between the allosterically activated and inhibited states.

However, as similar approaches have been used for other systems, the caution necessary for the interpretation of experiments of this kind deserves some discussion here. There is no question that more than two conformations are needed to explain this set of results, the question is as to the relevance of this to mechanism of allosteric action.

First, what counts for allosteric interactions of biological importance is the distribution of states in the presence of substrates. The thematic of hysteretic enzymes[36] has shown that for many enzymes, their state when stored in the absence of substrate is metastable and changes in the presence of substrate. This can be true whether or not the enzyme is observably hysteretic. That the states observable in the absence of substrate correspond to those in its presence is an assumption in the absence of experimental evidence.

The second problem is that all physical or chemical measurements like those above, and measurements of catalytic velocity too, are nothing but independent (low resolution) reporters on the state of an enzyme. Different news reporters file different reports; we do not expect them to agree in the weights they give to different things. So two conformations might have different optical properties, but identical interactions with substrate. Then what from the point of view of optical properties is two conformations, would be only one from the point of view of kinetics. From the physical point of view, different conformation reports might give equally interesting information. But from the biological point of view, it is those conformations that are kinetically distinct that are important.

Finally, when a difference in conformation induced by a ligand is demonstrated, it then needs to be known whether the conformational transition is co-operative or not. Consider the equilibrium:

$$T \rightleftharpoons R' \rightleftharpoons R''$$

in which only R' can bind a ligand. If however, the transition $R' \rightleftharpoons R''$ is noncooperative, then the set of conformations R', R'' can be treated as a single X binding conformation for the purposes of kinetics and equilibrium binding. This is true for an indefinite number of conformations. Even if a series of ligands x, y, z, . . . distribute themselves with different affinities on the different conformations R', R'', . . . , with noncooperative transitions between them and each conformation in the family has different catalytic properties, the whole set can still be treated as a single conformation R on which competitive and non-competitive interactions take place according to simple formulae.[22,27,37]

B. Conformationally Frozen Forms of the Enzyme

It has been possible to "freeze" the enzyme in either the activated or the inactivated conformation by reaction with the bifunctional reagent, glutaraldehyde (GA) in the presence of allosteric effectors or other modifiers.[17,33,38]

GA treatment in the absence of effectors inactivates the enzyme. Inclusion of dCTP during the GA treatment, however, protects against loss of activity, and the resulting enzyme has the properties expected for the enzyme frozen in the conformation induced by the activator,[17] although no activator remains bound to the protein. The properties are a nearly noncooperative substrate concentration-velocity curve with higher affinity than the native enzyme (i.e., a curve similar to what is observed with native enzyme when activated by dCTP) and insensitivity to dCTP activation and dTTP inhibition. (Some conformational flexibility remains — both triphosphates do inhibit the frozen enzyme, but at much higher concentrations than those of their normal effects on unmodified enzyme). The frozen enzyme no longer displays the activation by low dAMP concentrations discussed in a previous section, but is still inhibited by it. In other words, dAMP can still be bound at substrate sites, but then does not interact with other substrate sites.

GA treatment in the presence of dTTP likewise gives an enzyme with the properties expected for one frozen in the dTTP binding conformation.[33] This is distinguishable from merely inactivated enzyme since the dTTP binding conformation is catalytically active with the substrate dCMP-5-Hg-SCH$_2$CH$_2$OH. The properties of the enzyme obtained in this way are lack of catalytic activity toward dCMP but retention, indeed increase, of the activity toward the mercurated substrate, whose kinetics are essentially noncooperative and relative insensitivity toward dCTP and dTTP, both of which become inhibitory.

In view of the knowledge of the effects of dAMP and dGMP summarized in previous sections, it would be expected that these substances could also be used to stabilize the enzyme in forms similar to those stabilized by dCTP and dTTP during conformational fixation. In fact these monophosphates are usually included in the reaction mixture with GA and the triphosphates (dAMP with dCTP and dGMP with dTTP) to help stabilize the enzyme activity. dGMP does seem to do what is expected[17] though the properties of the resulting enzyme have not been investigated in detail; dAMP has not been investigated in this respect.

ACKNOWLEDGMENTS

The work of the reviewers has been supported by contract between the Commission of the European Communities and the University of Rome, and by the Consiglio Nazionale delle Ricerche. This is publication BIO 2313 of the Biology Division, D-G. XII, Commission of the European Communities.

Table 1
TABLE OF SYMBOLS

v	Steady-state velocity
V	Velocity at saturation by substrate
x	Substrate concentration in solution
X	Moles of bound substrate per mole of enzyme
y	Inhibitor concentration in solution
Y	Moles of bound inhibitor per mole of enzyme
h_x, h_y	Hill coefficients for substrate, inhibitor ($= d \log v/(V - v)/d \log, x,y$)
R	Substrate binding conformation of a subunit
T	Inhibitor binding concentration of a subunit
RX, TY, etc.	Subunits in R,T conformation with molecule of substrate, inhibitor bound, etc.
n	Number of effector sites

REFERENCES

1. **Geraci, G., Rossi, M., and Scarano, E.**, Deoxycytidylate aminohydrolase. I. Preparation and properties of the homogenous enzyme, *Biochemistry*, 6, 183, 1967.
2. **Keck, K., Mahler, H. R., and Fraser, D.**, Synthesis of deoxycytidine 5′-phosphate deaminase in *Eschericia coli* infected by T2 bacteriophage, *Arch. Biochem. Biophys.*, 86, 85, 1960.
3. **Weinberg, G., Buddy, U., and Martin, D. W. J.**, Mutator phenotypes in mammalian cell mutants with distinct biochemical defects and abnormal deoxyribonucleoside triphosphate pools, *Proc. Natl. Acad. Sci. U.S.A.*, 78, 2447, 1981.
4. **de Saint Vincent, B. R., Deschamps, M., and Buttin, G.**, The modulation of the thymidine triphosphate pool of Chinese hamster cells by dCMP deaminase and UDP reductase, *J. Biol. Chem.*, 255, 162, 1980.
5. **Giustl, G., Mangoni, C., De Petrocellis, B., and Scarano, E.**, Deoxycytidylate deaminase and deoxycytidine deaminase in normal and neoplastic human tissue, *Enzymol. Biol. Clin.*, 11, 375, 1970.
6. **Weymouth, L. A. and Loeb, L. A.**, Mutagenesis during *in vitro* DNA synthesis, *Proc. Natl. Acad. Sci. U.S.A.*, 75, 1925, 1978.
7. **Fersht, A. R.**, Fidelity of replication of phage φX174 DNA by DNA polymerase III holoenzyme: spontaneous mutation by misincorporation, *Proc. Natl. Acad. Sci. U.S.A.*, 76, 4946, 179.
8. **Hubner, U. and Alberts, B. M.**, Fidelity of DNA replication catalyzed *in vitro* on natural DNA template by the bacteriophage multienzyme complex, *Nature (London)*, 285, 300, 1980.
9. **Momparler, R. L., Rossi, M., Bouchard, J., Bartolucci, S., Momparler, L. F., Raia, C. A., Nucci, R., Vaccaro, C., and Sepe, S.**, 5-aza deoxycytidine synergystic action with thymidine on leukemic cells and interaction of 5-aza-dCMP with dCMP deaminase, in *Purine and Pyrimidine Metabolism in Man*, Nyhan, W. L., Thompson, L. F., and Watts, R. W. E., Eds., Plenum Press, New York, 1986, 157.
10. **Ellims, P. H. E., Kao, A. Y., and Chabner, B. A.**, Deoxycytidylate deaminase: purification and some properties of the enzyme isolated from human spleen, *J. Biol. Chem.*, 256, 6335, 1981.
11. **Maley, S. F. and Maley, F.**, Allosteric transition associated with binding of substrate and effector ligands to T2 phage-induced deoxycytidylate deaminase, *Biochemistry*, 21, 3780, 1982.
12. **Wyman, J.**, Linked functions and reciprocal effects in hemoglobin; a second look, *Adv. Protein Chem.*, 19, 223, 1964.
13. **Edsall, J. T. and Gutfreund, H.**, *Biothermodynamics*, John Wiley & Sons, New York, 1983.
14. **Whitehead, E. P.**, Subunit interactions and enzyme kinetics: the linkage approach, in *Biochemical Dynamics*, Ricard, J., and Cornish-Bowden, A., Eds., Plenum Press, New York, 1984, 45.
15. **Mastrantonio, S., Nucci, R., Vaccaro, C., Rossi, M., and Whitehead, E. P.**, Analysis of competition for substrate sites in an allosteric enzyme with co-operative kinetics. Effects of dAMP and dUMP on donkey spleen deoxycytidylate aminohydrolase, *Eur. J. Biochem.*, 137, 421, 1983.
16. **Scarano, E., Geraci, G., and Rossi, M.**, Deoxycytidylate aminohydrolase II: kinetic properties. The activatory effect of deoxycytidine triphosphate and the inhibitory effect of deoxythymidine triphosphate, *Biochemistry*, 6, 192, 1967.
17. **Nucci, R., Raia, C. A., Vaccaro, C., Sepe, S., Scarano, E., and Rossi, M.**, Freezing of dCMP aminohydrolase in the activated conformation by glutaraldehyde, *J. Mol. Biol.*, 124, 133, 1978.
18. **Carmardella, L., Raia, C. A., Cancedda, F., Nucci, R., Romano, M., Vaccaro, C., and Rossi, M.**, Studi sulla struttura primaria della dCMPasi da milza d'asino, *Congresso Società Italiana di Biochimica*, Laco Ameno d'Ischia, Abs., p. 312, 1984.

19. **Scarano, E., Geraci, G., and Rossi, M.,** Deoxycytidylate aminohydrolase IV: stoichiometry of binding of iososteric and allosteric effectors, *Biochemistry,* 6, 3645, 1967.
20. **Rossi, M., Raia, C. A., Nucci, R., and Vaccaro, C.,** manuscript in preparation, 1988.
21. **Monod, J., Wyman, J., and Changeux, J.-P.,** On the nature of allosteric transitions: a plausible model, *J. Mol. Biol.,* 12, 88, 1965.
22. **Whitehead, E. P., Raia, C. A., Nucci, R., and Rossi, M.,** submitted for publication, 1988.
23. **Poorman, R. A., Randolph, A., Kemp, R. G., and Henderson, R. A.,** Evolution of the phosphofructokinase gene and duplication and creation of a new effector site, *Nature (London),* 309, 467, 1984.
24. **Whitehead, E. P.,** Regulation of enzyme activity and the allosteric transition, *Prog. Biophys. Mol. Biol.,* 21, 321, 1970.
25. **Rossi, M., Geraci, G., and Scarano, E.,** Deoxycytidylate aminohydrolase III. Modifications of the substrate sites caused by allosteric effectors, *Biochemistry,* 6, 3640, 1967.
26. **Douglas, J., Haldane, J. S., and Haldane, J. B. S.,** The laws of combination of haemoglobin with carbon monoxide and oxygen, *J. Physiol. London,* 44, 275, 1912.
27. **Whitehead, E. P.,** Kinetics of enzymes with interacting effector molecules: tests of a configurational hypothesis in a quasi-equilibrium model, *Biochemistry,* 9, 1440, 1970.
28. **Whitehead, E. P.,** The mathematical formalism and the physical understanding of allosteric interactions in proteins, *Acta Biol. Med. Ger.,* 31, 227, 1973.
29. **Rossi, M., Momparler, R. L., Nucci, R., and Scarano, E.,** Studies on the analogs of isosteric and allosteric ligands of deoxycytidylate aminohydrolase, *Biochemistry,* 9, 2539, 1970.
30. **Momparler, R. L., Rossi, M., Bouchard, J., Vaccaro, C., Momparler, L. F., and Bartolucci, S.,** Kinetic interaction of 5-aza-deoxycytidine-5'-monophosphate and its 5'-triphosphate with deoxycytidylate aminohydrolase, *Mol. Pharmacol.,* 25, 436, 1984.
21. **Jones, P. A.,** Altering gene expression with 5-azacytidine, *Cell,* 40, 485, 1985.
32. **Momparler, R. L., Vesely, J., Momparler, R. F., and Rivard, G. E.,** Synergystic action of 5-aza-2'-deoxycytidine and 3-deazauridine on L1210 leukemic cells and EMY6 tumor cells, *Cancer Res.,* 39, 2041, 1982.
33. **Raia, C. A., Nucci, R., Vaccaro, C., Sepe, S., Rella, R., and Rossi, M.,** Reversal of the effect of the allosteric ligands of dCMP aminohydrolase and stabilisation of the enzyme in the T form, *J. Mol. Biol.,* 157, 557, 1982.
34. **Rossi, M., Dosseva, I., Pierro, M., Cacace, M., and Scarano, E.,** Studies on the conformational isomers of deoxycytidylate aminohydrolase, *Biochemistry,* 10, 3060, 1971.
35. **Cervone, F., Rossi, M., Vaccaro, C., Sepe, S., Scarano, E., and Rossi, M.,** The role of the sulphydryl groups of spleen deoxycytidylate aminohydrolase, *Eur. J. Biochem.,* 46, 401, 1974.
36. **Frieden, C.,** Kinetic aspects of regulation of metabolic processes: the hysteretic enzyme concept, *J. Biol. Chem.,* 245, 5788, 1970.
37. **Wyman, J.,** On allosteric models, *Curr. Top. Cell. Regul.,* 4, 209, 1972.
38. **Rossi, M., Raia, C. A., and Vaccaro, C.,** Chemical stabilisation of conformational states of deoxycytidylate aminohydrolase, *Methods Enzymol.,* 135(B), 577, 1987.

Chapter 10

RABBIT MUSCLE PHOSPHOFRUCTOKINASE

James C. Lee, Lyndal K. Hesterberg, Michael A. Luther, and Guang-Zuan Cai

TABLE OF CONTENTS

I. INTRODUCTION

Phosphofructokinase (ATP: D-fructose 6-phosphate 1-phosphotransferase, EC 2.7.1.11) catalyzes the transfer of the terminal phosphate of ATP to the C-1 hydroxy of fructose 6-phosphate to produce fructose 1,6-diphosphate. The catalyzed reaction represents a key regulatory point in the glycolytic pathway.[1-4] Hence, it is not surprising that phosphofructokinase (PFK) is well regulated by an apparently complicated set of mechanism(s) and exhibits unusual kinetic properties, e.g., ATP serves both as the substrate and an inhibitor, fructose 1,6-diphosphate is both a product and an activator. Since PFK plays a key role in metabolic regulation, it has been the subject of intensive investigation for the last 20 years. One of the basic aims of these studies is to identify the mode(s) of regulation. Significant progress has been made, although it was hampered by the instability of the protein (e.g., cold inactivation) and the apparently large variety of mechanisms that are involved in the regulation, including conformational changes, subunit assembly, ionization of amino acid side chains, and posttranslational modifications. The present review will only focus on the recent developments on the rabbit muscle enzyme. A molecular model for allosteric regulation for PFK will be proposed as a working hypothesis to highlight the various areas that are supported by experimental evidence available in the literature. Furthermore, it is hoped to identify the areas in which significantly more progress is still needed.

II. PHYSICAL PROPERTIES

One of the factors that prevents a quantitative characterization of the regulatory properties of rabbit muscle PFK is the instability of the enzyme prepared by the most commonly used purification procedures of Kemp and Ling et al.[5,6] PFK prepared by these procedures is proteolytically modified, rendering the enzyme unstable. Therefore, a significant amount of information reported in the literature pertaining to the physical properties of the enzyme may not reflect quantitatively the basic characteristics of the *native* enzyme, hence, some of the long established physical properties of the enzyme need to be clarified.

A. Subunit Molecular Weight

The purification procedures of Kemp[5] and Ling et al.[6] include differential heat denaturation and/or alcohol precipitation steps which are known to desensitize some allosteric enzymes.[7-9] On the other hand, the procedure adopted by this laboratory, designated as the Hesterberg procedure, is a combination of the procedures established by Hussey et al.[10,11] and Uyeda et al.[12] The potentially detrimental steps are omitted. PFK samples purified by all of these procedures are about 95% homogeneous, but the enzyme prepared by the Ling or Kemp procedures exhibits a faster migration rate in SDS polyacrylamide gel, as shown in Figure 1. The apparent molecular weights of PFK purified by the Ling, Kemp, and Hesterberg procedures are 75,000, 81,000, and 83,000, respectively. One-dimensional peptide mapping shows that the smaller peptide in the Ling preparation is a consequence of proteolytic cleavage of the larger peptide in the other enzyme preparation.[13]

B. Stability

The proteolytically modified PFK prepared by either the Kemp or Ling procedure is not stable, and a continuous conversion of active to an inactive form is observed. The inactive form sediments slowly and is not in dynamic equilibrium with other subunits.[13] In contrast, PFK isolated by the Hesterberg procedure is stable, active, and in rapid equilibrium with various aggregated forms of PFK. Frieden and co-workers studied the cold lability behavior of rabbit muscle PFK isolated by the Ling procedure.[14,15] After cold inactivation, the rate of regaining activity was reported to be a function of pH, temperature, and protein concen-

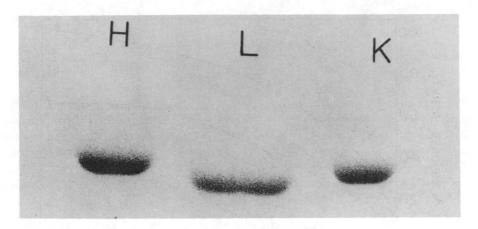

FIGURE 1. SDS-polyacrylamide gel electrophoresis results of PFK preparations. The samples in each lane are the following: H, L, and K are PFK samples purified by the Hesterberg, Ling, and Kemp procedures, respectively.

tration. The rate was reported to increase with decreasing pH, temperature, and protein concentration. PFK purified by the Hesterberg procedure exhibits a qualitatively similar behavior, as shown in Figure 2. However, native PFK is reactivated to its full activity within 60 min within the range of temperature of incubation, pH of solution, and protein concentration, whereas under similar conditions, PFK isolated by the Ling procedure can only be reactivated to less than 60% of activity after 240 min of incubation.[16] Hence, it can be concluded that *native* PFK is stable at low temperature. Its loss of activity at low temperature is rapidly reversible without apparent detrimental consequences. The cold lability phenomenon reported in the literature may only pertain to PFK isolated by the Ling procedure which has been shown to be a proteolytic modified enzyme.[13]

Similarly, native PFK is stable within the pH range of 6.0 to 10.0[17] Although the enzyme does not exhibit any significant amount of activity at pH values below 6.0 and above 9.5, normal activity can be observed once the solution pH is shifted to 7.0, again demonstrating the complete and rapid reversibility of this reaction. These observations are in contrast to those of Bock and Frieden carried out with PFK prepared by the Ling procedure.[16] Hence, their proposed mechanism of pH-dependent inactivation and reactivation of PFK appears applicable to the native enzyme, but the equilibrium and rate constants probably pertain only to the modified PFK and in the presence of phosphate buffer.

Most of the studies on PFK reported in the literature include phosphate in their buffer systems in order to stabilize the enzyme.[18-25] Native PFK purified by the Hesterberg procedure is stable in the absence of phosphate buffer. The identity of the buffer, e.g., Tris or glycylglycine, does not affect the stability of the enzyme. However, phosphate buffer can affect the regulatory behavior of PFK, since it can serve as an activator of the enzyme.[26] In the phosphate buffer, PFK exhibited a hyperbolic relation between its relative specific activity and protein concentration, whereas in the glycylglycine buffer, such relation was altered and become sigmoidal. Consequently, to explore the mode of regulation on PFK activity, the experimental condition should be carefully chosen so that it amplifies the regulatory behavior of the enzyme.

The isolation of PFK in a native and stable form facilitated a systematic search for the modes of regulation and identification of components which participate in these various modes of regulation.

III. ACTIVE FORM OF PFK

There are reports suggesting that PFK aggregates larger than tetramer are enzymatically active and have the same specific activity as tetramer.[27] It was also suggested that PFK

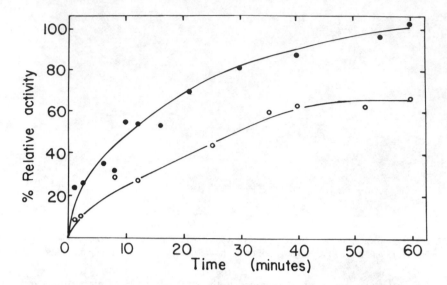

FIGURE 2.   Kinetics of reactivation at pH 8.0 and 23°C. The final enzyme concentration was 0.5 μg/ml, and it was inactivated at pH 6.5, 12°C. At zero time, the inactive enzyme was diluted to pH 8.0 in (●) 25 mM Imidazole, 1 mM EDTA, 6 mM MgCl$_2$, 3 mM (NH$_4$)$_2$SO$_4$, or (○) 100 mM phosphate, 1 mM EDTA. Activity of enzyme kept at pH 8.0 was used as reference for 100% activity.

dimer and monomers are either inactive or have lower specific activities.[23] All of these studies, however, suffer in varying degrees from technical difficulties. For investigation of the identity of active enzyme-substrate complexes, density gradient centrifugation and gel filtration would not be the methods of choice because these methods rely on a physical separation of the different aggregates of the enzyme and a subsequent test for catalytic activity.[23,24,27,28] Such procedures cannot rule out possible changes in the aggregation state after completion of separation or in the assay mixture. The reacting enzyme sedimentation technique employed by Hesterberg and Lee overcomes these technical difficulties.[10,29,30] The advantages of the method are as follows: (1) the hydrodynamic properties of the enzyme-substrate complex can be determined while it is fully active; (2) sedimentation can be observed at the same dilute enzyme concentrations used in kinetic studies; therefore, the determined hydrodynamic properties can be directly compared with the kinetic data.[10]

Detailed studies to identify the physical characteristics of active PFK forms were performed at pH 7.0, 8.55, and 23°C.[10,31] Two assay systems were employed, and these include a coupled enzyme and a pH-dependent dye-linked system. These systems were chosen to maximize the chance of detecting the slow-moving components, such as monomeric PFK, that is enzymatically active. The active PFK species sediments as a single component with a sedimentation coefficient ($s_{20,w}$) of 12.4 ± 0.5S. Although PFK undergoes association-dissociation, there is no observable change in the value of $s_{20,w}$ for the active species over a 57-fold range of protein concentration. Throughout this range, only a single active species of PFK was observed, and within an experimental uncertainty of ±10%, the enzymatic activity observed in the sedimentation studies accounts for the total enzymatic activity determined in the steady-state kinetics.

In order to define more precisely the structural properties of the active form of PFK, boundary sedimentation studies of the enzyme in the presence of 1.0 mM fructose 6-phosphate and 0.1 mM adenylyl-imidodiphosphate was performed at pH 7.0 and 23°C. The results showed that the sedimentation coefficient of PFK remains constant up to 100 μg/ml and assumes a value of 12.4S. The molecular weights of the subunit and the 12.4S component

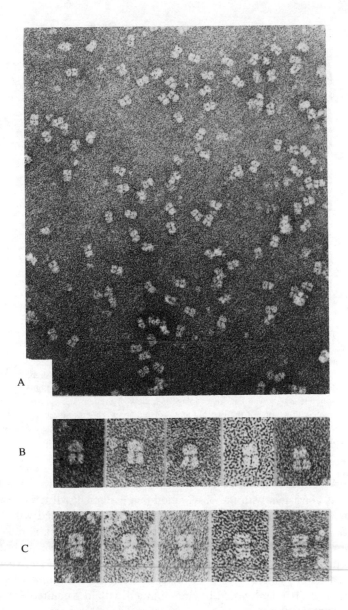

FIGURE 3. Images of negatively stained phosphofructokinase at pH 7.0. (A) A field from a specimen is shown at × 250,000; (B and C) individual particles showing two different characteristic views, are presented both at × 500,000. (From Hesterberg, L. K., Lee, J. C., and Erickson, H. P., *J. Biol. Chem.*, 256, 9724, 1981. With permission.)

were measured by sedimentation equilibrium yielding values of 83,000 and 330,000. Hence, the active form of PFK is a tetrameric species with a sedimentation coefficient of 12.4S.

The structure of the PFK tetramer was also studied by electron microscopy of negatively stained specimens.[31] Particles identified as tetramers measured approximately 9 nm in diameter by 14 nm in length, as shown in Figure 3. The arrangement and dimensions of subunits deduced from hydrodynamic parameters agree reasonably well with those observed in the electron micrographs. All of the images of PFK in Figure 3 are consistent with D_2, dihedral, symmetry. Although C_4 symmetry cannot be excluded by the results available at present, a model with D_2 symmetry is a much simpler and more satisfactory interpretation

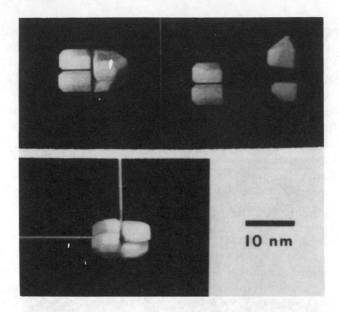

FIGURE 4.   A model of the phosphofructokinase molecule constructed from approximately globular subunits with D_2 symmetry. The view of the tetramer at the upper left corresponds to the images of Figure 3 revealing a cleft between the subunits of one dimer and conical shape for the other. At the upper right, a dimer and two views of individual subunits are shown. The lower image is a view of the tetramer approximately down one of the twofold axis, rotated about 15° from the upper model. The skewed conical shape of the monomers and the 30° rotation of one dimer with respect to the other give views consistent with the images, but these features are not definitely established.

of the images. A similar model was proposed for porcine liver PFK.[32] Furthermore, the proposed model is consistent with the observation that essentially all tetrameric proteins of known structure display D_2 symmetry.[33] A model for PFK constructed with four identical subunits with D_2 symmetry is shown in Figure 4. Each subunit is approximately 4 × 6 × 6 nm (tapered to a trapezoidal shape). An earlier electron microscopic study by Telford et al.[34] concluded that the individual subunit was a prolate ellipsoid with an axis of 6.7 × 2.5 × 2.5 nm. The subunit size proposed is certainly in error, as the volume of such a particle is only 22 nm³, while 101 nm³ would be the minimum volume required for a polypeptide chain of 83,000 with a partial specific volume of 0.730.[10] Furthermore, it was proposed that dimers consisted of two monomers lying side by side, and tetramers are formed from two dimers in a planar end to end array, with total dimensions of 13 × 8 × 2.5 nm. The model of a tetrameric enzyme 2.5 nm thick is also too small for a protein of 330,000 in molecular weight. Since the images published by Telford et al.[34] are very similar to those in Figure 4, the differences in models must be attributed to interpretation.

IV. INACTIVE TETRAMER

Based mainly on kinetic observations, the currently accepted model for the regulation of PFK is that it exists in two conformations, namely, an active R form and an inactive T form. The model assumes that the R and T forms represent two conformational isomers of the tetrameric enzyme.[24,35-38] Frieden and co-workers further propose that the activation or inactivation of the enzyme depends on either the protonation or deprotonation of certain ionizable groups on this tetramer.[16,25,38-40] Mindful of the fact that most of their studies were conducted on PFK purified by the Ling procedure and that the physical state of an inactive tetramer has not been identified, Luther et al.[41] reported a characterization of an inactive tetramer of PFK. The study was initiated as a consequence of a report by Gilbert that PFK can be reversibly inactivated by oxidized glutathione and greater than 95% of the original activity can be recovered upon the addition of reducing agent.[42]

Sedimentation velocity and equilibrium studies on inactivated PFK in the presence of 1.0 mM fructose 6-phosphate and 0.1 mM adenylyl-imidodiphosphate show the presence of an inactive tetrameric PFK which sediments at 13.5S.[41] Hence, the inactive form assumes a more compact structure whereas the active form is more relaxed or asymmetric. An important observation is that the inactivation reaction is completely reversible, and the enzyme shows no evidence of irreversible modifications as indicated by the identical results obtained by steady-state kinetic measurements and from sedimentation experiments on the native and reactivated PFK.

The presence of a 13.5S tetrameric PFK has also been inferred as a consequence of computer fitting of sedimentation velocity experiments.[43,44] A value of 13.5S is required to best fit the data in the presence of activators, inhibitors, or buffer alone. However, in the presence of either fructose 6-phosphate or ATP, or a combination of both fructose 6-phosphate and adenylyl imidodiphosphate, a value of 12.4S is essential for obtaining the best fit of the data, thus demonstrating a high degree of specificity in the ability of substrates or substrate analogs to induce such structural changes. At present, one can at least conclude that the hydrodynamic properties of this chemically inactivated form of PFK are similar to those inferred in the native unmodified protein. Definitive characterization is still not available to establish a direct correspondence of a chemically modified and unmodified 13.5S tetramer. However, all the metabolites tested in this study interact with the oxidized, inactive PFK and induce the enzyme to associate or dissociate in a manner quantitatively very similar to that observed in the native enzyme. Hence, these observations further support the assignment of the 13.5S tetramer as one of the inactive species.

V. SUBUNIT SELF-ASSOCIATION

Having established the presence of two conformers of PFK tetramer, one active and the other inactive, the issue of subunit self-association and its role in enzyme regulation will be addressed.

Hofer reported that for rabbit muscle PFK, the values of $K_{m,app}$, the Michaelis-Menten constant, for fructose 6-phosphate and ATP, respectively, are dependent on enzyme concentration.[45] The fructose 6-phosphate concentrations required for half-saturation increase at lower enzyme concentrations, as shown in Figure 5, whereas with decreasing PFK concentration, the ATP inhibition is enhanced, as shown in Figure 6. Similar results were reported on beef heart PFK.[46] These results imply strongly that subunit self-association plays a significant role in enzyme regulation. On the other hand, Frieden proposed that self-association plays no role in the mechanism of regulation due to the apparently slow rate constant governing the association reaction.[16,38,39] Hence, PFK subunit self-association becomes the subject of intensive investigation by many investigators using such techniques as electron microscopy,[34,47] X-ray diffraction,[48] elastic light scattering,[49] sedimentation equilibrium,[18] gel chromatography,[23,27] and sedimentation velocity.[6,18,27,47] The consensus is that PFK undergoes self-association in a complicated fashion. Until recently, neither the mode of association nor the thermodynamic parameters governing the association-dissociation equilibrium have been systematically determined.

Leonard and Walker reported that at pH 8.0, 20°C, and protein concentration greater than 1 mg/ml, the sedimentation patterns showed three peaks with $s_{20,w}$ values of 12, 18, 30S.[18]

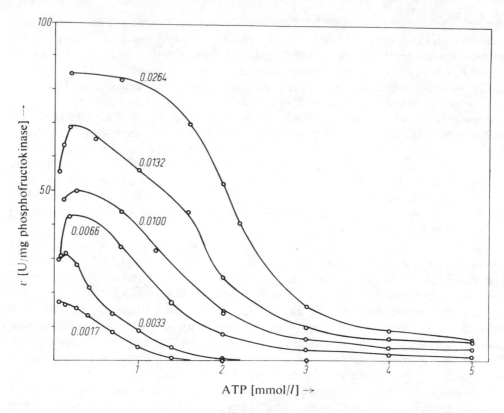

FIGURE 5.   Specific activity of phosphofructokinase at variable concentrations of ATP and at different concentrations of the enzyme. The assay system contained in the volume of 1 ml: fructose 6-phosphate 1.0 mM; MgCl$_2$ 4.0 mM; KCl 10.0 mM; phosphoenolpyruvate 1.0 mM; NADH 0.33 mM; pyruvate kinase 2 mg; lactate dehydrogenase 3 mg and imidazole HCl 50 mM. The pH was adjusted to 6.0. The assays were initiated by addition of 40 μl of the enzyme solution. The final enzyme concentrations in the assay are given at the curves in mg/ml. (From Hofer, H. W., *Hoppe Seyler's Z. Physiol. Chem.*, 352, 997, 1971. With permission.)

These peaks were not completely resolved under their experimental conditions. With increasing protein concentration, the amount of protein in the fastest peak increased, whereas the amount of protein in the two slower peaks remained constant. They concluded that the sedimentation behavior was typical of a system of several components in a rapid reversible equilibrium of undetermined stoichiometry. At the same time, Aaronson and Frieden also reported the observation of three peaks at pH 8.0 and 10°C.[27] With increasing protein concentration, however, these authors reported that the areas encompassed by these peaks were linear fractions of the total protein concentration, implying that these peaks represent noninteracting components. An attempt was made to separate the various polymeric forms of PFK by column chromatography. Fractions from the leading and trailing edges of the protein peak eluted from the column were subjected to sedimentation velocity analysis. The results revealed that these fractions exhibit different sedimentation profiles. On the basis of these observations, Aaronson and Frieden concluded that PFK appears to be a mixture of three components not in a rapid reversible equilibrium.[27] These components were interpreted to represent three different monomeric forms of the enzyme with different propensity to self-associate.

Pavelich and Hammes,[23] utilizing frontal gel chromatography at 5°C, studied the self-association of PFK over a pH range of 6.0 to 8.0 and at protein concentrations below 1 mg/ml. These authors reported that PFK undergoes dissociation from tetramers to dimers and probably to monomers. Due to the length of the experiment and the instability of PFK, up

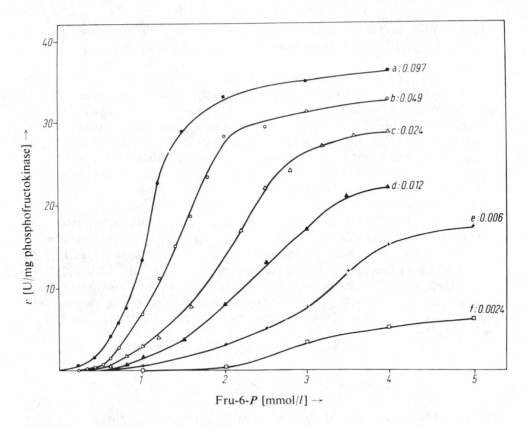

FIGURE 6.   Specific activity of phosphofructokinase at different enzyme concentrations and variable fructose 6-phosphate concentrations. The test cuvettes contained: ATP 3.5 mM; MgCl$_2$ 4.0 mM; KCl 10.0 mM; phosphoenolypyruvate 1.0 mM; NADH 0.33 mM; pyruvate kinase 2 mg; lactate dehydrogenase 3 mg and imidazole HCl 50 mM. The pH was 6.8 and the volume 1 ml. The fructose 6-phosphate concentrations were variable as indicated in the figure. The reactions were started with 40 μl of phosphofructokinase solution. The final enzyme concentrations in the assay are given at the curves in mg/ml. (From Hofer, H. W., Hoppe-Seyler's Z. *Physiol. Chem.*, 352, 997, 1971. With permission.)

to 20% loss of enzyme activity was reported, thus preventing an accurate assessment of the mode of association and the equilibrium constants of the reaction.

Contrary to these literature reports, studies from this laboratory have established that PFK is capable of undergoing association-dissociation in a rapid, dynamic equilibrium when studied by sedimentation velocity.[13,41,43,44] Since the question of whether PFK oligomers are in rapid, dynamic equilibrium is intimately linked to the role of subunit interaction in enzyme regulation, we shall examine the results and validity of tests employed to establish these conclusions.

Five different tests were conducted to establish the dynamic nature of PFK subunit interaction:

1.   Rotor speed dependence of weight-average sedimentation coefficient, $\bar{s}_{20,w}$. Whether a reaction is considered rapid or slow, it is in reference to the rate of separating the components in the reaction. If the half-time of reaction is faster than the rate of fractionation, then the reaction is considered rapid. Whereas, if the half-time of reaction is slower than the rate of separation of components, then the reaction is classified as slow. If a self-associating system is in a slow equilibrium or if noninteracting due to the presence of denatured components, then $\bar{s}_{20,w}$ would be expected to change as a function of the rotor speed. Results showed that under all experimental conditions tested, PFK solutions purified by the Hesterberg procedure yield identical values of

$\bar{s}_{20,w}$ at either 40,000 rpm or 60,000 rpm in the analytical ultracentrifuge. Samples purified by the Kemp or Ling procedure do exhibit speed dependence in their values of $\bar{s}_{20,w}$.

2. $\bar{s}_{20,w}$ as a function of time after dilution. If the association-dissociation is governed by a slow equilibrium, the relaxation of the system to reach a new equilibrium state after dilution will take time and a decrease in $\bar{s}_{20,w}$ is expected with increasing time after dilution. The value of $\bar{s}_{20,w}$ should eventually reach a value characteristic of the low protein concentration. On the other hand, if the system is in a rapid equilibrium, the $\bar{s}_{20,w}$ at zero time should be characteristic of the lower protein concentration, and its value will not change over time. The sedimentation properties of the enzyme purified by the Hesterberg procedure were monitored from 0.5 to 6.5 h after the initial dilution. The results show that $\bar{s}_{20,w}$, measured at 0.5 h after dilution, decreases to the value expected of PFK at the low concentration. This value of $\bar{s}_{20,w}$ remains constant throughout the duration of the experiment of 6.5 h, indicating that a slow equilibrium cannot exist under the experimental conditions. In contrast, PFK samples prepared by the Kemp or Ling procedures do show a time dependence in $\bar{s}_{20,w}$ after dilution.

3. $\bar{s}_{20,w}$ of diluted and reconcentrated PFK sample. If a self-association system is at equilibrium, the value of $\bar{s}_{20,w}$ is governed by the stoichiometry, equilibrium constants, and total protein concentration under a specific set of experimental conditions, since

$$\bar{s} = \sum_i s_i^\circ (1 - g_i C) K_i C_1^i / \sum_i K_i C_1^i$$

where S_i° is the sedimentation coefficient of the i th species at infinite dilution, g_i is the nonideality coefficient, $C = \sum_i K_i C_1^i$, K_i is the equilibrium constant between any i-mer and the monomer, and C_1 is the monomer concentration. Hence, at a given protein concentration a specific value of $\bar{s}_{20,w}$ should be observed regardless of whether the protein concentration is established by dilution of a stock concentrated solution or by concentrating a diluted sample. Luther et al.[13] reported sedimentation velocity studies of PFK sample prepared by the Hesterberg method at 25 μg/ml. Also, aliquot was concentrated and subsequently subjected to sedimentation velocity measurement. Results from seven experiments showed that PFK at 25 μg/ml sediments with a $\bar{s}_{20,w}$ of 12.3 ± 0.3S. These values are in exact agreement with PFK samples diluted from a stock solution of 5 mg/ml. Hence, again PFK purified by the Hesterberg procedure undergoes reversible association-dissociation.

4. Sedimentation profile as a function of protein concentration. When native PFK in 25 mM Tris-CO$_3$, 1 mM EDTA, 6.0 mM MgCl$_2$, and 3.4 mM (NH$_4$)$_2$SO$_4$ at pH 7.0 and 23°C is centrifuged at 60,000 rpm, the sedimentation profile, shown in Figure 7 as derivative scans, is a function of the initial loading concentration of PFK. At a concentration of 150 μg/ml, PFK sediments as a slightly skewed peak with $\bar{s}_{20,w}$ = 12.3S. At 300 μg/ml, the peak sediments at about the same rate although the leading edge is now skewed. The formation of a second peak is apparent at 500 μg/ml, and the sedimentation coefficients of the slow and fast peaks are 12.2 and 18.7S, respectively. When bimodality sets in, the area encompassed by the slow peak does not change with increasing protein concentration, but most of the protein is added to the faster peak as shown clearly by the sedimentation profile at 750 μg/ml. Such sedimentation behavior is characteristic of a self-associating system in rapid equilibrium relative to the length of the experiment.[50] The presence of a bimodal sedimentation pattern implies a cooperative self-association with stoichiometry of the association ≥ 3, as described by Gilbert.[50-52]

5. Fractionation and reanalysis. In a rapidly reversible association-dissociation system, fractions isolated from any part of the sedimentation profile, such as the one shown

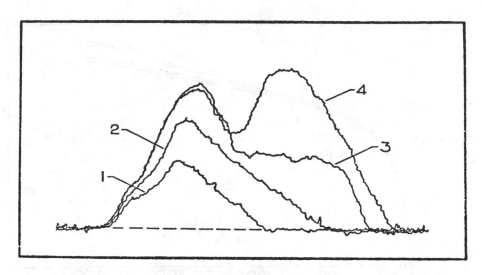

FIGURE 7.　Sedimentation velocity profiles of PFK expressed as derivative scans in 25 mM Tris-CO$_3$, 1 mM EDTA, 6 mM MgCl$_2$, and 3.4 mM (NH$_4$)$_2$SO$_4$ buffer at pH 7.00, 23°C. Protein concentrations (in μg/ml) were the following: 1, 150; 2, 300; 3, 500; 4, 750. Profiles were traced with the derivative mode of UV scanner at 30, 32, 28, and 28 min for profiles 1 to 4, respectively, at 52,000 rpm. (From Hesterberg, L. K. and Lee, J. C., *J. Biol. Chem.*, 20, 2974, 1981. With permission.)

in Figure 6, should behave identically when adjusted to the same concentration and subjected to analysis under the same experimental conditions. Luther et al.[13] isolated two fractions, the leading and trailing peaks of a PFK sample in velocity sedimentation by using an aluminum partition centerpiece. The two fractions were diluted to the same protein concentration as was the unfractionated PFK. These three samples were then subjected to sedimentation velocity studies. The leading, trailing, and unfractionated samples of PFK purified by the Hesterberg procedure all show similar sedimentation patterns. The observed values for $\bar{s}_{20,w}$ and specific activities are identical. These results are indicative of a system undergoing rapid and fully reversible association-dissociation. However, identical experiments yielded different results for PFK purified by the Ling or Kemp procedures. The values of $\bar{s}_{20,w}$ of the trailing fractions are lower than that of the unfractionated samples which in turn are lower than that of the leading fractions. Furthermore, the specific activity of each fraction is different, the trailing fractions exhibiting the lowest specific activity and the leading fractions the highest. These results indicate that these samples contain components that are not in rapid equilibrium with active PFK. It was shown subsequently that these noninteracting inactive components are denatured PFK. It was demonstrated that the *active* fractions of PFK also undergo rapid dynamic association-dissociation under tests 1 and 2. Additional sedimentation velocity experiments of freshly prepared active PFK were conducted as a function of enzyme concentration. Figure 8 shows the relation between $\bar{s}_{20,w}$ and PFK concentration. It is evident that *active* PFK self-associates in a similar manner.

Having established that *active* PFK undergoes rapid reversible association-dissociation, the sedimentation velocity data can thus be analyzed quantitatively to determine the stoichiometry and thermodynamic parameters characteristic of the self-associating reaction. It is evident that the value of $\bar{s}_{20,w}$ is a function of the protein concentraiton; $\bar{s}_{20,w}$ decreases with decreasing PFK concentration as shown in Figure 8. The extrapolated value of $\bar{s}_{20,w}$ at infinite dilution is about 5S, indicating that the smallest species under the experimental

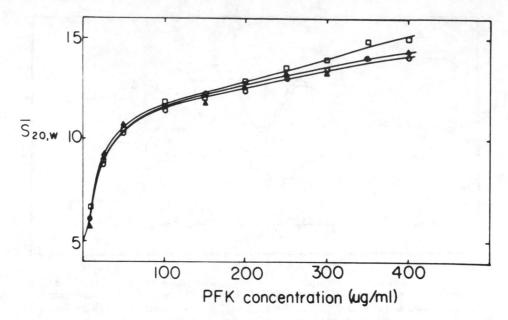

FIGURE 8. $\overline{S}_{20,w}$ as a function of PFK concentration for the PFK preparations in TEMA buffer at pH 7.0, 23°C. The symbols and preparations of PFK are: (□) Hesterberg PFK; (▲) Kemp PFK; (○) Ling PFK. The lines represent the theoretical fit of the experimental data using the association model of $M_1 \rightleftharpoons M_4 \rightleftharpoons M_{16}$ and other parameters shown as follows:

Preparation	s_4° (s)	Stoichiometry	[ml/mg)³]	[(ml/mg)¹⁵]	σ
H-PFK	13.5	1:4:16	5.0×10^5	2.5×10^{23}	0.16
K-PFK	13.5	1:4:16	4.08×10^5	7.18×10^{22}	0.19
L-PFK	13.2	1:4:16	3.44×10^5	5.00×10^{22}	0.05

(From Luther, M. A., Hesterberg, L. K., and Lee, J. C., *Biochemistry*, 24, 2463, 1985. With permission.)

condition is probably the monomeric unit of PFK at 83,000 mol w. In order to determine the physical constants characteristic of the self-association of PFK, a quantitative analysis was initiated by theoretical fitting of the concentration dependence of $\overline{s}_{20,w}$ using a nonlinear least-squares method.[53] The fitness of calculated and experimental data is judged by the standard root mean square deviation, σ. Typical results of the calculations for relevant combinations of the stoichiometry of self-association of PFK are shown in Table 1. It is evident that the simplest model of self-association at pH 7.0 and 23°C is a sequential tetramerization of monomer $\rightleftharpoons$ tetramer $\rightleftharpoons$ 16-mer with s_4° assuming a value of 13.5S, in excellent agreement with the inactive PFK tetramer. It is interesting to note that the assignment of s_4° 12.4S leads to a very poor fit with a σ-value of 2.18.

PFK self-association at 5°C cannot be fitted well by an association stoichiometry of monomer $\rightleftharpoons$ tetramer $\rightleftharpoons$ 16-mer. By including in the dimers in the model, i.e., monomer $\rightleftharpoons$ dimer $\rightleftharpoons$ tetramer $\rightleftharpoons$ 16-mer, the theoretical fit of the experimental data improves dramatically. Thus, it appears that the primary effect of temperature below 23°C is on the dimerization constant. The presence of dimer at 5°C as determined by sedimentation velocity is consistent with the reports by Hammes and co-workers.[43] Pavelich and Hammes reported that at 5°C and concentrations less than 0.2 mg/ml, PFK is in a pH-dependent equilibrium between a dimer and a tetramer as studied by frontal gel chromatography.[23] Such a conclusion was further substantiated by the results of a study at 4°C employing cross-linkage of PFK with a bifunctional reagent.[54]

Table 1
SUMMARY OF FITTING FOR WEIGHT-AVERAGE
SEDIMENTATION VELOCITY DATA, IN 25 mM TRIS-
CO$_3$, 1 mM EDTA, 6.0 mM MgCl$_2$, and 3.4 mM (NH$_4$)$_2$SO$_4$
BUFFER AT pH 7.0 AND 23° C

Model	K_2^{app}	K_4^{app}	diK^{app}_n	$s_4^{°a}$	σ
1-4-16	0	5.06×10^5	3.25×10^{23}	13.5	0.31
	0	2.98×10^6	5.23×10^{26}	12.4	2.18
1-4-8	0	2.00×10^6	7.04×10^{12}	13.5	5.80
2-4-16	1.0^b	468.7	2.17×10^{11}	13.5	19.2
	1.0^b	733.6	1.90×10^{12}	12.4	13.8
1-2-4-16	8.00×10^{-10}	5.03×10^5	3.34×10^{23}	13.5	0.31
	18.62	2.80×10^6	4.87×10^{26}	12.4	1.78

a   The values of $s_i^°$ remained constant for all species except $s_4^°$. These values are
    the following: $s_1^° = 4.95S$, $s_2^° = 7.6S$, $s_8^° = 19.7S$, and $s_{16}^° = 34.0S$.
b K_2^{app} assumes a value of 1.0 by definition.

From Hesterberg, L. K., and Lee, J. C., *Biochemistry*, 20, 2974, 1981. With
permission.

It has been clearly demonstrated qualitatively that the equilibria of PFK self-association
can be affected by the presence of ligands.[27,47,55,56] In general, substrates and activators were
reported to favor PFK self-association, while inhibitors enhance the formation of smaller
aggregates. There is, however, no quantitative analysis of the effects of ligands on the
association. Hesterberg and Lee initiated a quantitative study on the effects of ligands.[44] In
the presence of a substrate, fructose 6-phosphate or ATP, the self-association of PFK is
enhanced although the stoichiometry remains the same. An interesting observation is that
using $s_4^°$ equals 13.5S the data do not fit well, but instead a value of 12.4S is required to
achieve a better fit of data. The obvious conclusion is that the substrate is capable of inducing
a strong shift in the equilibrium favoring the formation of larger aggregates and to induce
a structural change in PFK so that the tetrameric PFK is in its active form with a $s_4^°$ of
12.4S.

The self-association of PFK under conditions that are reported to activate the enzyme was
also studied. These conditions include the presence of activator ADP, phosphate, and high
pH. Hesterberg and Lee reported that ADP and phosphate enhance the self-associatin reaction.[44] The stoichiometry of the reaction remains the same and $s_4^°$ assumes a value of
13.5S. Similar results are obtained by decreasing concentrations of proton.

The presence of citrate, an allosteric inhibitor of PFK, favors the formation of dimers
with $s_4^°$ assuming a value of 13.5S. Further analysis of the data by the linked function theory
derived by Wyman led to the conclusion that the formation of each dimer involves the
binding of one additional citrate molecule per PFK subunit.[57]

On the basis of the results, the simplest mode of self-association for the rabbit muscle
PFK system is

$$M_1 \rightleftharpoons M_2 \rightleftharpoons M_4 \rightleftharpoons M_{16}$$

Ligands and temperature would perturb the various equilibrium constants without altering
the basic scheme of association. One of the effects unique to the substrates is the induction
of a significant structural change in PFK so that the tetrameric form assumes a hydrodynamic
structure of the active form.

VI. LIGAND BINDING

The most extensively studied ligand is the nucleotide ATP and its derivatives. Since ATP can serve both as a substrate and as an inhibitor, the conclusions from these ligand binding studies are not unequivocal and straightforward. This is further complicated by the observations that ADP, AMP, or cyclic AMP can serve as activators of muscle PFK and that PFK undergoes ligand-induced subunit association. At present, there is no consensus in the literature concerning either the equilibrium constants governing these reactions or the mechanism with which other metabolites modulate the interaction of PFK with ATP. Some of these results are summarized in Table 2. Kemp and Krebs conducted equilibrium binding studies at pH 6.95 and 23°C using the chromatographic method of Hummel and Dryer.[58,59] It was reported that 3 mol of ATP are bound to 90,000 g of PFK. Double reciprocal plots were curvilinear, indicating heterogeneity in the binding sites whose dissociation constants (K_D) vary from 4 to 13 μM. In the presence of cyclic AMP, binding of ATP to the site with the highest affinity is reduced in a noncompetitive manner. Whereas in the presence of 0.5 μM citrate, there is an apparent enhancement of affinity for ATP. The double reciprocal plot is more linear with an apparent K_D of 1.5 to 3 μM. From these results, it is evident that the binding of ATP to PFK is complex and is modulated by both activators and inhibitors of the enzyme. More recently, Pettigrew and Frieden monitored the binding of ATP and AMP-PNP to rabbit muscle PFK with intrinsic protein fluorescence and sedimentation.[40] At pH 6.9, two AMP-PNP binding sites per subunit were found by the fluorescence technique; however, only one site was detected by sedimentation. At pH 8.0, there is also a discrepancy in the stoichiometries determined by these methods. These results are also in contrast to those reported by Wolfman et al.[60] who monitored binding of AMP-PNP to PFK at 4°C with both fluorescence and equilibrium dialysis techniques. In the presence of 10 mM phosphate, a binding stoichiometry of two was reported and a cooperative binding at the second site was invoked to fit the data. Furthermore, fructose 6-phosphate was shown to further complicate the interaction of AMP-PNP-PFK by inducing cooperativity in both binding sites leading to a total of four binding constants to fit the data. It is evident that the current knowledge on nucleotide-PFK interaction is still very limited, and there is discrepancy even in the stoichiometry of the ligand.

The interaction between fructose 6-phosphate and PFK was studied by Kemp and Krebs,[58] as well as by Hill and Hammes.[56] At pH 6.95 and 23°C, there is no apparent cooperativity in fructose 6-phosphate binding.[58] The binding is modulated by a variety of ligand effectors, as summarized in Table 2. In general, activators of PFK enhance fructose 6-phosphate binding, whereas inhibitors lower the affinity. At pH 8.0 and 5°C, Hill and Hammes reported that PFK interacts with fructose 6-phosphate with a negative cooperativity,[56] the degree of which decreases at pH 7.0. It was proposed that dimeric PFK exhibits an extreme negative cooperativity in its interaction with fructose 6-phosphate, while the higher aggregates of PFK exhibit less or very little cooperativity. In a ligand induced self-associating system, such as the rabbit muscle PFK system, Cann and Himman have elegantly demonstrated in a theoretical treatment that depending on the mode of ligand-macromolecular interaction both apparent negative or positive cooperativity can be observed, even though there is in reality no direct coupling among these binding sites. Hence, the presence and absence of cooperative interactions as reported by the authors of these two studies may actually reflect the behavior of PFK subunit interaction under different experimental conditions. Regardless of the actual mechanism governing fructose 6-phosphate-PFK interaction, it is clear that a systematic study taking into the accounts of subunit interaction and induced conformational changes is essential in elucidating the thermodynamic linkages that control PFK-substrate interactions.

The interactions of PFK with activators and inhibitors are apparently less complicated.

Table 2

SUMMARY OF LIGAND BINDING STUDIES TO RABBIT MUSCLE PHOSPHOFRUCTOKINASE

Ligand	pH	Temp.	Method	Effector	Stoichiometry	Observation	Ref.
ATP	6.95	23	Equilibrium — Hummel-Dreyer	—	3	Curvature in double reciprocal plots	58
				Citrate		Reduced curvature in double reciprocal plots	
				cAMP		Reduced ATP binding at the site with the highest affinity	
ATP	6.9	25	Fluorescence	—	2	Two sites with 15-fold difference in binding constant	25
AMP-PNP	8.00	20-25	Fluorescence	—	1	Binding constant >1 mM	
	6.9	20	Fluorescence	—	2	>2000-fold difference in binding constant	
	6.9	20	Sedimentation	—	1	Only one strong site	
	8.0	20	Fluorescence	—	0	—	
	8.0	20	Sedimentation	—	1	—	
AMP-PNP	7.0	4	Fluorescence and equilibrium dialysis	Phosphate	2	Two sites with one assuming two affinities as a consequence of cooperative interaction	60
				Phosphate-cAMP	1		
				FDP	2	Two independent sites with 40-fold difference in binding constant	
				F6P	2	Two independent sites each of which assumes two affinities as a consequence of cooperative interaction	
cAMP	6.95	23	Equilibrium — Hummel-Dreyer	—	1	Binding constant is not significantly changed as a function of pH or protein concentration	58
cAMP	8.0	20	Sedimentation	—	1	No apparent cooperativity	25
F6P	6.95	23	Equilibrium — Hummel-Dreyer	ADP/NH$_4^+$/phosphate	1	All of these effectors enhance F6P binding	58
				AMP	1		
				Citrate/40 μM ATP	1	Inhibit F6P binding	
F6P	8.0	5	Equilibrium	—	1	Concave Scatchard plots	56

Table 2 (continued)
SUMMARY OF LIGAND BINDING STUDIES TO RABBIT MUSCLE PHOSPHOFRUCTOKINASE

Ligand	pH	Temp.	Method	Effector	Stoichiometry	Observation	Ref.
	7.0	5	Equilibrium	—	1	Less pronounce concave Scatchard plots	
FDP	8.0	5	Equilibrium	—	1	Degree of curvature in Scatchard plots is a function of protein concentration	
Citrate	6.9	23	Equilibrium — Hummel-Dreyer	—	1		63

Kemp and Krebs reported a binding stoichiometry of one ligand per PFK subunit for cAMP, AMP, or ADP.[58] The apparent affinity is affected by changing Mg^{++}, but not by variation in pH. Foe et al.[62] reported that the binding of fructose 2,6-diphosphate to PFK shows negative cooperativity. The ligand induces PFK to aggregate. Citrate, an allosteric inhibitor, also binds PFK with a stoichiometry of one with no apparent cooperativity.[63]

In summary, the current knowledge on ligand-PFK interactions is still quite limited. The general consensus is that these reactions are complicated and are intimately linked to the exact experimental conditions, namely, temperature, pH, protein concentration, and the presence or absence of other ligands. Hence, an establishment of quantitative linkages awaits more systematic studies.

VII. POSTTRANSLATIONAL MODIFICATION

Besides being regulated by a variety of metabolites in a complex manner, the enzyme activity of PFK may also be regulated by posttranslational modifications. Since the initial report by Höfer and Furst on the *in vivo* phosphorylation of PFK,[64] the role of phosphorylation-dephosphorylation on the regulation of muscle PFK activity has been the subject of active investigation.

Höfer and Sorensen-Ziganke reported that the incorporation of ^{32}P-phosphate into mouse muscle PFK is not affected by the presence of cycloheximide.[65] The amount of phosphate incorporated is a function of muscle activity. It is shown that the phosphorylation of PFK increases with muscle contraction and up to eight phosphates per PFK tetramer were reported. Kemp et al.[66] identified the site of phosphorylation as the serine residue located at the sixth amino acid from the carboxyl terminus. Furthermore, the same authors reported that rabbit muscle PFK can be phosphorylated *in vitro* by the catalytic subunit of cyclic AMP-dependent protein kinase and evidently, at least 80% of the *in vivo* phosphorylation takes place at the same site as that introduced by the kinase. The carboxyl terminal of rabbit muscle PFK is particularly susceptible to proteolysis and its exposure is influenced by a number of ligands.[67] The presence of ATP, $MgCl_2$, or citric acid protects the segment from proteolysis, but AMP and fructose-1,6-diphosphate induce the opposite effect. In addition, phosphorylated PFK can be dephosphorylated *in vitro* by treatment with alkaline phosphatase.[68]

Luther and Lee reported a possible functional role of phosphorylation in regulating PFK-actin interaction.[69] Liou and Anderson reported earlier that F-actin acts as a positive effector of PFK.[70] The formation of the PFK-actin complex was monitored by sedimentation. The binding isotherms for both phosphorylated and dephosphorylated PFK are sigmoidal, although the binding isotherm for the phosphorylated PFK is located to the left of that for the dephosphorylated form. The data were further analyzed by the Hill plots to determine the Hill coefficient, n, and $\overline{Y}_{0.5}$, the concentration of PFK subunit where half-saturation is observed.[71] For the phosphorylated forms, the averaged value of $\overline{Y}_{0.5}$ is (1.2 ± 0.2) μM, whereas the dephosphorylated form of PFK assumes an averaged value of (3.8 ± 0.2) μM, as summarized in Table 3. The *in vitro* phosphorylated PFK yielded very similar results as the *in vivo* phosphorylated sample. The apparent values of $\overline{Y}_{0.5}$ for both the phosphorylated and dephosphorylated forms are lower in the presence of both substrates — namely 2 mM fructose 6-phosphate and 1 mM of AMP-PNP, as shown in Table 3, and the maximum amount of PFK bound increases by a factor of 1.7 ± 0.1. The binding isotherm show little or no change in the presence either of AMP-PNP or fructose 6-phosphate alone, thus implying that the two substrates act synergistically. Control experiments show that the PFK-actin interaction is specific and not due to simple nonspecific macromolecular trapping or ionic interaction.

Having established that phosphorylation of PFK enhances the affinity of the enzyme for F-actin, the effect of phosphorylation on the steady-state kinetics of PFK was monitored.

Table 3
SUMMARY OF BINDING DATA OF PFK TO F-ACTIN[a]

Preparation	$\overline{Y}_{0.5}$(PFK μM)		n	
	With substrates	Without substrates	With substrates	Without substrates
In vivo				
Phosphorylated	1.1 ± 0.2	1.2 ± 0.2	1.3 ± 0.1	1.8 ± 0.1
In vitro				
Phosphorylated		1.2 ± 0.2		1.9 ± 0.1
Dephosphorylated	2.9 ± 0.4	3.8 ± 0.2	1.9 ± 0.1	2.3 ± 0.2

[a] Binding experiments were conducted in 65 μM glycylglycine, 10 mM KCl, 15 mM MgCl$_2$, 1 mM EDTA, and 1 mM DTT, pH 7.0, 23°C. F-actin final concentration was 1.0 mg/ml. $\overline{Y}_{0.5}$ is expressed as concentration of the PFK subunit.

From Luther, M. A. and Lee, J. C., *J. Biol. Chem.*, 261, 1753, 1986. With permission.

Luther and Lee reported that at pH 7.0 and 23°C the phosphorylated PFK, be it formed *in vivo* or *in vitro*,[69] exhibits a distinct sigmoidal relationship between its activity and the fructose 6-phosphate concentration, whereas the same relationship for the dephosphorylated form is shifted to the left. Hence, dephosphorylated PFK has a higher apparent affinity for fructose 6-phosphate and the interaction between the enzyme and the substrate is apparently less cooperative. The inhibitory behavior of ATP on PFK was also probed, and there is no significant difference in the apparent inhibition constants, k_I^{app}, for ATP among the various phosphorylated and dephosphorylated forms of PFK. The overall kinetic observations reported by Luther and Lee are consistent with previously published reports,[69] despite the fact that the enzyme was prepared by different methods.[12,68,72]

The effect of F-actin on the regulatory behavior of the phosphorylated and dephosphorylated forms of PFK was monitored. At pH 7.0 and 23°C, F-actin does not affect the regulatory behavior of dephosphorylated PFK. For the phosphorylated forms of PFK, F-actin acts as a positive effector by shifting an otherwise sigmoidal curve to the left, rendering the relationship between enzyme activity and fructose 6-phosphate less sigmoidal. In addition, the presence of F-actin increases the values of K_I^{app} of ATP for both the phosphorylated and dephosphorylated forms of PFK. Thus, it can be concluded that F-actin renders the enzyme less sensitive to ATP inhibition.[69]

Based on these observations, a scheme is proposed by Luther and Lee to correlate the phosphorylation state of PFK with muscle activity and cellular location of the enzyme. In contracting muscle, the need for energy is high. One way of enhancing the efficiency of the system is to have the energy source at the point where it is needed, i.e., by localizing the glycolytic enzymes. Thus, if the enzymes are bound to the muscle matrix near the site where ATP is utilized, then the ATP produced can be more efficiently delivered and used. Being a key regulatory enzyme in glycolysis, it is logical that the location of PFK be controlled. Enzyme phosphorylation could be the control mechanisms of the process. In contracting muscle, PFK is phosphorylated, and its apparent affinity for F-actin is increased. There may be a time lag between enzyme phosphorylation and the formation of a PFK-actin complex. Hence, to conserve energy resources, the kinetic behavior of phosphorylated PFK not complexed with actin is more sigmoidal than that of the dephosphorylated form, i.e., its activity at the same fructose 6-phosphate concentration is lower. Once the complex is formed, the enzymic activity is increased so that the energy flux can be increased. In contrast to the contracting muscle, the need for energy is lower for resting muscle. Therefore, the physical contact of glycolytic enzymes with the muscle matrix is not needed. Under these

Table 4
PROPOSED RELATIONSHIPS AMONG MUSCLE ACTIVITY, PHOSPHORYLATION STATE, AND CELLULAR LOCATION OF PFK

	Resting muscle	Contracting muscle	Ref.
Need for energy in muscle	Low	High	
Phosphorylation state of PFK	Dephosphorylated	Phosphorylated	65
Affinity for F-actin	Low	High	69
Location of glycolytic enzyme	Cytoplasm	Particulate	80, 81, 87—93
Kinetic behavior	Low cooperativity and low K_m^{app}	High cooperativity and high K_m^{app} before complex formation;	69, 12
		Low cooperativity and low K_m^{app} after formation of complex	68

From Luther, M. A. and Lee, J. C., *J. Biol. Chem.*, 261, 1753, 1986. With permission.

conditions, PFK is dephosphorylated, and less PFK is associated with the muscle matrix. Although a high energy flux through the muscle matrix is not required in resting muscle, some energy is still required for other cellular activities. It is logical for the dephosphorylated PFK not complexed with actin to exhibit less sigmoidal kinetic properties, i.e., a higher enzyme activity at the same concentration of substrate. The evidence which support the proposed scheme is summarized in Table 4.

In conclusion, evidence is provided that phosphorylation of PFK not only affects its kinetic properties, but also regulates the formation of an enzyme-actin complex. This complex formation is specific and may serve as a means to compartmentalize glycolytic enzymes so as to provide energy to the muscle cells where it is needed. Such a hypothesis has also been advanced by the recent work of Anderson and co-workers.[73]

The formation of F-actin-enzyme complex is not specific only to PFK. There is ample evidence in the literature indicating that several glycolytic enzymes, especially aldolase, glyceraldehyde 3-phosphate dehydrogenase, and pyruvate kinase, have significant affinities for the myofibrillar apparatus, i.e., actin, myosin, troponin, and tropomyosin.[74-82] Moreover, the binding to F-actin is preferred over the other muscle matrix proteins.[75,77,83] PFK differs from the other glycolytic enzymes in that phosphorylation can regulate its affinity toward F-actin. This distinction may be relevant. At present, the time sequence of absorption of glycolytic enzymes to F-actin, if there is one, is not known. It is conceivable that the complex between the phosphorylated PFK- and F-actin can serve as a nucleation site for the subsequent absorption of other glycolytic enzymes. There is credence to this proposal since a preference was reported in the manner that glycolytic enzymes form complexes with the myofibrillar apparatus.[84] Hence, more work is required to probe this interaction between glycolytic enzymes and muscle proteins and the role of PFK in the formation of these complexes.

VIII. MOLECULAR MECHANISM OF REGULATION

Currently, there is an agreement among the various models proposed that PFK exists in two tetrameric conformations,[24,38,85] and the recent physical studies have unequivocally defined the specific hydrodynamic features of these two forms — an inactive 13.5S tetramer and an active 12.4S tetramer.[41,43,44] However, there is disagreement on the role of subunit interaction in the regulation of PFK activity. Frieden et al.[38] contended that subunit assembly plays no role. On the other hand, the models proposed by the Kemp and Hammes laboratories include dimeric PFK in equilibrium with one of the tetrameric forms. Based on the hydrodynamic and thermodynamic studies from this laboratory, the presence of not only tetrameric PFK, but also dimeric, monomeric, and other higher aggregates of PFK are defined under specific experimental conditions.

These physical measurements are essential in establishing the identities of the various components participating in the mechanism of regulating PFK. Based on all of the hydrodynamic and thermodynamic data, a model for the regulation of PFK can be proposed as a working hypothesis.

$$\text{Active} \rightleftharpoons \text{Inactive}$$

$$M_4 \ (12.4s) \underset{S}{\rightleftharpoons} M_4 \ (13.5s) \overset{I}{\underset{A/S}{\rightleftharpoons}} M_2 \overset{I}{\underset{A/S}{\rightleftharpoons}} M_1$$

The significant features of the proposed model are (1) activators and inhibitors perturb the equilibria between the inactive species; (2) only substrates favor the formation of the active form, in addition to shifting the association-dissociation equilibria between the inactive species. Hence, association-dissociation of PFK subunits is linked to conformational changes between active and inactive tetrameric PFK. Since these events are linked by reversible equilibrium processes, perturbation on any part of this network of equilibria will ultimately be reflected in the amount of active tetrameric PFK present, and thus the enzymatic activity. The quantitative significance of association-dissociation and conversion between the active and inactive tetrameric PFK in the regulation of enzymatic activity remains to be defined; however, from the thermodynamic viewpoint, the association-dissociation of PFK subunits must play a role in the regulation of PFK activity.

The proposed model can serve as a first approximation to qualitatively describe some of the experimental data from this laboratory and that in the literature. The observation that the activators tested do not induce the formation of an active PFK is consistent with the results of Gottschalk and Kemp that both ADP-ribose and nicotinimide dinucleotides occupy the AMP binding site, but do not promote the active conformation.[86]

It is clear that a more definitive quantitative description of the thermodynamics of allosteric regulation of rabbit muscle PFK awaits for further precise definition of quantitative linkages among ligand binding, structural changes, and steady-state kinetics. It is also clear that the molecular mechanisms employed by nature to regulate PFK include ramifications of the basic thermodynamic concepts of linkage between the binding of two ligands to the same macromolecule. Let us review the general knowledge on linkages as revealed by the study on rabbit muscle PFK.

In a reaction which defines the formation of a complex MX from its components M and X, such as

$$A + M + X \rightleftharpoons A + MX \rightleftharpoons AMX$$

and if the reaction is modulated by A, then this is an allosterically regulated reaction. The chemical nature of A can be as simple as H^+ or as complex as another macromolecule. In the case of PFK, the sigmoidicity of a velocity *vs.* substrate concentration plot can be shifted by a change in pH or the presence of F-actin.[26,69]

The facts that A can alter the interaction of X with M are due to the possibility of M existing in multiple states which are in equilibrium. The presence of A and X can perturb the equilibria among these states. The physical identities of these states can represent M and M′, conformational states which differ in their secondary and tertiary structures or in their aggregational state. In the case of PFK, the conversion of the inactive 13.5S tetramer is an active 12.4S tetramer by the substrates and ligand induced association and dissociation of PFK subunits are examples of such changes in states.

In summary, within the system of rabbit muscle phosphofructokinase, a variety of mechanisms are employed to regulate the enzyme activity. The apparent complexity of the mechanism can be resolved into identifiable and potentially isolated steps so that systematic and quantitative elucidation of the regulatory mechanism is feasible.

ACKNOWLEDGMENTS

The authors deeply appreciate the critical and constructive review of this manuscript by Dr. George Na and Thomas G. Consler. This work was supported by NIH grants AM-21489 and NS-14269.

REFERENCES

1. **Krebs, H. A.,** The Pasteur effect and the relations between respiration and fermentation, in *Essays in Biochemistry,* Vol. 8, Academic Press, New York, 1972, 1.
2. **Heinrich, R., Rapoport, S. M., and Rapoport, T. A.,** Metabolic regulation and mathematical models, *Prog. Biophys. Mol. Biol.,* 32, 1, 1977.
3. **Hofmann, E.,** The significance of phosphofructokinase to the regulation of carbohydrate metabolism, *Rev. Physiol. Biochem. Pharm.,* 75, 1, 1976.
4. **Hess, B.,** Oscillating reactions, *TIBS,* 2, 193, 1977.
5. **Kemp, R. G.,** Phosphofructokinase from rabbit skeletal muscle, *Methods Enzymol.,* 42C, 71, 1975.
6. **Ling, K. H., Marcus, F., and Lardy, H. A.,** Purification of some properties of rabbit skeletal muscle phosphofructokinase, *J. Biol. Chem.,* 240, 1893, 1965.
7. **Kurganov, B. I.,** *Allosteric Enzymes,* John Wiley & Sons, New York, 1982, 26.
8. **Stadtman, E. R.,** Allosteric regulation of enzyme activity, *Adv. Enzymol.,* 28, 41, 1966.
9. **Seya, T. and Nagasawa, S.,** Limited proteolysis of the third component of human complement, C_3, by heat treatment, *J. Biochem. (Tokyo),* 89, 659, 1981.
10. **Hesterberg, L. K. and Lee, J. C.,** Sedimentation study of a catalytically active form of rabbit muscle phosphofructokinase at pH 8.55, *Biochemistry,* 20, 2974, 1980.
11. **Hussey, G. R., Liddle, P. F., Ardon, P., and Kellett, G. L.,** The isolation and characterization of differentially phosphorylated fractions of phosphofructokinase from rabbit skeletal muscle, *Eur. J. Biochem.,* 80, 497,1977.
12. **Uyeda, K., Miyatake, A., Luby, L. J., and Richards, E. G.,** Isolation and characterization of muscle phosphofructokinase with varying degrees of phosphorylation, *J. Biol. Chem.,* 253, 8319, 1978.
13. **Luther, M. A., Hesterberg, L. K., and Lee, J. C.,** Subunit interaction of rabbit muscle phosphofructokinase: effects of purification procedures, *Biochemistry,* 24, 2463, 1985.
14. **Bock, P. E. and Frieden, C.,** pH-Induced cold lability of rabbit skeletal muscle phosphofructokinase, *Biochemistry,* 13, 4191, 1974.
15. **Bock, P. E., Gilbert, H. R., and Frieden, C.,** Analysis of the cold lability behavior of rabbit muscle phosphofructokinase, *Biochem. Biophys. Res. Commun.,* 66, 564, 1975.
16. **Bock, P. E. and Frieden, C.,** Phosphofructokinase. I. Mechanism of the pH-dependent inactivation and reactivation of the rabbit muscle enzyme, *J. Biol. Chem.,* 251, 5630, 1976.
17. **Cai, G.-Z., Lee, L. L.-Y., and Lee, J. C.,** unpublished data, 1986.
18. **Leonard, K. R. and Walker, I. O.,** The self-association of rabbit muscle phosphofructokinase, *Eur. J. Biochem.,* 26, 442, 1972.
19. **Paradies, H. H. and Vettermann, W.,** On the quaternary structure of native rabbit muscle phosphofructokinase, *Biochem. Biophys. Res. Commun.,* 71, 520, 1976.
20. **Bloxham, D. P. and Lardy, H. A.,** Phosphofructokinase, *Enzymes,* 8A, 1973, 238.
21. **Jones, R., Dwek, R. A., and Walker, I. O.,** Spin-labelled phosphofructokinase, *Eur. J. Biochem.,* 60, 187, 1975.
22. **Kee, A. and Griffin, C. C.,** Kinetic studies of rabbit muscle phosphofructokinase, *Arch. Biochem. Biophys.,* 149, 361, 1972.

23. **Pavelich, M. J. and Hammes, G. G.**, Aggregation of rabbit muscle phosphofructokinase, *Biochemistry,* 12, 1408, 1973.

24. **Goldhammer, A. R. and Hammes, G. G.**, Steady-state kinetic study of rabbit muscle phosphofructokinase, *Biochemistry,* 17, 1818, 1978.

25. **Pettigrew, D. W. and Frieden, C.**, Rabbit muscle phosphofructokinase — a model for regulatory kinetic behavior, *J. Biol. Chem.,* 254, 1896, 1979.

26. **Hofer, H. W. and Pette, D.**, Aktive and inaktive formen der phosphofructokinase des kananchenskelet muskels, *Hoppe-Seyler's Z. Physiol. Chem.,* 349, 1105, 1968.

27. **Aaronson, R. P. and Frieden, C.**, Rabbit muscle phosphofructokinase: studies on the Polymerization, *J. Biol. Chem.,* 247, 7502, 1972.

28. **Hofmann, E., Kurganov, B. I., Schellenberger, W., Schultz, J., Sparmann, G., Wenzel, K. W., and Zimmermann, G.**, Association-dissociation behavior of erythrocyte phosphofructokinase and tumor pyruvate kinase, *Adv. Enzymol. Regul.,* 13, 247, 1975.

29. **Cohen, R., Giraud, B., and Messiah, A.**, Theory and practice of the analytical centrifugation of an active substrate-enzyme complex, *Biopolymers,* 5, 203, 1967.

30. **Cohen, R. and Mire, M.**, Analytical-band centrifugation of an active enzyme-substrate complex. I. Principle and practice of the centrifugation, *Eur. J. Biochem.,* 23, 267, 1971.

31. **Hesterberg, L. K., Lee, J. C., and Erickson, H. P.**, Structural properties of an active form of rabbit muscle phosphofructokinase, *J. Biol. Chem.,* 256, 9724, 1981.

32. **Foe, L. G. and Trujillo, J. L.**, Quaternary structure of pig liver phosphofructokinase, *J. Biol. Chem.,* 255, 10537, 1980.

33. **Klotz, I. M., Darnall, D. W., and Langerman, N. R.**, Quaternary structure of proteins, in *The Proteins,* Vol. 1, 3rd ed., Academic Press, New York, 1975, 294.

34. **Telford, J. N., Lad, P. M., and Hammes, G. C.**, Electron microscope study of native and cross-linked rabbit muscle phosphofructokinase, *Proc. Natl. Acad. Sci. U.S.A.,* 72, 3054, 1975.

35. **Roberts, D. and Kellett, G. L.**, The kinetics of effector binding to phosphofructokinase — the allosteric conformational transition induced by $1,N^6$-ethenoadenosine triphosphate, *Biochem. J.,* 183, 349, 1979.

36. **Roberts, D. and Kellett, G. L.**, The kinetics of effector binding to phosphofructokinase. The binding of Mg^{++}-$1,N^6$-ethenoadenosine triphosphate to the catalytic site, *Biochem. J.,* 189, 561, 1980.

37. **Roberts, D. and Kellett, G. L.**, The kinetics of effector binding to phosphofructokinase. The influence of effectors on the allosteric conformational transition, *Biochem. J.,* 189, 568, 1980.

38. **Frieden, C., Gilbert, H. R., and Bock, P. G.**, Phosphofructokinase. III. Correlation of the regulatory kinetic and molecular properties of the rabbit muscle enzyme, *J. Biol. Chem.,* 251, 5644, 1976.

39. **Bock, P. G. and Frieden, C.**, Phosphofructokinase. II. Role of ligands in pH-independent structural changes of the rabbit muscle enzyme, *J. Biol. Chem.,* 251, 5637, 1976.

40. **Pettigrew, D. W. and Frieden, C.**, Binding of regulatory ligands to rabbit muscle phosphofructokinase, *J. Biol. Chem.,* 254, 1887, 1979.

41. **Luther, M. A., Gilbert, H. F., and Lee, J. C.**, Self-association of rabbit muscle phosphofructokinase. Role of subunit interaction in regulation of enzymatic activity, *Biochemistry,* 22, 5494, 1983.

42. **Gilbert, H. F.**, Biological disulfides: the third messenger?, *J. Biol. Chem.,* 257, 12086, 1982.

43. **Hesterberg, L. K. and Lee, J. C.**, Self-association of rabbit muscle phosphofructokinase at pH 7.0: stoichiometry, *Biochemistry,* 20, 2974, 1981.

44. **Hesterberg, L. K. and Lee, J. C.**, Self-association of rabbit muscle phosphofructokinase: effects of ligands, *Biochemistry,* 21, 216, 1982.

45. **Hofer, H. W.**, Influence of enzyme concentration on the kinetic behavior of rabbit muscle phosphofructokinase, *Hoppe-Seyler's Z. Physiol. Chem.,* 352, 997, 1971.

46. **Hulme, E. C. and Tipton, K. F.**, The dependence of phosphofructokinase kinetics upon protein concentration, *FEBS Lett.,* 12, 197, 1971.

47. **Parmeggiani, A., Lutt, J. H., Love, D. S., and Krebs, E. G.**, Crystallization and properties of rabbit skeletal muscle phosphofructokinase, *J. Biol. Chem.,* 241, 4625, 1966.

48. **Paradies, H. H.**, Structure of cross-linked rabbit muscle phosphofructokinase in solution, *J. Biol. Chem.,* 254, 7495, 1979.

49. **Goldhammer, A. R. and Paradies, H. H.**, Phosphofructokinase: structure and function, *Curr. Top. Cell. Regul.,* 15, 109, 1979.

50. **Gilbert, G. A.**, *Faraday Discuss. Chem. Soc.,* 20, 68, 1955.

51. **Gilbert, G. A.**, Sedimentation and electrophoresis of interacting substances. I. Idealized boundary shape for a single substance aggregating reversibly, *Proc. R. Soc., London, Ser. A.,* 250, 377, 1959.

52. **Gilbert, G. A.**, Sedimentation and electrophoresis of interacting substances. III. Sedimentation of a reversibly aggregating substance with concentration dependent sedimentation coefficients, *Proc. R. Soc., London, Ser. A,* 276, 354, 1963.

53. **Magar, M. E.**, *Data Analysis in Molecular Biology,* Academic Press, New York, 1973.

54. **Lad, P. M. and Hammes, G. G.**, Physical and chemical properties of rabbit muscle phosphofructokinase cross-linked with dimethyl suberimidate, *Biochemistry*, 13, 4530, 1974.

55. **Lad, P. M., Hill, D. E., and Hammes, G. G.**, Influence of allosteric ligands on the activity and aggregation of rabbit muscle phosphofructokinse, *Biochemistry*, 12, 4303, 1973.

56. **Hill, D. E. and Hammes, G. G.**, An equilibrium binding study of the interaction of fructose 6-phosphate and fructose 1,6-bis phosphate with rabbit muscle phosphofructokinase, *Biochemistry*, 14, 203, 1975.

57. **Wyman, J.**, Linked function and reciprocal effects in hemoglobin: a second look, *Adv. Protein Chem.*, 19, 223, 1964.

58. **Kemp, R. G. and Krebs, E. G.**, Binding of metabolites by phosphofructokinase, *Biochemistry*, 6, 423, 1967.

59. **Hummel, J. P. and Dreyer, W. J.**, Measurement of protein-binding phenomena by gel filtration, *Biochim. Biophys. Acta*, 63, 530, 1962.

60. **Wolfman, N. M., Thompson, W. R., and Hammes, G. G.**, Study of the interaction of adenylyl imidodiphosphate with rabbit muscle phosphofructokinase, *Biochemistry*, 17, 1813, 1978.

61. **Cann, J. R. and Hinman, N. D.**, Hummel-Dreyer gel chromatographic procedure as applied to ligand-mediated association, *Biochemistry*, 15, 4614, 1976.

62. **Foe, L. G., Latshaw, S. P., and Kemp, R. G.**, Binding of hexose biphosphates to muscle phosphofructokinase, *Biochemistry*, 22, 4601, 1983.

63. **Colombo, G., Tate, P. W., Girotti, A. W., and Kemp, R. G.**, Interaction of inhibitors with muscle phosphofructokinase, *J. Biol. Chem.*, 250, 9404, 1975.

64. **Höfer, H. W. and Furst, M.**, Isolation of a phosphorylated form of phosphofructokinase from skeletal muscle, *FEBS Lett.*, 62, 118, 1976.

65. **Hofer, H. W. and Sorensen-Ziganke, B.**, Phosphorylation of phosphofructokinse from skeletal muscle: correlations between phosphorylation and muscle function, *Biochem. Biophys. Res. Commun.*, 90, 199, 1979.

66. **Kemp, R. G., Foe, L. G., Latshaw, S. P., Poorman, R. A., and Heinrikson, R. L.**, Studies on the phosphorylation of muscle phosphofructokinase, *J. Biol. Chem.*, 256, 7282, 1981.

67. **Riquelme, P. T. and Kemp, R. G.**, Limited proteolysis of native and *in vitro* phosphorylated muscle phosphofructokinase, *J. Biol. Chem.*, 255, 4367, 1980.

68. **Foe, L. G. and Kemp, R. G.**, Properties of phospho and dephospho forms of muscle phosphofructokinase, *J. Biol. Chem.*, 257, 6368, 1982.

69. **Luther, M. A. and Lee, J. C.**, The role of phosphorylation in the interaction of rabbit muscle phosphofructokinase with F-actin, *J. Biol. Chem.*, 261, 1753, 1986.

70. **Liou, R.-S. and Anderson, S.**, Activation of rabbit muscle phosphofructokinase by F-actin and reconstituted thin filaments, *Biochemistry*, 19, 2684, 1979.

71. **Hill, A. V.**, A new mathematical treatment of changes of ionic concentration in muscle and nerve under the action of electric currents, with a theory as to their mode of excitation, *J. Physiol. (London)*, 40, 190, 1910.

72. **Kitajima, S., Sakakibara, R., and Uyeda, K.**, Significance of phosphorylation of phosphofructokinase, *J. Biol. Chem.*, 258, 13292, 1983.

73. **Kuo, H.-J., Malencik, D. A., Liou, R.-S., and Anderson, S. R.**, Factors affecting the activation of rabbit muscle phosphofructokinase by actin, *Biochemistry*, 25, 1278, 1986.

74. **Hofer, H. W. and Pette, D.**, Verfahren einer standardisierten extration und reinigung der phosphofructokinase aus kaninelenakclctmuscle, *Hoppe-Seyler's Z. Physiol. Chem.*, 349, 995, 1968.

75. **Arnold, H. and Pette, D.**, Binding of glycolytic enzymes to structure proteins of the muscle, *Eur. J. Biochem.*, 6, 163, 1968.

76. **Arnold, H. and Pette, D.**, Binding of aldolase and triosephosphate dehydrogenase to F-actin and modification of catalytic properties of aldolase, *Eur. J. Biochem.*, 15, 360, 1970.

77. **Arnold, H., Henning, R., and Pette, D.**, Quantitative comparison of the binding of various glycolytic enzymes to F-actin and the interaction of aldolase with G-actin, *Eur. J. Biochem.*, 22, 121, 1971.

78. **Melnick, R. L. and Hultin, H. O.**, Studies on the nature of the subcellular localization of lactate dehydrogenase and glyceraldehyde-3-phosphate dehydrogenase in chicken skeletal muscle, *J. Cell. Physiol.*, 81, 139, 1973.

79. **Clarke, F. M., Masters, C. J., and Winzor, D. J.**, Interaction of aldolase with the troponin-tropomyosin complex with bovine muscle, *Biochem. J.*, 139, 785, 1974.

80. **Clarke, F. M. and Masters, C. J.**, On the association of glycolytic enzymes with structural proteins of skeletal muscle, *Biochim. Biophys. Acta*, 381, 37, 1975.

81. **Clarke, F. M. and Masters, C. J.**, Interactions between muscle proteins and glycolytic enzymes, *Int. J. Biochem.*, 7, 359, 1976.

82. **Dagher, S. M. and Hultin, H. O.**, Association of glyceraldehyde-3-phosphate dehydrogenase with the particulate fraction of chicken skeletal muscle, *Eur. J., Biochem.*, 55, 185, 1975.

83. **Walsh, T. P., Clarke, F. M., and Masters, C. J.,** Modification of the kinetic parameters of aldolase on binding to the actin-containing filaments of skeletal muscle, *Biochem. J.,* 165, 165, 1977.
84. **Masters, C. J.,** Interactions between glycolytic enzymes and components of the cytomatrix, *J. Cell. Biol.,* 99, 2225, 1984.
85. **Kemp, R. G. and Foe, L. G.,** Allosteric regulatory properties of muscle phosphofructokinase, *Mol. Cell. Biochem.,* 57, 147, 1983.
86. **Gottschalk, M. E. and Kemp, R. G.,** Interaction of dinucleotides with muscle phosphofructokinase, *Biochemistry,* 20, 2245, 1981.
87. **Pette, D. and Brandau, H.,** Intracellular localization of glycolytic enzymes in cross-striated muscles of locusta migratoria, *Biochem. Biophys. Res. Commun.,* 9, 367, 1962.
88. **Sigel, P. and Pette, D.,** Intracellular localization of glycogenolytic and glycolytic enzymes in white and red rabbit skeletal muscle, *J. Histochem. Cytochem.,* 17, 225, 1969.
89. **Arnold, H., Nolte, J., and Pette, D.,** Quantitative and histochemical studies on the desorption and readsorption of aldolase in cross-striated muscle, *J. Histochem. Cytochem.,* 17, 314, 1969.
90. **Dolken, G., Leisner, E., and Pette, D.,** Immunofluorescent localization of glycogenolytic and glycolytic enzyme proteins and of malate dehydrogenase isozymes in cross-striated skeletal muscle and heart of the rabbit, *Histochemistry,* 43, 113, 1975.
91. **Walsh, T. P., Winzor, D. J., Clarke, F. M., Masters, C. J., and Morton, D. J.,** Binding of aldolase to actin containing filaments, *Biochem. J.,* 186, 89, 1980.
92. **Walsh, T. P., Masters, C. J., Morton, D. J., and Clarke, F. M.,** The reversible binding of glycolytic enzymes in ovine skeletal muscle in response to tetanic stimulation, *Biochim. Biophys. Acta,* 675, 29, 1981.
93. **Clarke, F. M., Shaw, F. D., and Morton, D. J.,** Effect of electrical stimulation post mortem of bovine muscle on the binding of glycolytic enzymes, *Biochem. J.,* 186, 105, 1980.

Chapter 11

YEAST PHOSPHOFRUCTOKINASE

Michel Laurent and Jeannine M. Yon

TABLE OF CONTENTS

I. INTRODUCTION

Phosphofructokinase (PFK) or ATP D-fructose-6-P-1 phosphotransferase (EC 2.7.1.11) is an enzyme of the glycolytic pathway which catalyzes the transfer of the terminal phosphate of ATP to the C-1 hydroxyl of the fructose 6-phosphate according to the following reaction:

$$\text{Fru-6-P} + \text{ATP} \xrightarrow{\text{Mg}^{++}} \text{Fru-1,6-P}_2 + \text{ADP}$$

This enzyme is highly specific for the hexose phosphate, but can react with a wide variety of nucleoside triphosphates such as phosphoryl donors. Its molecular characteristics as well as its regulation by various effectors depend on the source of the enzyme.

The chemical reaction catalyzed by PFK is exergonic, and therefore practically irreversible, and gluconeogenesis proceeds through another reaction catalyzed by fructose 1,6-bisphosphatase:

$$\text{Fru-1,6-P}_2 + \text{H}_2\text{O} \rightarrow \text{Fru-6-P} + \text{P}_i$$

which is also exergonic. Multiple regulations ensure the energetics of the cell. In some cells, glycolysis and gluconeogenesis occur in separate compartments. In other cells, such as liver cells, the activity of each enzyme depends on antagonistic regulations, the inhibitors of one enzyme being activators of the other. Phosphofructokinase is regulated by various metabolites in a manner that controls glycolysis according to the energy requirements of the cells.

The earliest studies of yeast phosphofructokinase by a long time precede the development of allosteric concepts. However, there is at least a semantic relationship between phosphofructokinase and allostery. Sols[1] relates as follows his first meeting with phosphofructokinase: "Phosphofructokinase . . . has long been known as a difficult enzyme. When I first encountered it, while studying hexokinase in Cori's laboratory in 1951, I heard comments about it being an "unreliable enzyme". A decade later, Carl Cori remarked that perhaps it was an "all hysteric enzyme". The term "hysteric" made its way into the tiny world of enzymologists. However, it is difficult to know if it designated either the enzymes which were claimed to be allosteric or the unfortunate ones who tried to study such an enzyme. Happily, as William James wrote:[2] "When a thing was new, people said 'It is not true'. Later, when the truth became obvious people said 'Anyway, it is not important', and when the importance could not be denied, people said 'Anyway, it is not new'." Today, since allostery is no longer new, specialists can quietly talk about it to examine if the concepts are realistic.

The question deserves our attention since the original Koshland-Néméthy-Filmer (KNF)[3] and Monod-Wyman-Changeux (MWC)[4] models fail for any enzyme as soon as a sufficient amount of experimental data are available. For instance, the only extensively studied enzyme for which the MWC model seems to satisfactorily describe the kinetic behavior, is phosphofructokinase from *Escherichia coli*.[5] However, in spite of the fact that reports on crystallographic studies[6] are not of much use to elucidate the overall mechanism of the allosteric transition (no X-ray diffraction data for the T conformation are yet available), earlier binding studies[7] suggested a much more complex mechanism than a simple two-state T-R transition.

Evidently, one cannot expect that the studies on yeast phosphofructokinase will provide answers on a putative universal mechanism for the allosteric transitions. Moreover, the knowledge of yeast phosphofructokinase is much more elementary than that for homologuous enzymes, especially bacterial enzymes. The primary structure of yeast phosphofructokinase is unknown and no crystallographic data are available. Furthermore, the oligomeric structure of yeast phosphofructokinase is much more complex than that of the bacterial enzyme.

Despite this lack of information, yeast phosphofructokinase may be considered as a good model for the study of the dynamical aspects of the allosteric transition. Yeast phosphofructokinase allosteric behavior has been characterized by means of several physical methods, and correlation may be tentatively established between functional and structural aspects of the conformational transition.

II. STRUCTURE OF YEAST PHOSPHOFRUCTOKINASE

Phosphofructokinase (PFK) displays a wide variety of structures in prokaryotic and eukaryotic organisms. The enzymes from prokaryotic organisms such as *E. coli* or *Bacillus stearothermophilus* are generally tetrameric and made up of four identical subunits; their molecular weight is around 150K.[8,9] The molecular weight of PFK from *E. coli* is 140K, but the active form of the isoenzyme PFK 2 is dimeric.[10] In mammals, PFKs are generally tetrameric with molecular weights ranging between 300 and 500K, according to the tissues. However, these enzymes undergo association or dissociation processes which depend on the pH, temperature, enzyme, or effector concentration. Polymeric species have been described with PFK from various origins: sheep heart,[11] chicken liver,[12] rabbit liver,[13] rat liver,[14,15] pig liver[16] and heart,[17] rabbit skeletal muscle,[18] and red cells,[19] and human red cells.[20] In rabbit muscle, the smallest active species have a molecular weight around 320 to 380K[21] and seem tetrameric with subunits of identical molecular weight.[22] By contrast, human red cells PFK contains two different types of subunits and has complex high molecular weight quaternary structures.[20] Aggregation of mammalian PFK can reach 20 times the molecular weight of the tetramer without any significant loss of activity; this phenomenon is only observed in the case of purified enzymes. Thus, polymerization does not seem to be involved in the regulation of PFK. Dissociation to dimeric and monomeric species has been reported for several PFKs, those from rabbit skeletal muscle,[21,23] pig liver,[24] and rabbit red cells.[19] These species are inactive.[17]

Like human erythrocyte enzyme, yeast phosphofructokinase also contains two types of subunits, α and β, differing slightly in their electrophoretic mobility.[25] They differ also in their immunological properties,[26] in their susceptibility to proteases in the presence of specific ligands,[27] and in their amino acid composition.[28] No aggregation, or conversely, no dissociation phenomena have been reported with yeast PFK. In contrast with the other PFKs, the yeast enzyme is not a tetramer. Although it was well established that the native enzyme is made up of an equal number of α and β-subunits, the number of subunits constituting the native enzyme was for a while matter of controversy.[29,30]

A. Oligomeric Structure of Yeast Phosphofructokinase

The discrepancies in the values of PFK molecular weight and quaternary structure can be explained, on the one hand, by the susceptibility of the enzyme to yeast proteases,[27] and the difficulties encountered in obtaining a "true native enzyme", and, on the other hand, by the difficulties in evaluating accurately the molecular weight of the oligomer. The enzyme is very sensitive to endogeneous proteases. Fragmentation products are easily detectable;[27] two minor bands α' and β' have been identified as corresponding, respectively, to proteolyzed forms of α- and β-subunits. For Tamaki et Hess,[29] α and β are converted into smaller molecular weight species (92 and 87K, respectively) upon hyaluronidase, triacylglycerol lipase, or neuraminidase treatment. However, there is neither carbohydrate nor lipid in the yeast enzyme phosphofructokinase.[25] Furthermore, it was shown that the enzyme contains no neutral sugar.[30] Therefore, the enzymatic treatment was supposed only to "desensitize" the α- and β-subunits toward dodecyl suflate. The subunit molecular weights of the hyaluronidase treated enzyme correspond to the molecular weight of the α'- and β'-species.[30] From this data, it was difficult to know whether the α'- and β'-components correspond to partially proteolyzed polypeptide chains[31] or to species which behave normally on sodium

Table 1
MOLECULAR WEIGHT OF YEAST
PHOSPHOFRUCTOKINASE

Method	Molecular weight (thousands)
X-ray scattering[30]	800 ± 50
Light scattering[30]	730 ± 30
Gel filtration on ultrogel AcA 22[30]	730 ± 50
Sepharose 6B[25]	720
X-ray scattering[38]	740
Equilibrium sedimentation[31]	835 ± 32

Table 2
APPARENT MOLECULAR WEIGHT OF YEAST PHOSPHOFRUCTOKINASE
SUBUNITS

Method	α	β	α′[a]	β′[a]
Electrophoresis 0,1% SDS[30]				
−7% Acrylamide	116 ± 7	110 ± 7	92 ± 6	88 ± 5
−5% Acrylamide	121 ± 7	113 ± 7	93 ± 6	88 ± 5
−4% Acrylamide	124 ± 7	113 ± 7	95 ± 5	88 ± 5
Gel filtration on Sepharose CL-6B in 6 M Gdn HCl[21]		110 ± 10[b]		
Electrophoresis 10% Acrylamide[29]	140	130	92	87
Electrophoresis 7% Acrylamide[31]	118 ± 3	112 + 3		

[a]   α′ and β′ are partially proteolyzed species.
[b]   α and β are not resolved in this case.

dodecyl sulfate gel electrophoresis upon hyaluronidase treatment.[29] On the other hand, analytical gel filtration on a Sepharose CL-6B column in the presence of 6 M Gdn HCl allowed the determination of an average molecular weight of 110 ± 10K for the α- and β-subunits which are not resolved under these conditions.[30] In order to minimize proteolytic degradations, procedures of purification eliminating any delay have been described.[30,32,34]

A hexameric structure has been suggested by several authors.[33,34,35] However, Kopper-schläger et al.[31] showed later that the molecular weight of the α- and β-subunits determined by gel electrophoresis in the presence of SDS, was overestimated by about 15% compared with the values obtained by equilibrium sedimentation. In contrast, using this last method, the molecular weight of the oligomeric enzyme (830K) was overestimated compared with the values obtained by both gel filtration and light scattering (720K),[29,34,35] This later data suggested an octameric structure. However, several ambiguities remained in the interpretation. More recently, the molecular weight of PFK was reinvestigated by Chaffotte et al.[30] The molecular weight of the oligomeric enzyme was estimated using several methods, analytical gel filtration, analytical ultracentrifugation, light scattering, and X-ray scattering measurements.[30] Table 1 summarizes the data. The molecular weight of the α- and β-subunits was also evaluated using gel electrophoresis in SDS, and analytical gel filtration in 6 M Gdn HCl. The results are given in Table 2. Thus, assuming an equal statistical weight for each kind of experiment, an average value of 775 ± 75K can be estimated for the molecular weight of the oligomer. Gel filtration and light scattering experiments certainly gave underestimated values. In the case of gel filtration, the lack of high molecular weight protein markers required a linear extrapolation which may be questionable. For light scat-

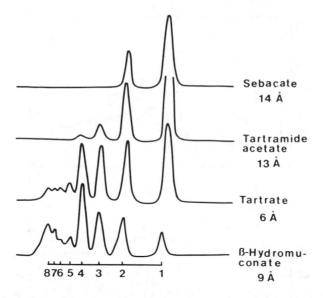

FIGURE 1.   Densitometric recordings of SDS polycrilamide gel electrophoresis of yeast phosphofructokinase cross-linked with disuccinimide esters of variable chain length. Cross-linking reaction was carried out for 20 min; longer time reaction time does not produce qualitative modifications of the corresponding densitograms.[30,37]

tering, the molecular weight estimation relied heavily on the determination of the dn/dc value. As several methods lead to underestimated values, the molecular weight of the native enzyme was considered to be slightly higher than 800K. An average molecular weight around 110K determined for the two types of subunits was compatible with an octameric quaternary structure. But taking into account the uncertainty, it was not possible to conclude to an oligomeric structure only from the molecular weight determinations.

Cross-linking experiments gave the most significant results. However, in an earlier report, Kopperschläger et al.[36] using dimethylsuberimidate, only obtained polymers up to tetrameric species, but not with higher molecular weight. In more recent papers,[30,37] the same kind of experiments were carried out with different cross-linking reagents; β-hydromuconate (9 Å) derivatives produce species ranging from the monomer to the octamer of yeast phosphofructokinase (Figure 1). Thus cross-linking patterns obtained from chemical reaction of yeast phosphofructokinase with β-hydromuconate demonstrate the octameric structure of the enzyme. This data differed significantly from that previously reported by Kopperschläger et al.[36] with dimethylsuberimidate. From X-ray diffusion data, Plietz et al.[38] proposed that the arrangement of the subunits within the oligomer is $\alpha_2 \beta_4 \alpha_2$, the radius of gyration increasing from 70 to 73 Å upon ATP binding.

B. Physicochemical Differentiation of the α- and β-Subunits of Yeast Phosphofructokinase

The two kinds of subunits have been differentiated by several of their properties. They differ slightly in their electrophoretic mobility in polyacrylamide gel in the presence of SDS.[29,31,34] The α- and β-subunits have distinct antigenic properties as shown by bidimensional immunoelectrophoresis.[25] Proteolytic degradation of the enzyme in the presence of specific ligands by yeast proteinase B or chymotrypsin, led to distinct α'- and β'-fragments, and attains a plateau corresponding to the formation of these fragments which are stable for several hours; proteinase A or subtilisin causes more complete proteolysis.[27] There is also

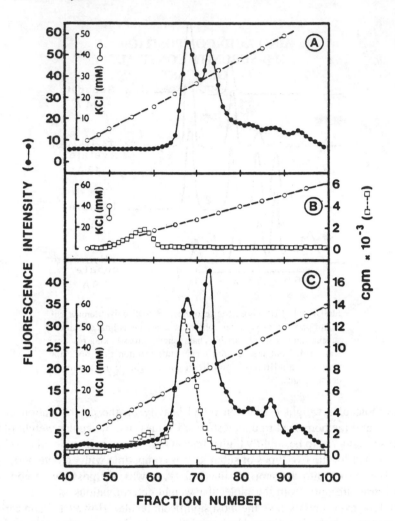

FIGURE 2.   Separation of yeast phosphofructokinase subunits by ion exchange
chromatography on DEAE-52 cellulose (A). Specific localization on the β-subunits
of -SH groups protected by fructose 6-phosphate from Mal NEt chemical modification (C). (A) Chromatography of chemically unmodified phosphofructokinase
under urea-denaturing conditions. (B) Chromatography of a solution of radioactive Mal NEt indicating the presence of a small amount of radioactive contam-
ination. (C) Separation of α- and β-subunits of yeast phosphofructokinase after
differential labeling of fructose 6-phosphate binding sites. See Reference 41 for
experimental details.

some genetic evidence indicating that the α- and β-chains are not coded by the same gene[40]
(see below). These subunits have been isolated by ion exchange chromatography under
conditions of denaturation.[28] The phosphofructokinase denatured in 8 *M* urea was chromatographied on a DEAE-52 cellulose column; α-subunits, more acidic than β-subunits,
are eluted with higher ionic strength. The two kinds of subunits are partially separated by
such a treatment (Figure 2); a second chromatography of the α- and β-pool on the DEAE-
52 cellulose column allowed a complete separation of the α- and β-subunits.[28]

This highly purified material was used for amino acid analysis and peptide mapping. The
amino acid composition of the two types of subunits reveals significant differences, especially
in the content of Ala, Phe, Ser, His, Ile, Leu, Met, and Arg (Table 3). The peptide maps
obtained after tryptic hydrolysis differ not only in the localization of the peptides, but also

Table 3
AMINO ACID COMPOSITION OF α
AND β-SUBUNITS CONSTITUTIVE
OF YEAST
PHOSPHOFRUCTOKINASE

Amino acid	α-Subunit	β-Subunit
Asp(Asn)	93.2	94.3
Glu(Gln)	91.6	92.7
Gly	74.1	71.5
Ala	74.7	95.2
Phe	36.8	25.5
His	22.4	17.7
Lys	58.5	54.9
Arg	46.2	51.7
Ile	62.1	68.8
Leu	83.8	77.5
Val	63.5	64.9
Thr	55.7	60.6
Ser	71.7	63.5
Tyr	23	22.7
Cys	11.1	8.9
Met	9.7	16.3
Trp	6.9	5.1
Pro	26.1	26.4

From Tijane, M. N., Chaffotte, A. F., Yon, J. M., and Laurent, M., *FEBS Lett*. 148, 267, 1982. With permission.

in the intensity of the ninhydrine staining of the major spots (Figure 3).[28] Greater differences were reported between the two polypeptide chains M and E constitutive of human erythrocyte PFK. While the most significant differences reside in the Ala content of the α- and β-chains of the yeast enzyme, it mainly concerns the (Asp + Asn) content of the two chains of the human erythrocyte PFK.[20]

All this data has revealed structural differences in α- and β-chains. The two types of subunits also have different functional properties. Using selective differential labeling of the SH groups in yeast PFK, it was possible to demonstrate that the substrate fructose 6-phosphate binding sites are localized exclusively on β-subunits.[41] The oligomer contains 83 ± 2 cysteinyl groups titratable when the protein is unfolded; in the native enzyme, these groups exhibit different reactivities toward 5-5′ dithiobis 2 nitrobenzoate or *N*-ethylmaleimide. In a first set of experiments, PFK was labeled by *N*-ethyl maleimide (Mal NEt) in the presence of fructose 6-phosphate. Under these conditions, the enzyme remained fully active. After removal of the substrate, 12 SH groups per oligomer were titratable either by NbS2 or by Mal NEt. The enzyme specifically labeled with radioactive Mal NEt in its fructose 6-phosphate binding sites was denatured further and chromatographied as previously described,[28] in order to separate the α- and β-subunits. The labeled reagent was found only in β-subunits (Figure 2). Thus, these results permit us to conclude that the substrate binding sites are localized exclusively on β-subunits. Although there are no direct indications concerning the localization of effector (ATP and AMP) binding sites on each type of subunits, the data constitute strong evidence for the absence of a direct role of α-subunits in the catalysis and suggest the existence of distinct catalytic and regulatory subunits in yeast phosphofructokinase.

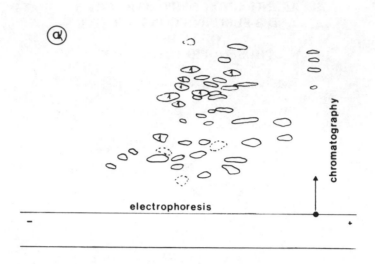

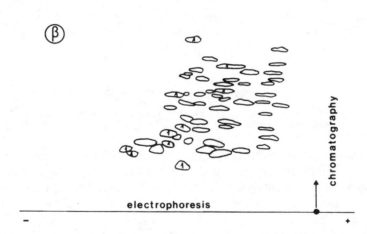

FIGURE 3.   Tryptic peptide maps of α- and β-subunits constitutive of yeast PFK. (According to Tijane, M. N., Chaffotte, A. F., Yon, J. M., and Laurent, M., *FEBS* Lett., 148, 267, 1982. With permission.)

C. Genetic Evidence for the Existence of Distinct Catalytic and Regulatory Subunits in Yeast Phosphofructokinase

Mutant studies[42] have shown that *Saccharomyces cerevisiae* possesses a second ATP: D-fructose 6-phosphate 1-phosphotransferase activity. This activity was attributed to a "particulate" phosphofructokinase or PFK 2. The "particulate" phosphofructokinase is a membrane-bound enzyme, and its regulatory properties are distinct from those of the well-known cytosolic phosphofructokinase: PFK 2 is not inhibited by ATP but is strongly inhibited by fructose 1,6-bisphosphate. The kinetic saturation curves are hyperbolic with respect to fructose 6-phosphate or ATP. Fructose 2,6-bisphosphate does not activate the "particulate" phosphofructokinase.

Lobo and Maitra[40] have isolated glucose-negative double mutants pfk1-pfk2 having neither the soluble nor the particulate phosphofructokinase. These mutants can be suppressed by the gene *PFK 1*. Two distinct genes (*PFK 2* and *PFK 3*) may be involved in the second mutation, so that the glucose negative strains may have either the genotype pfk1⁻ pfk2⁻ or pfk1⁻ pfk3⁻. By crossing a double mutant pfk1⁻ pfk2⁻ to a wild-type strain, spores growing on glucose and having the genotype pfk1$^+$ pfk2$^-$ are found. The phosphofructo-kinase activity of these spores is distinct from *PFK 1* or *PFK 2* activities, since it is neither inhibited by ATP nor activated by fructose 2,6-bisphosphate on the one hand, nor inhibited by fructose 1,6-biphosphate on the other hand. The allosteric properties of the enzyme isolated from spores pfk1$^+$ pfk2$^-$ can be restored by mixing it with the extract obtained from a pfk1$^-$ pfk2$^+$ strain. Experiments using a temperature-sensitive allele of the gene *PFK 2* indicate that it is the product of the *PFK 2* gene itself that confers on the catalytic subunit the property of ATP inhibition.

These studies support the conclusion that the allosteric properties of yeast phosphofruc-tokinase derive from the interaction of catalytic subunits specified by the gene *PFK 1* with regulatory subunits specified by the gene *PFK 2*.

Independent mutant studies[43] also conclude that one gene controls or specifies β-subunits while another gene specifies α-subunits, the mutations in the second gene being missense. These results would also accord with β having a catalytic function in the yeast phospho-fructokinase catalyzed reaction. It should be noted that the nomenclature of cytosolic phos-phofructokinase subunits is inversed in Lobo and Maitra's papers[40,42,44] compared to the conventions adopted by the other groups.

It is surprising to note that the gene which specifies regulatory subunits in the soluble yeast phosphofructokinase seems to be an essential determinant for the synthesis of the membrane-bound enzyme.[40] This observation was partially confirmed by Hüse and Kopperschläger[45] who have shown that antibodies obtained from rabbits against soluble yeast phosphofructokinase also react with the particulate yeast phosphofructokinase.

III. ALLOSTERIC REGULATION OF YEAST PHOSPHOFRUCTOKINASE

A. The Main Effectors of the Yeast Enzyme. Comparision with Phosphofructokinases from Other Sources

The activity of phosphofructokinase is regulated by a variety of intracellular metabolites, depending on the source. Unlike the bacterial enzymes, yeast phosphofructokinase appears to be "self-regulated", i.e., its effectors are produced by reactions which are very close to the catalyzed reaction in the metabolic pathway.

For instance, *E. coli* phosphofructokinase is activated by ADP and is strongly inhibited by phosphoenolpyruvate.[5] On the other hand, ATP and AMP have no effect on its activity. The allosteric behavior of yeast phosphofructokinase is different: phosphoenolpyruvate has no significant effect on the properties of the enzyme (at least in the physiological concen-tration range) and ADP is only a poor activator. But other adenylates (AMP and ATP) are strong effectors of yeast phosphofructokinase. Unlike the muscle enzyme, yeast phospho-fructokinase is insensitive to fructose 1,6-bisphosphate and citrate.[46] The fact that yeast phosphofructokinase is self-controlled has been proposed as a factor which may play a central role in the triggering of metabolic oscillations observed on yeast extracts (see Section IV).

The regulation of yeast phosphofructokinase essentially involves the following metabolites: fructose 6-phosphate (allosteric substrate); AMP and fructose 2,6-biphosphate (activators); ATP (substrate and inhibitor).

The molecular characteristics of these controls, for each metabolite, are examined below.

B. Molecular Mechanism of the Allosteric Regulation of Yeast Phosphofructokinase
1. Fructose 6-Phosphate as an Allosteric Substrate

The binding isotherm of fructose 6-phosphate to yeast phosphofructokinase is typical of a positive cooperative phenomenon.[47,48] However, there is a discrepancy for the binding stoichiometries obtained under distinct experimental conditions (four[48] or eight to ten[47] molecules of ligand bound per enzyme oligomer). The differential chemical labeling ex-

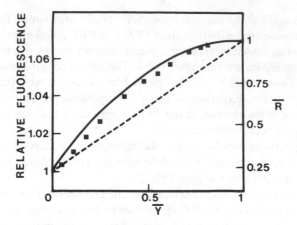

FIGURE 4. Nonlinear relationship between the state function
$\overline{R}$ and the degree of fructose 6-phosphate binding to yeast phosphofructokinase.[48] The state function $\overline{R}$ which is proportional
to the extent of conformational changes was calculated from
fluorimetric data.[48] The saturation function $\overline{Y}$ corresponds to
data obtained in binding experiments.[48] Solid curve corresponds
to a concerted transition assuming a nonexclusive coefficient c
= 0.02 and four interacting protomers (see Reference 48 for
experimental details). The dashed line illustrates the results
expected in the case of a sequential allosteric model.[3]

periments (see Section II) are in favor of four binding sites, unless we suppose that: (1)
either each β-subunit has two binding sites for fructose 6-phosphate; or (2) positive interactions exist between β-subunits (positive cooperative binding) and that strong negative
interactions do occur between α- and β-subunits (half-site reactivity).

Anyway, all the experiments indicate that the maximum Hill coefficient is significantly
lower than the number of total subunits. The Hill coefficient does not exceed 2.5, whether
for binding experiments or for kinetic experiments. The average Hill number calculated for
any fructose 6-phosphate kinetic saturation depends on the concentration of effector.[49] However, the dependence of the observed Hill number with respect to the half-saturation parameter
is adequately described[49] on the basis of the bell-shaped relationship[50] which characterizes
a K system in the concerted allosteric model of Monod, Wyman, and Changeux.[4]

The concerted character of the conformational transition which involves the subunits
sharing the substrate binding sites is further evidenced by plotting the extent of conformational
change as indicated by fluorimetric measurements as a function of the degree of fructose 6-
phosphate binding (Figure 4). This is a typical $\overline{R}$ (the state function) vs. $\overline{Y}$ (the saturation
function) plot. The nonlinearity of this relationship for the allosteric substrate may be taken
as an indication of the concerted character of the allosteric transition. However, this criteria
does not allow a much more complex sequential mechanism, such as the Haber and Koshland's model,[51] to be excluded. In agreement with the previous interpretation, the kinetic
constant for the fructose 6-phosphate mediated T-R transition, as monitored by fluorimetric
stopped-flow measurements,[48] is invariant with respect to substrate and enzyme concentrations, as could be expected for a concerted transition involving only two quaternary states.

2. ATP as a Michaelian Substrate and an Allosteric Inhibitor

ATP has two functions in the phophofructokinase catalyzed reaction. First, it acts as a
Michaelian substrate. The maximum velocity of the fructose 6-phosphate saturation curves
exhibits a Michaelian dependence with respect to the ATP concentration.[49] In the same

manner, the ATP saturation curves are biphasic. The lowest substrate concentrations correspond to the Michaelian saturation. An increase of the ATP concentration leads to a decrease in phosphofructokinase activity; this second phenomenon corresponds to the saturation of the regulatory ATP sites.[49] The two classes of ATP sites may be also characterized in binding experiments.

The binding of ATP is heterogeneous and is adequately described by assuming the existence of two independent classes of binding sites.[35,52] Kinetic data, as well as displacement ex periments in binding studies,[35] indicates that the tight ATP binding sites (K_d = 3 μM) corresponds to the regulatory sites. The dissociation constant for the second class of sites (K_d = 36 μM) compares well with the K_m value for ATP (31 μM).

However, the inhibition of yeast phosphofructokinase by ATP cannot be described in terms of the concerted transition model of Monod, Wyman, and Changeux.[4] The progressive displacement of the kinetic saturation curves for fructose 6-phosphate occurs in a large range of ATP concentrations.[48] On the contrary, Monod's model predicts that a steeper variation of the allosteric constant must occur in a narrower range of effector concentrations. In agreement with this observation, a local conformational change is evidenced by fluorescence experiments. A part of the quenching of the intrinsic fluorescence of the protein which occurs upon ATP binding to the regulatory sites[48] cannot be attributed to a shift of the allosteric equilibrium (R $\leftrightarrow$ T transition), but is related to a local conformational change (T $\leftrightarrow$ T' transition). The kinetic constant of the T $\leftrightarrow$ T' transition is much greater than the corresponding one for the T $\leftrightarrow$ R transition.[48]

The existence of a T' conformational state distinct from the T-state is further evidenced by X-ray scattering experiments[39] in solution. This technique is specifically sensitive to change in the overall structure of the object studied. In absence of any ligands, the allosteric T $\leftrightarrow$ R equilibrium is significantly shifted toward the T-state (L_o = T_o/R_o = 3 to 6), as indicated by steady-state kinetic data, binding studies, and fluorescence experiments.[48] The scattering curve obtained in an extended angular range for the ATP-saturated enzyme is significantly different from the curve obtained for the native enzyme or that obtained for the fructose 6-phosphate saturated enzyme (Figure 5). The two latter curves are quite similar, indicating that X-ray scattering fails to provide informations on the T $\leftrightarrow$ R transition. The differences between the two quaternary states are probably too small to be detected by this approach, at least in the angular range studied. On the other hand, the curve obtained for the ATP-saturated enzyme clearly demonstrates the existence of a T'-induced form distinct in quaternary structure from the T conformation: in the presence of ATP, an increase of the radius of gyration (of about 2 to 3 Å) of yeast phosphofructokinase is observed and the secondary minima and maxima of the scattering curves are shifted to lower angles.

Although authors agree for the qualitative interpretation of the complex mechanism of ATP regulation of yeast phosphofructokinase, the same discrepancy as for fructose 6-phosphate holds for the number of ATP binding sites. For Nissler et al.,[52] yeast phosphofructokinase binds two molecules of ATP per subunit. For Laurent et al.,[35] there are only four sites in each of the two classes of ATP binding sites.

3. AMP as an Allosteric Activator

5'-AMP is one of the two most efficient allosteric activators of yeast phosphofructokinase. In the absence of other effector, AMP binds monophasically to yeast phosphofructokinase, and the stoichiometry of binding is estimated to four molecules.[35,53] AMP may displace the ATP molecules bound to the regulatory binding sites, but it cannot at all displace ATP bound to the substrate binding site.[35] Fructose 2,6-bisphosphate does not influence the stoichiometry of AMP binding.[53] The stoichiometry of four AMP binding sites per enzyme oligomer and the hyperbolicity of the binding function have both been evidenced by the use of different experimental techniques (ultrafiltration, gel filtration, flow dialysis) in different laboratories.[30,35,53]

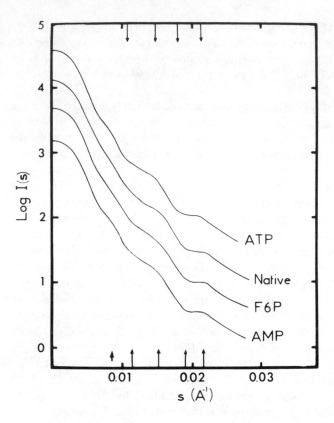

FIGURE 5. X-ray scattering curves of yeast phosphofructokinase in the presence or absence of ligands. The scattered
intensities are corrected for background and collimations distortions and vertically shifted to facilitate comparisons. The
secondary minima and maxima observed with native, Fru 6-P
and AMP-enzyme around s = 0.0119, 0.0149, 0.188, and
0.214 Å⁻¹ (fine arrows, bottom of the figure) are shifted to
lower angles (fine arrows, top of the figure) with the ATP
saturated enzyme. For s = 0.0086 A⁰⁻¹, a shoulder is clearly
visible (thick arrow, bottom of the figure) with the ATP and
AMP-enzyme. However, this difference is at the limit of statistical accuracy (see Reference 39 for experimental details).
(From Laurent, M., Tijane, M. N., Roucous, C., Seydoux,
F. J., and Tardieu, A., *J. Biol. Chem.*, 259, 3124, 1984. With
permission.)

As in the case of ATP, the effect of the AMP on yeast phosphofructokinase is complex.
For Nissler et al.,[54] AMP does not exhibit any effect on the allosteric equilibrium mediating
the cooperativity in fructose 6-phosphate binding (T ↔ R transition) and on the fructose 6-
phosphate affinity for the T- and R-states of the enzyme. The interdependence between AMP
activation and ATP inhibition has been kinetically investigated.[48] The activating effect of
AMP is strongly dependent on the ATP concentration. At low ATP concentration, the
activation by AMP is small and occurs in a range of concentrations (20 to 40 μM) which
is consistent with the value estimated for its dissociation constant (about 16 μM[35]). At higher
ATP concentrations, AMP activates the reaction up to a 100-fold. At nearly physiological
ATP concentrations, activation by AMP shows positive apparent cooperativity (apparent
Hill number = 1.75) but the inhibition of the reaction by ATP cannot be completely reversed
by AMP.[48]

Nissler et al.[54] interpreted similar data by assuming that AMP acts by inducing a R′ subconformation distinct from the initial R-state. However, no significant difference is evidenced between the X-ray scattering curves corresponding to the fructose 6-phosphate induced conformation and the AMP-induced conformation.[39] This observation does not mean that a R ↔ R′ transition does not occur, but only that if there is a local effect on the R-state, its amplitude is to small to be detected by solution X-ray scattering studies.

4. Fructose 2,6-Bisphophate as an Allosteric Activator

Our understanding of glycolysis has been questioned since Van Schaftingen et al.[55] and Furuya and Uyeda[56] discovered that the cytosol of various kinds of cells, including yeast cells, contains an important regulatory metabolite which was later identified as fructose 2,6-bisphosphate.

Fructose 2,6-bisophosphate is synthesized by 6-phosphofructo-2-kinase, an enzyme which was recently purified from yeast[57] and which seems to exist in two distinct forms. By analogy with 6-phosphofructo, 2-kinase from rat liver,[58] it is supposed that the two forms of the yeast enzyme correspond to a phosphorylated form and a nonphosphorylated one, respectively.

Fructose 2,6-bisphosphate appears to be the most powerful activator of yeast phosphofructokinase. The nature of the activation is of some controversy. For Avigad,[59] fructose 2,6-bisphosphate does not affect the maximum velocity of the catalyzed reaction with respect to fructose 6-phosphate. On the contrary, it does for Nissler et al.[60] and Bartrons et al.[61] For all the authors, however, fructose 2,6-bisphosphate acts as an allosteric activator by decreasing the half-saturation parameter of the enzyme for fructose 6-phosphate. The activation constant for fructose 2,6-bisphosphate is in the range 0.6 to 1.7 μM.[60] That means that yeast phosphofructokinase is at least one order of magnitude more sensitive to fructose 2,6-bisphosphate than to AMP. The reactivation by fructose 2,6-bisphosphate of the ATP-inhibited enzyme is cooperative[60] and the apparent Hill coefficient (about 1.5 at 3 mM ATP) is in the range of that observed for the reactivation by AMP. Of course, the value of the Hill coefficient is dependent on the ATP concentration. Until now, no mechanistic interpretation of the activating effect of fructose 2,6-bisphosphate to yeast phosphofructokinase has been given. However, it does not seem, unlike what happens with the rat liver enzyme,[62,63] that AMP and fructose 2,6-bisphosphate act synergistically in releasing the ATP-inhibition. At the cellular level, the regulation of phosphofructokinase by AMP is assumed to contribute to the homeostasis of ATP,[64] while fructose 2,6-bisphosphate seems to be involved in the coordination between glycolysis and gluconeogenesis.[65]

C. Models for the Allosteric Regulation of Yeast Phosphofructokinase

The following scheme was proposed[48] for the regulation of yeast phosphofructokinase:

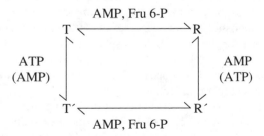

where R and R′, T and T′ are distinct quaternary states of the enzyme oligomer with low and high affinity for fructose 6-phosphate, respectively.

In this model, the transition from T to R (or from T′ to R′) is concerted, but the transition T to T′ (or R to R′) is sequential, i.e., proportional to ligand binding. The T′ form may be distinguished from the T and R conformations by X-ray scattering measurements. The concerted character of the allosteric transition mediated by fructose 6-phosphate results from the fact that the affinity of this substrate for the R and R' forms (or for the T and T' forms) are supposed to be identical. On the other hand, the inhibitory effect of ATP acts through a local "site by site" mechanism, i.e., it involves essentially the sequential T ↔ T' transition. The activating effect of AMP may involve simultaneously sequential (R ↔ R') and concerted (T ↔ R) transitions. For the derivation of the equations corresponding to this model, the number of interacting protomers is equal to four, according to binding studies. In the absence of any ligand, the allosteric constant L_o is around 3 to 6, i.e., the allosteric equilibrium is slightly, but significantly, shifted toward the T conformation. Some uncertainties remain concerning this model, particularly concerning the integration of the activating effect of fructose 2,6-bisphophate. The specific activities of the various forms are unknown.

This model accounts for the major characteristics of the regulation of yeast phosphofructokinase, i.e., it associates concerted homotropic interactions (for the allosteric substrate) and local heterotropic effects (ATP inhibition). This feature may be correlated to structural data, especially to the existence of two types of subunits in the enzyme oligomer. A series of experimental evidences (see Section I) suggest the existence of distinct catalytic and regulatory subunits in yeast phosphofructokinase. X-ray scattering experiments[38] and cross-linking patterns analysis[37] agree with an $\alpha_2\beta_4\alpha_2$ structure, i.e., the regulatory subunits would be located toward the periphery of the enzyme with catalytic subunits forming some kind of core. Such a model would allow tight interactions between β-subunits and looser ones between α-subunits. This could also explain why the T ↔ T' conformational changes are more easily detected by X-ray scattering measurements than those involving the T ↔ R transition: for a given displacement of a subunit, the further the subunit is from the center of gravity, the bigger the change in the radius of gyration.[66]

Although similar in its formalism, Reuter et al.[46] proposed a different four-state model for the regulation of yeast phosphofructokinase. In this model, two inactive subconformations R2 and T2 are degenerated from the active R and T conformational states, respectively. The inhibitor ATP would only bind to the R2 and T2 forms, while fructose 6-phosphate would bind only to the active R- and T-states. AMP would have the same affinity for the R and T conformations, but it would have no affinity for the R2- and T2-states. This model was built in order to explain why AMP and ATP shift the half-saturation parameter of the fructose 6-phosphate kinetic saturation curves without modifying the associated Hill number.[67]

This point is of particular interest concerning the problem (and the limits) of kinetic modeling. We argued in the introduction that the simple KNF and MWC allosteric models fail as soon as a sufficient amount of data is available. The two models proposed for the regulation of yeast phosphofructokinase both agree with the kinetic behavior of the enzyme under a given set of experimental conditions. These conditions especially concern the presence or the absence of phosphate in the kinetic medium, since it has been shown[49] that the presence of phosphate abolishes the invariance of the Hill number with respect to the ATP and AMP concentrations, for the fructose 6-phosphate saturation curves. This influence of phosphate on yeast phosphofructokinase was also reported independently by Banuelos et al.[68] Thus, probably the two models proposed for yeast phosphofructokinase represent two particular cases of a more general model.

The observation of distinct kinetic behavior under different experimental conditions is not really surprising, especially in the case of yeast phosphofructokinase. In fact, since the intrinsic allosteric constant has a value around 3 to 6 (at pH 6.8, in the presence of phosphate), the allosteric equilibrium is not drastically shifted toward one or other basic conformational state, in the absence of any ligand. Some experimental factors (controlled or uncontrolled) may contribute to modify this equilibrium and to lead to apparent discrepancies between results. This point is supported by Johansson and Kopperschläger's[69] observation that the partition of yeast phosphofructokinase between the liquid phases of an aqueous biphasic system changes drastically in the pH interval 7 to 8, in contrast to other proteins. A conformational change in the enzyme oligomer is supposed to be associated with the pH modification.[69]

IV. REGULATION OF YEAST PHOSPHOFRUCTOKINASE AND GLYCOLYTIC OSCILLATIONS

In their basic paper[4] on the allosteric model, Monod, Wyman, and Changeux concluded: "It is clear that the sigmoidal shape of the saturation curve (of an allosteric enzyme) may in itself offer a physiological advantage, since it provides the possibility of threshold effects in regulation". The concept of amplification upon response to a signal, such as ligand binding, is one of the two main properties which are invoked for the explanation of the molecular basis of glycolytic periodicities. We will discuss at the end of this section the limit of the relationship between allostery and amplification. The other important principle involved in most of the regulatory mechanisms of cell metabolism (and particularly in glycolytic oscillations) is control by feedback. Today, numerous patterns of metabolic control interactions are believed to result from a concerted, cooperative, and cumulative feedback inhibition.

In an enzymatic system, the enzyme reacts to the binding of substrate and effectors by inducing conformational changes of its own structure. In an open system, the enzyme conformation and its activity are adjusted according to the input and output conditions. The input is given by the concentration of a substrate or control metabolite, and the output is the rate of change of the concentration of a product of the regulated enzyme.

The discovery of periodic changes in the concentration of metabolites in various cells and cell-free extracts was of considerable influence on the development of dynamic conceptions about metabolic regulation.[70,71] Oscillatory phenomena in metabolic pathways are due either to the regulation of protein synthesis (epigenetic oscillations) or to the regulation of the activity of the enzyme itself (metabolic oscillations). From a thermodynamical point of view, oscillations represent a dissipative structure, i.e., a process of temporal self-organization beyond instability for a nonequilibrium stationary state. Among several examples, glycolytic oscillations are the best studied system, since research has been extended toward the elucidation of the molecular control mechanism of a single enzyme, such as control by allosteric interactions. There are many valuable reviews on oscillations, particularly on yeast glycolytic oscillations.[72-74] In the present paper devoted to allostery, we only focus our attention on the following particular point: how can we explain, from a mechanistic point of view, the occurrence of periodic changes in metabolic concentrations upon allosteric regulation of a single enzyme? This question is relevant for the illustration of the coupling between an elementary enzymatic reaction (which occurs at the molecular level) and a physiological phenomenon which is observed at the macroscopic level.

The fundamental principles of the relevant models[75,76] are identical. They suppose that phosphofructokinase obeys Monod, Wyman, and Changeux's[4] concerted transition theory and catalyzes the irreversible formation of a product which behaves as a positive effector of the enzyme. The system is open as the substrate is injected at a constant rate (or at a periodic rate in some experiments) while the product disappears according to first order kinetics. Quasi steady-state is assumed, i.e., the time scale for the catalytic or transconformational steps is supposed to be fast compared to the changes in metabolic concentrations. In these conditions, substrate and product concentrations are governed by a set of two nonlinear differential equations which can be solved by numerical methods of integration. The instability of the stationary state of the system may be studied by normal mode analysis, i.e., by determining whether infinitesimal perturbations in substrate or product concentrations

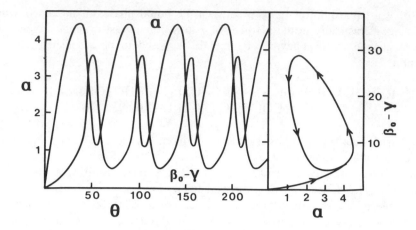

FIGURE 6. Simulation of the oscillating behavior of an open system containing yeast phosphofructokinase. α represents the normalized concentration of fructose 6-phosphate. β_0-γ is proportional to the quantity (AMP) + (ADP)/2. The normalized time corresponds to tV_M/K_{F6P}, where V_M is the maximum velocity of the phosphofructokinase catalyzed reaction and K_{F6P} is the dissociation constant of the R form of the enzyme for its allosteric substrate. This model takes into account some of the regulatory properties of yeast phosphofructokinase. However, the ratio of the dissociation constants of the R conformation of the enzyme for the allosteric substrate and the allosteric activator is overestimated by one order of magnitude (see text and Reference 77 for further details). The figure on the left shows the temporal behavior of the system while the figure on the right represents the corresponding phase plane analysis. The period of the simulated oscillations is 2.1 min with a phosphofructokinase activity of 10 mM/min. (From Demonqeot, J. and Seydoux, F. J., *Applications en Biologie,* Vol. 2, Delattre, P. and Thellier, M., Eds., Maloine, Paris, 1979, 519. With permission.)

from the steady-state will decay or grow in time. The dynamic behavior of the system may be visualized in a phase plane, i.e., a plane in which axes represent substrate and product concentrations, respectively (Figure 6).

Analysis of such models shows that in a large range of parameters values, the system undergoes sustained oscillations corresponding to a limit cycle around a nonequilibrium, instable stationary state.[76,77] At the molecular level, metabolic oscillations are thus supposed to result from the oscillations of the conformation of the enzyme between T- and R-states. There is a good agreement between model predictions and experimental observations of oscillations on yeast extracts or yeast cells. The correlation may be satisfactorily established for the oscillatory range of substrate injection rate, the period and the amplitude of the metabolite concentration changes, the periodic change in phosphofructokinase activity, and the phase-shift observed upon addition of the product ADP.[78,79] Recently, investigations have been carried out on chaotic dynamics in yeast glycolysis, i.e., on the randomness in time patterns of a nonlinear system resulting from periodic perturbations. This approach is of particular interest since the system may constitute an experimental model to study physiological processes which occur in cellular dynamic coupling, for instance in a neural system.[80] Again in this case, experiments[80] are in agreement with model predictions based on the regulatory properties of phosphofructokinase.[81]

Despite the fact that there is no doubt about the central role of phosphofructokinase in the triggering of glycolytic oscillations, some points remain to be discussed:

1.   The enzymatic model used for theoretical studies on glycolytic oscillations is oversimplified. For instance, the allosteric properties of *E. coli* phosphofructokinase were taken into account in order to explain yeast glycolytic oscillations.[76] This observation

is not relevant if we consider only the principle of the phenomenon. It is no longer true if quantitative correlations must be established between experimental and theoretical studies. For instance, we may note that Demongeot and Seydoux[77] concluded that it is necessary to take into account some parallel phenomena (such as the Fru 6-P $\leftrightarrow$ G6P equilibrium) in order to have a good agreement between model predictions on the conditions of oscillatory behavior and the values of yeast phosphofructokinase kinetic parameters as determined under *in vitro* conditions. Conversely, although model studies demonstrate that phosphofructokinase from *E. coli* must induce glycolytic oscillations, there is no experimental report on the occurrence of such a phenomenon in the bacterial cell.

2. Until now, experimental studies were unsuccessful in obtaining sustained oscillations in a soluble reconstituted enzyme system containing only phosphofructokinase as a controlling enzyme. For instance, Eschrich et al.[82] concluded that a reconstituted homogeneous system operating under open conditions and composed of phosphofructokinase, pyruvate kinase, adenylate kinase, and glucose 6-phosphate isomerase, was unable to produce sustained oscillations. On the contrary, the same authors[83] demonstrated that oscillations may be experimentally observed in a reconstituted enzyme system containing both yeast phosphofructokinase and fructose 1,6-bisphosphatase. However, fructose 1,6-bisphosphatase is not present in a large amount in yeast cells operating under glycolytic conditions, since it is an inducible enzyme which is synthetized under gluconeogenesis.[84]

3. If glycolytic oscillations are a periodic phenomenon and if circadian rhythms are another periodic phenomenon, there has been no evidence, until now, that the origin of circadian rhythms may be reduced to a coupling between some elementary (enzymatic or non-enzymatic) oscillators. Such an affirmation is often considered as obvious in papers dealing with the meaning of glycolytic oscillations observed in yeast cells and cell-free extracts.[85] Probably it is more useful to consider that glycolytic oscillations are only a predictable consequence of a regulated open pathway.

In the same way, we mentioned earlier in this chapter that the concept of amplificatory response to a signal was generally associated to the allosteric behavior of regulated enzymes. We saw later that the amplification properties (i.e., the nonlinear response) of allosteric enzymes are considered to be the microscopic event which allows metabolic oscillations. This supposes implicitly that the regulatory properties of enzymes determined under *in vitro* conditions are representative of enzyme behavior under physiological conditions. This is probably not often the case. For instance, one of the major differences between studies *in vivo* and *in vitro* is the enormous gap between enzyme concentrations used under these distinct conditions. It has been shown recently[86] that at physiological enzyme concentration, apparent cooperativity cannot be ascribed only to the structural properties of allosteric enzymes. The relative distribution between free and bound species must also be taken into account. A concentrated allosteric enzyme does not amplify substrate fluctuations much more than an equally concentrated Michaelian enzyme with the same number of independent substrate binding sites. Thus, is allostery the molecular event which allows oscillations? Of course, the concentration of phosphofructokinase is not particulary high in yeast cells (compared, for instance, to the intracellular concentration of glyceraldehyde 3-phosphate dehydrogenase). But the question of enzyme concentration constitutes only one example of the differences in conditions prevaling under *in vivo* and *in vitro* environments. The problem is the same for other different questions: important phenomena such as compartmentalization, channelling, etc. . . . may occur in the cell, and they prevent the enzyme from exhibiting identical kinetic behavior under *in vitro* and *in vivo* conditions.

In a recent report in which oscillations were presented as an example of a dynamic self-organization phenomena, Westerhoff,[87] argued that "the understanding of the spatiotemporal organization of metabolism arises from the breakdown of the common assumption that enzymes are dissolved in a homogeneous solution of metabolites". Such a remark is not new,[1,88] but it seems that now enzymologists are able not only to raise the question, but also to begin to provide some answers to the problems of cellular enzymology. Weber and Bernhard's[89] work is of particular interest in this respect. They studied experimentally the mechanisms of metabolic transfer between consecutive enzymes in a metabolic pathway. Their kinetic investigations on glycolysis clearly demonstrate that 1,3-diphosphoglycerate is directly transferred between glyceraldehyde 3-phosphate and 3-phosphoglycerate kinase via an enzyme-substrate-enzyme complex. Thus, the released intermediate metabolite 1,3-diphosphoglycerate does not disperse freely in the cytosol before proceeding along the glycolytic pathway. This observation is not independent of the problem of enzyme concentrations in the cell, since it is known that the molar concentrations of sites of glycolytic enzymes are in excess of the molar concentrations of the three-carbon intermediates of glycolysis.[89]

V. CONCLUSION

Yeast phosphofructokinase is structurally distinct from all other homologous enzymes isolated from various sources. While prokaryotic phosphofructokinases (*E. coli, B. stearothermophilus*) are tetrameric, the yeast enzyme is an octamer. In opposition to the mammalian phosphofructokinases, it does not undergo dissociation association processes upon variations of pH, temperature, or effector concentration. However, the molecular weight of the subunit (100K) is the same for the yeast phosphofructokinase as for the eukaryotic enzymes, the molecular weight of bacterial enzyme subunits being close to 40K.

As in the case of human phosphofructokinase, the yeast enzyme results from the association of two different types of subunits. Both chemical and genetic evidence seem to indicate that the two types of phosphofructokinase polypeptide chains correspond to distinct catalytic and regulatory subunits, the β-subunits being identified as the catalytic ones.

All phosphofructokinases are allosteric enzymes regulated by a wide variety of metabolites depending on the source. Unlike bacterial enzymes, yeast phosphofructokinase seems to be "self-regulated", its effectors being produced by reactions very similar to the reaction catalyzed by PFK in the metabolic pathway. For example, the enzyme from *E. coli* is activated by ADP and phosphoenol pyruvate; these effectors have no effect on yeast phosphofructokinase. Whereas the functional properties of *E. coli* phosphofructokinase fit well with Monod-Wyman-Changeux's concerted model, the main characteristics of yeast phosphofructokinase are the coupling between concerted homotropic interactions (with respect to the substrate) and local heterotropic modulations (regulatory effects). Such a mechanism is similar to that described for asparate transcarbamylase from *E. coli*,[90] an enzyme in which catalytic and regulatory subunits are unambiguously characterized. However, it should be noted that even for enzymes in which the catalytic and regulatory sites are located on the same polypeptide chain, such a dichotomy between homotropic and heterotropic effects may exist.[23,91] Hervé[92] suggests that this phenomenon may be significant in the evolution of proteins. Proteins are believed to evolve from molecules in which catalytic and regulatory sites are located on different subunits which are associated in an oligomeric structure to proteins in which these two kinds of sites are on the same polypeptide chain. The integrated structures represent an economy for the cell; furthermore they are more stable to the fluctuations of the environment. Such an organization would allow more complex interactions to occur including positive or negative interactions between the regulatory sites, in rabbit muscle phosphofructokinase[23] or glycogen phosphorylase.[91] However, in the case of phosphofructokinase, the reference to evolution is not evident since *E. coli* enzyme as well as muscle enzymes are made of a unique kind of subunit, whereas the yeast enzyme has two different kinds of subunits.

Because of its regulatory properties, and the different ways by which the enzyme activity can be modulated, phosphofructokinase has been considered to have a central role in the regulation of the glycolytic pathway and to be probably responsible for the oscillatory behavior under certain conditions. Since 1970, the metabolic step catalyzed by phosphofructokinase has been subject of extensive investigation, both in theoretical and experimental fields for researchers concerned with the mechanisms of metabolic regulation. The yeast enzyme has been a privileged target for the researchers interested in oscillations. We discussed earlier in this paper the limits of such an approach: the microenvironment in which metabolic activity actually occurs in cells is not equivalent to the conditions used to determine the properties of enzymes *in vitro*; nevertheless, these properties are extensively used to modelize metabolism and to evaluate key steps in metabolic regulation. Such a view seems to be now widely accepted and a general opinion may be reflected in this statement:[2] "Local domains and microenvironments in the aqueous compartments (and their surroundings) are not only important, but represent conditions that must first be studied and described before an understanding of metabolism and its regulation is possible".

Moreover, in the 5 last years, our knowledge of the metabolic reaction which transforms fructose 6-phosphate has been profoundly questioned, especially for yeast cells; until 1980, it was believed that this glycolytic step involved a complex but unique enzyme, phosphofructokinase. Several new facts, however, underline the involvment of other enzymes. In a number of cells, an enzyme other than phosphofructokinase also reacts with fructose 6-phosphate and ATP to produce fructose 2,6-bisphosphate. The fructose 2,6-bisphosphate is the most powerful activator of the phosphofructokinase from various sources. In yeast cells, a second ATP D-fructose 6-phosphate 1-phosphotransferase activity is present. This activity is due to a membrane-bound enzyme which has distinct regulatory properties compared to the cytosolic phosphofructokinase. Although their role has not yet been clearly elucidated, these factors cannot be ignored for the understanding of glycolytic regulation. The mechanisms which are responsible of such a regulation are certainly more complex than it was assumed until now.

REFERENCES

1. **Sols, A.**, Multimodulation of enzyme activity, *Curr. Top. Cell. Regul.*, 19, 77, 1981.
2. **Clegg, J. S.**, Properties and metabolism of the aqueous cytoplasm and its boundaries, *Am. J. Physiol.*, 246, 133, 1984.
3. **Koshland, D. E., Néméthy, G., and Filmer, D.**, Comparison of experimental binding data and theoretical models in proteins containing subunits, *Biochemistry*, 5, 365, 1966.
4. **Monod, J., Wyman, J., and Changeux, J. P.**, On the nature of allosteric transitions: a plausible model, *J. Mol. Biol.*, 12, 88, 1965.
5. **Blangy, D., Buc, H., and Monod, J.**, Kinetics of the allosteric interactions of phosphofructokinase from *E. Coli*, *J. Mol. Biol.*, 31, 13, 1968.
6. **Evans, P. R. and Hudson, P. J.**, Structure and control of phosphofructokinase from *Bacillus stearothermophilus*, *Nature (London)*, 279, 500, 1979.
7. **Blangy, D.**, Propriétés allostériques de la phosphofructokinase d'E. coli. Etude de la fixation des ligands par dialyse à l'équilibre, *Biochimie*, 53, 135, 1971.
8. **Blangy, D.**, Phosphofructokinase from *E. coli*. Evidence for a tetrameric structure of the enzyme, *FEBS Lett.*, 2, 109, 1968.
9. **Hudson, F., Hengartner, H., Kolb, E., and Harris, J. I.**, The primary structure of phosphofructokinase from *B. stearothermophilus*, *Proc. 12th FEBS Symp.*, 52, 341, 1979.
10. **Kotlarz, D. and Buc, H.**, Regulatory properties of phosphofructokinase from *E. coli*, *Eur. J. Biochem.*, 117, 569, 1981.

11. **Mansour, T. E., Wakid, N., and Sprouse, H. M.,** Studies on heart phosphofructokinase. Purification, crystallisation and properties of sheep heart phosphofructokinase, *J. Biol. Chem.*, 241, 1512, 1966.

12. **Kono, N. and Uyeda, K.,** Chicken liver phosphofructokinase. I. Crystallization and physiochemical properties, *J. Biol. Chem.*, 248, 8592, 1973.

13. **Kemp, R. G.,** Rabbit liver phosphofructokinase, *J. Biol. Chem.*, 246, 245, 1971.

14. **Brand, I. A. and Söling, H. D.,** Rat liver phosphofructokinase. Purification and characterization of its reaction mechanism, *J. Biol. Chem.*, 249, 7824, 1974.

15. **Dunaway, G. A. and Weber, G.,** Rat liver phosphofructokinase isozymes, *Arch. Biochem. Biophys.*, 82, 70, 1974.

16. **Trujillo, J. L. and Deal, W. C.,** Pig liver phosphofructokinase: asymmetry properties, proof of rapid association-dissociation equilibria, and effect of temperature and protein concentration on the equilibria, *Biochemistry*, 16, 3098, 1977.

17. **Mansour, J. E.,** Phosphofructokinase, *Curr. Top. Cell. Regul.*, 5, 1, 1972.

18. **Aaronson, R. P. and Frieden, C.,** Rabbit muscle phosphofructokinase: studies on the polymerization, *J. Biol. Chem.*, 247, 7502, 1972.

19. **Tarui, S., Kono, N., and Uyeda, K.,** Purification and properties of rabbit erythrocyte phosphofructokinase, *J. Biol. Chem.*, 247, 1138, 1972.

20. **Karadsheh, N. S., Uyeda, K., and Oliver, R. M.,** Studies on structure of human erythrocyte phosphofructokinase, *J. Biol. Chem.*, 252, 3515, 1977.

21. **Paetkau, V. and Lardy, H. A.,** Phosphofructokinase. Correlation of physical and enzymatic properties, *J. Biol. Chem.*, 242, 2035, 1967.

22. **Leonard, K. R. and Walker, I. O.,** The self-association of rabbit muscle phosphofructokinase, *Eur. J. Biochem.*, 26, 442, 1972.

23. **Lad, P. M. and Hammes, G. G.,** Physical and chemical properties of rabbit muscle phosphofructokinase cross linked with dimethyl suberimidate, *Biochemistry*, 13, 4530, 1974.

24. **Massey, T. H. and Deal, W. C.,** Unusual metabolite dependent solubility of phosphofructokinase, *J. Biol. Chem.*, 248, 56, 1973.

25. **Tamaki, N. and Hess, B.,** Purification and properties of phosphofructokinase (EC 2.7.1.11) of *Saccharomyces carlbergensis*, *Hoppe Seyler's Z. Physiol. Chem.*, 356, 399, 1975.

26. **Hermann, K., Diezel, W., Kopperschläger, G., and Hofmann, E.,** Immunological evidence for nonidentical subunits in yeast phosphofructokinase, *FEBS Lett.*, 36, 190, 1973.

27. **Hüse, K., Kopperschläger, G., and Hofmann, E.,** Differences in the degradation of yeast phosphofructokinase by proteinases A and B from yeast, *Biochem. J.*, 155, 721, 1976.

28. **Tijane, M. N., Chaffotte, A. F., Yon, J. M., and Laurent, M.,** Separation and chemical differentiation of α and β subunits in yeast phosphofructokinase, *FEBS Lett.*, 148 267, 1982.

29. **Tamaki, N. and Hess, B.,** Subunit structure of 6-phosphofructokinase from brewer's yeast, *Hoppe Seyler's Z. Physiol. Chem.*, 356, 1663, 1975.

30. **Chaffotte, A. F., Laurent, M., Tijane, M. N., Tardieu, A., Roucous, C., Seydoux, F., and Yon, J. M.,** Studies on the structure of yeast phosphofructokinase, *Biochimie*, 66, 49, 1984.

31. **Kopperschläger, G., Bär, J., Nissler, K., and Hofmann, E.,** Physiochemical parameters and subunit composition of yeast phosphofructokinase, *Eur. J. Biochem.*, 81, 317, 1977.

32. **Kopperschläger, G. and Johansson, G.,** Affinity partitioning with polymer-bound Cibacron blue F3G-A for rapid, large-scale purification of phosphofructokinase from baker's yeast, *Anal. Biochem.*, 124, 117, 1982.

33. **Kopperschläger, G., Lorenz, I., Diezel, W., Marquardt, I., and Hofmann, E.,** *Acta Biol. Med. Ger.*, 29, 561, 1972.

34. **Diezel, W., Böhme, H. J., Nissler, K., Freyer, R., Heilman, W., Kopperschläger, G., and Hofmann, E.,** A new purification procedure for yeast phosphofructokinase minimizing proteolytic degradation, *Eur. J. Biochem.*, 38, 479, 1973.

35. **Laurent, M., Chaffotte, A. F., Roucous, C., Tenu, J. P., and Seydoux, F. J.,** Binding of nucleotides AMP and ATP to yeast phosphofructokinase: evidence for distinct catalytic and regulatory subunits, *Biochem. Biophys. Res. Comm.*, 80, 646, 1978.

36. **Kopperschläger, G., Usbeck, E., and Hofmann, E.,** Studies on the oligomeric structure of phosphofructokinase by means of cross linking with diimido esters, *Biochem. Biophys. Res. Comm.*, 71, 371, 1976.

37. **Tijane, M. N., Seydoux, F. J., Hill, M., Roucous, C., and Laurent, M.,** Octameric structure of yeast phosphofructokinase as determined by cross linking experiments with dissuccinimidyl β-hydromuconate, *FEBS Lett.*, 105, 249, 1979.

38. **Plietz, R., Damashun, H., Kopperschläger, G., and Müller, J. J.,** Small angle X-ray scattering studies on the quaternary structure of phosphofructokinase from baker's yeast, *FEBS Lett.*, 91, 230, 1978.

39. **Laurent, M., Tijane, M. N., Roucous, C., Seydoux, F. J., and Tardieu, A.,** Solution X-ray scattering studies of the yeast phosphofructokinase allosteric transition, *J. Biol. Chem.*, 259, 3124, 1984.

40. **Lobo, Z. and Maitra, P. K.**, Genetic evidence for distinct catalytic and regulatory subunits in yeast phosphofructokinase, *FEBS Lett.*, 139, 93, 1982.
41. **Tijane, M. N., Chaffotte, A. F., Seydoux, F. J., Roucous, C., and Laurent, M.**, Sulfhydril groups of yeast phosphofructokinase. Specific localization on β subunits of fructose-6-phosphate binding sites as demonstrated by a differential chemical labeling study, *J. Biol. Chem.*, 255, 10188, 1980.
42. **Lobo, Z. and Maitra, P. K.**, A particulate phosphofructokinase from yeast, *FEBS Lett.*, 137, 279, 1982.
43. **Clifton, D. an Fraenkel, D. G.**, Mutant studies of yeast phosphofructokinase, *Biochemistry*, 21, 1935, 1982.
44. **Nadkarni, M., Lobo, Z., and Maitra, P. K.**, Particulate phosphofructokinase of yeast: physiological studies, *FEBS Lett.*, 147, 251, 1982.
45. **Hüse, K. and Kopperschläger, G.**, Interactions of antibodies against soluble phosphofructokinase with the soluble and particulate enzymes from yeast, *FEBS Lett.*, 155, 50, 1983.
46. **Reuter, R., Eschrich, K., Schellenberger, W., and Hofmann, E.**, Kinetic modelling of yeast phosphofructokinase, *Acta Biol. Med. Ger.*, 38, 1067, 1979.
47. **Nissler, K., Kessler, R., Schellenberger, W., and Hofmann, E.**, Binding of fructose 6-phosphate to phosphofructokinase from yeast, *Biochem. Biophys. Res. Commun.*, 79, 973, 1977.
48. **Laurent, M., Seydoux, F. J., and Dessen, P.**, Allosteric regulation of yeast phosphofructokinase. Correlation between equilibrium binding, spectroscopic and kinetic data, *J. Biol. Chem.*, 254, 7515, 1979.
49. **Laurent, M. and Seydoux, F. J.**, Influence of phosphate on the allosteric behavior of yeast phosphofructokinase, *Biochem. Biophys. Res. Commun.*, 78, 1289, 1977.
50. **Watts-Tobin, R. J.**, The oxidation of haemoglobin, *J. Mol. Biol.*, 23, 305, 1967.
51. **Haber, J. E. and Koshland, D. E.**, Relation of protein subunit interactions to the molecular species observed during cooperative binding of ligands, *Proc. Natl. Acad. Sci., U.S.A.*, 58, 2087, 1967.
52. **Nissler, K., Schellenberger, W., and Hofmann, E.**, Substrate binding to yeast phosphofructokinase: modelling of the Mg ATP binding, *Stud. Biophys.*, 2, 125, 1976.
53. **Kessler, R., Nissler, K., Schellenberger, W., and Hofmann, E.**, Fructose 2,6-bisphosphate increases the binding affinity of yeast phosphofructokinase to AMP, *Biochem. Biophys. Res. Commun.*, 107, 506, 1982.
54. **Nissler, K., Kessler, R., Schellenberger, W., and Hofmann, E.**, Effects of AMP and Cibacron blue F3G-A on the fructose 6-phosphate binding of yeast phosphofructokinase, *Biochem. Biophys. Res. Commun.*, 91, 1462, 1979.
55. **Van Schaftingen, E., Hue, L., and Hers, H. G.**, Control of the fructose 6-phosphate/fructose 1,6-bisphosphate cycle in isolated hepatocytes by glucose and glucagon, *Biochem. J.*, 192, 887, 1980.
56. **Furuya, E. and Uyeda, K.**, An activation factor of liver phosphofructokinase, *Proc. Natl. Acad. Sci. U.S.A.*, 77, 5861, 1980.
57. **Yamashoji, S. and Hess, B.**, Yeast 6-phosphofructo-2-kinase, *FEBS Lett.*, 172, 51, 1984.
58. **Van Schaftingen, E., Davies, D. R., and Hers, H. G.**, Inactivation of phosphofructokinase 2 by cyclic AMP-dependent protein kinase, *Biochem. Biophys. Res. Commun.*, 103, 362, 1981.
59. **Avigad, G.**, Stimulation of yeast phosphofructokinase activity by fructose 2,6-bisphosphate, *Biochem. Biophys. Res. Commun.*, 102, 985, 1981.
60. **Nissler, K., Otto, A., Schellenberger, W., and Hofmann, E.**, Similarity of activation of yeast phosphofructokinase by AMP and fructose 2,6-bisphosphate, *Biochem. Biophys. Res. Commun.*, 111, 294, 1983.
61. **Bartrons, R., Van Schaftingen, E., Vissers, S., and Hers, H. G.**, The stimulation of yeast phosphofructokinase by fructose 2,6-bisphosphate, *FEBS Lett.*, 143, 137, 1982.
62. **Uyeda, K., Furuya, E., and Luby, L. J.**, The effect of natural and synthetic D-Fructose 2,6-bisphosphate on the regulatory kinetic properties of liver and muscle phosphofructokinases, *J. Biol. Chem.*, 256, 8394, 1981.
63. **Van Schaftingen, E., Jett, M. F., Hue, L., and Hers, H. G.**, Control of liver 6-phosphofructokinase by fructose 2,6-bisphosphate and other effectors, *Proc. Natl. Acad. Sci. U.S.A.*, 78, 3483, 1981.
64. **Schellenberger, W., Eschrich, K., and Hofmann, E.**, Self-stabilisation of the energy charge in a reconstituted enzyme system containing phosphofructokinase, *Eur. J. Biochem.*, 118, 309, 1981.
65. **Hers, H. G. and Van Schaftingen, E.**, Fructose 2,6-bisphophate 2 years after its discovery, *Biochem. J.*, 206, 1, 1982.
66. **Luzzati, V. and Tardieu, A.**, Recent developments in solution X-ray scattering, *Annu. Rev. Biophys. Bioeng.*, 9, 1, 1980.
67. **Freyer, R., Eschrich, K., and Schellenberger, W.**, Kinetic modelling of yeast phosphofructokinase, *Stud. Biophys.*, 57, 123, 1976.
68. **Banuelos, M., Gancedo, C., and Gancedo, J. M.**, Activation by phosphate of yeast phosphofructokinase, *J. Biol. Chem.*, 252, 6394, 1977.
69. **Johanson, G. and Kopperschläger, G.**, Indications of pH-induced conformational changes in phosphofructokinase from baker's yeast, *Acta Biol. Med. Ger.*, 38, 1639, 1979.

70. **Hess, B. and Boiteux, A.,** Oscillatory phenomena in biochemistry, *Annu. Rev. Biochem.,* 40, 237, 1971.
71. **Heinrich, R., Rapoport, S. M., and Rapoport, T. A.,** Metabolic regulation and mathematical models, *Prog. Biophys. Mol. Biol.,* 32, 1, 1977.
72. **Goldbeter, A. and Caplan, S. R.,** Oscillatory enzymes, *Annu. Rev. Biophys. and Bioeng.,* 5, 449, 1976.
73. **Hess, B., Goldbeter, A., and Lefever, R.,** Temporal, spatial and functional order in regulated biochemical and cellular systems, *Adv. Chem. Phys.,* 38, 363, 1978.
74. **Schellenberger, W., Eschrich, K., and Hofmann, E.,** Self-organization of a glycolytic reconstituted enzyme system: alternate stable stationary states, hysteretic transitions and stabilisation of the energy charge, *Adv. Enzyme Regul.,* 19, 257, 1981.
75. **Sel'kov, E.,** Self-oscillations in glycolysis. I. A simple kinetic model, *Eur. J. Biochem.,* 4, 79, 1968.
76. **Goldbeter, A. and Nicolis, G.,** An allosteric enzyme model with positive feedback applied to glycolytic oscillations, *Prog. Theor. Biol.,* 4, 65, 1976.
77. **Demongeot, J. and Seydoux, F. J.,** Oscillations glycolytiques: modélisation d'un système minimum à partir des données physiologiques et moléculaires, in *Elaboration et Justification des Modèles. Applications en Biologie,* Vol. 2, Delattre, P. and Thellier, M., Eds., Maloine, Paris, 1979, 519.
78. **Boiteux, A., Goldbeter, A., and Hess, B.,** Control of oscillating glycolysis of yeast by stochastic, periodic and steady source of subtrate: a model and experimental study, *Proc. Natl. Acad. U.S.A.,* 72, 3829, 1975.
79. **Puissant, H.,** Rythmes Biologiques. Phénomènes de Regulation, Medecine thesis, University of Angers, France, 1979.
80. **Markus, M., Kuschmitz, D., and Hess, B.,** Chaotic dynamics in yeast glycolysis under periodic substrate input flux, *FEBS Lett.,* 172, 235, 1984.
81. **Decroly, O. and Goldbeter, A.,** Biorhythmicity, chaos and other patterns of temporal self-organization in a multiply regulated biochemical system, *Proc. Natl. Acad. Sci. U.S.A.,* 79, 6917, 1982.
82. **Eschrich, K., Schellenberger, W., and Hofmann, E.,** In vitro demonstration of alternate stationnary states in an open system containing phosphofructokinase, *Arch. Biochem. Biophys.,* 205, 114, 1980.
83. **Eschrich, K., Schellenberger, W., and Hofmann, E.,** Sustained oscillations in a reconstituted enzyme system containing phosphofructokinase and fructose 1,6-biphosphate, *Arch. Biochem. Biophys.,* 222, 657, 1983.
84. **Pontremoli, S. and Horecker, B. L.,** Fructose 1,6-diphosphatase, in *The Enzymes,* Vol. 8, Boyer, P.D., Ed., Academic Press, New York, 1971, 611.
85. **Hess, B.,** The glycolytic oscillator, *J. Exp. Biol.,* 81, 7, 1979.
86. **Laurent, M. and Kellershohn, N.,** Apparent cooperativity for highly concentrated michaelian and allosteric enzymes, *J. Mol. Biol.,* 174, 543, 1984.
87. **Westerhoff, H. V.,** Organization in the cell soup, *Nature (London),* 318, 106, 1985.
88. **Srere, P. A.,** Enzyme concentration in tissues, *Science,* 158, 936, 1967.
89. **Weber, J. P. and Bernhard, S. A.,** Transfer of 1,3-disphosphoglycerate between glyceraldehyde-3-phosphate dehydrogenase and 3-phosphoglycerate kinase via an enzyme-substrate-enzyme complex, *Biochemistry,* 21, 4189, 1982.
90. **Thiry, L. and Hervé, G.,** The stimulation of Escherichia coli aspartate transcarbamylase activity by adenosine triphosphate, *J. Mol. Biol.,* 125, 515, 1978.
91. **Wang, J. H. and Tu, J. I.,** Allosteric properties of glutaraldehyde-modified glycogen phosphorylase b, *J. Biol. Chem.,* 245, 176, 1970.
92. **Hervé, G.,** Is the association of concerted homotropie cooperative interactions and local heterotropie effects a general basic feature of regulatory enzymes?, *Biochimie,* 63, 103, 1981.

Chapter 12

CONCLUDING REMARKS

Guy Hervé

Nearly 25 years have elapsed since the original formulations of the theoretical models aimed at explaining the behavior of allosteric enzymes were published.[1,2] During this period, these models have undergone converging modifications which were imposed by experimental observations.[3-5] They were originally restricted to the explanation of cooperative binding of ligands to proteins. The recent development of formalisms aimed at describing the influence of subunit interactions on the catalytic parameters of enzymes is an important new step.[6] About a dozen regulatory enzymes have now been studied in sufficient detail for meaningful comparison with each other and with the theoretical models. However, it is clear that precise conclusions can be drawn mainly when detailed structural information is available.

From this landscape, the impression emerges that each of these enzymatic systems has its own logic and that, as in other aspects of biological phenomena, nature has found a variety of solutions to resolve similar problems. However, some allosteric enzymes do share common properties, that others do not have. Some of these properties are in accordance with the theoretical models and others are not. Extreme situations are encountered, such as the fact that monomeric enzymes can be sensitive to effectors (lactobacillus ribonucleotide reductase,[7] *Escherichia coli* carbamylphosphate synthetase) or show sigmoidal saturation curves (mnemonic enzymes[8]). Some oligomeric enzymes such as glutamine-PRPP aminotransferase[9] or particular aspartate transcarbamylases[10] exhibit homotropic cooperative interactions between the catalytic sites only in the presence of effectors. In other cases, heterotropic interactions operate without any manifestation of cooperativity between the catalytic sites (oligomeric ribonucleotide reductase[7]). Structural studies have shown that most oligomeric regulatory enzymes exist in several different states of aggregation, although it is only in rare cases that this phenomenon is likely to be of physiological significance (carbamylphosphate synthetase, muscle phosphofructokinase,[11] *E. coli* ribonucleotide reductase[7]). In some cases, this property might be part of the mechanism involved in the modulation of activity by effectors (*E. coli* carbamylphosphate synthetase).

According to early predictions,[1] it is a general observation that catalytic sites and regulatory sites are distinct; however, when substrates and effectors have similar structures, additional regulatory processes through "squatting" phenomenon[12] must be of physiological significance (ribonucleotide reductase,[7] aspartokinase[15]).

Sophisticated situations are encountered when allosteric regulation is associated to modulation of enzyme activity through covalent modification (muscle phosphofructokinase,[11] glycogene phosphorylase,[14] glutamine synthetase[16]).

It is more and more frequently observed in oligomeric enzymes that catalytic sites are located in the region of contact between adjacent subunits and involve amino acid side chains belonging to both the neighboring subunits. In some allosteric enzymes, the quaternary structure change might increase the efficiency of the catalytic sites through influencing such intersubunit contacts (glutamine synthetase,[16] aspartate transcarbamylase,[17] citrate synthase[18]). The same phenomenon applies to domain interactions and is evoked in the description of the homotropic cooperative interactions between the catalytic sites in aspartate transcarbamylase.[19]

The importance of the quaternary structure changes in the cooperative binding of substrates to allosteric enzymes is extensively documented. However, in this regard, it is interesting to note that the three-dimensional structure of yeast phosphofructokinase[13] and *Acinetobacter* citrate synthase[18] is more extensively affected by the effectors than by the substrates.

It is striking that the catalytic and regulatory properties of virtually all allosteric enzymes are sensitive to phosphate, an observation which must be of significance under the physiological conditions. This is probably an "environmental" property as it was previously defined.[20]

Finally, it appears as a general conclusion that a simple two-state equilibrium is too simple to account satisfactorily for the allosteric properties of the regulatory enzymes described here.

As far as the heterotropic interactions between regulatory and catalytic sites are concerned, it seems that the "primary-secondary effects" model which was proposed to account for the influences of effectors on aspartate transcarbamylase[17] is of a more general significance than solely explaining these interactions in this enzyme. A similar conclusion was reached in the case of the regulation of yeast phosphofructokinase activity by AMP and ATP.[13-21] In a more general way, this mechanism seems to apply also to muscle phosphofructokinase,[22] glycogene phosphorylase,[23] and Acinetobacter citrate synthase.[24] This conclusion is suggested by the fact that when the extreme T and R conformations of these enzymes are "frozen" through the use of chemical cross-linking reagents, the noncooperative extreme conformations of these enzymes which are obtained are still sensitive to their effectors. This phenomenon is especially impressive in the case of Acinetobacter citrate synthase whose native form exhibits homotropic cooperative interactions not only between the catalytic sites, but also between the regulatory sites for effectors binding. The treatment of this enzyme with the cross-linking reagent abolishes both these two types of interactions, but the "frozen" extreme conformations obtained still show the heterotropic interactions.[24]

The elegant experiment reported recently by Tat-Kwong Lau and Fersht[25] strongly suggests that this mechanism also applies to *E. coli* phosphofructokinase. The inhibition of this enzyme by phosphoenolpyruvate was converted to activation as the result of changing a single amino acid residue through site-directed mutagenesis. The important point is that this modified form of enzyme is activated by phosphoenolpyruvate in spite of the fact that this effector still has a higher affinity for the T-state of the enzyme. This feature shows unambiguously that the effector does not act through shifting the $T \rightleftharpoons R$ equilibrium involved in the homotropic cooperative interactions between the catalytic sites.

It has been concluded in the case of *E. coli* aspartate transcarbamylase that the primary effects of its two effectors ATP and CTP do not correspond to opposite directions of the same transition.[17] The uncoupling of AMP and NADH influences on the activity of *Pseudomonas* citrate synthase suggests the same conclusion in the case of this enzyme.[18] Interestingly, the same conclusion is reached concerning the mechanisms of action of AMP and IMP, two activators of glycogene phosphorylase *b*.[14]

It is also a general observation that the polarity of the "primary effects" in heterotropic interactions lies on very subtle structural features. A small modification in the structure of the effector, of the enzyme, or of the substrate can either abolish or, more impressively, inverse the influence of the effector. Among the numerous examples of this phenomenon are the following:

AMP is a strong inhibitor of glutamine-phosphoribosylpyrophosphate amidotransferase, but 3'-iso-AMP is an inhibitor.[9] UMP is a very potent inhibitor of *E. coli* carbamylphosphate synthetase, while dialdehyde-UMP and Ara-UMP are activators. As a consequence of a single amino acid replacement in *E. coli* phosphofructokinase, phosphoenolpyruvate becomes an activator of this enzyme instead of an inhibitor.[25] The forward reaction catalyzed by *E. coli* aspartate transcarbamylase is stimulated by ATP; the reverse reaction is inhibited by this nucleotide.[27] When the activity of this enzyme is measured in the presence of acetylphosphate as pseudosubstrate instead of carbamylphosphate, ATP has no effect while CTP still behaves as an inhibitor[29] In the presence of a particular pseudosubstrate, dCMP, aminohydrolase activity is activated by dTTP and inhibited by dCTP instead of the contrary.[28]

The precise answers to the questions dealing with the mechanisms of allosteric regulation lie in the domain of the physical chemistry of proteins, and the results to be obtained in the coming years will benefit both this field and that of enzymology of metabolic regulation. Among the interesting aspects which are under development, the implication of the fast movements in proteins in their regulatory properties,[26] and the mechanisms of long-range regulatory effects which do not involve the quaternary structure change[17] are especially attractive. The year 2000 version of this book should be fascinating.

REFERENCES

1. **Monod, J., Wyman, J., and Changeux, J. P.,** On the nature of allosteric transitions: a plausible model, *J. Mol. Biol.,* 12, 88, 1965.
2. **Koshland, D. E., Nemethy, G., and Filmer, D.,** Comparison of experimental binding data and theoretical models in proteins containing subunits, *Biochemistry,* 5, 365, 1966.
3. **Haber, J. E. and Koshland, D. E., Jr.,** Relation of protein subunit interactions to the molecular species observed during cooperative binding of ligands, *Proc. Natl. Acad. Sci. U.S.A.,* 58, 2087, 1967.
4. **Buc, H., Johannes, K. J., and Hess, B.,** Allosteric kinetics of pyruvate kinase of *Saccharomyces carlbergenis, J. Mol. Biol.,* 76, 199, 1973.
5. **Viratelle, O. and Seydoux, F. J.,** Pseudo-conservative transition: a two state model for the cooperative behavior of oligomeric proteins, *J. Mol. Biol.,* 92, 193, 1975.
6. **Ricard, J.,** Concepts and models of enzyme cooperativity, in *Allosteric Enzymes,* Hervé, G., Ed., CRC Press, Boca Raton, FL, 1989, chap. 1.
7. **Eriksson, S. and Sjöberg, B. M.,** Ribonucleotide reductase, in *Allosteric Enzymes,* Hervé, G., Ed., CRC Press, Boca Raton, FL, 1989, chap. 8.
8. **Ricard, J., Meunier, J. C., and Buc, J.,** Regulatory behavior of monomeric enzymes. I. The mnemonical enzyme concept, *Eur. J. Biochem.,* 49, 195, 1974.
9. **Switzer, R. L.,** Regulation of bacterial glutamine phosphoribosyl-pyrophosphate amidotransferase, in *Allosteric Enzymes,* Hervé, G., Ed., CRC Press, Boca Raton, FL, 1989, chap. 5.
10. **Bethell, M. R. and Jones, M. E.,** Molecular size and feed-back regulation characteristics of bacterial aspartate transcarbamylases, *Arch. Biochem. Biophys.,* 134, 352, 1969.
11. **Lee, J. C., Hesterberg, L. K., Luther, M. A., and Cai, G. Z.,** Rabbit muscle phosphofructokinase, in *Allosteric Enzymes,* Hervé, G., Ed., CRC Press, Boca Raton, FL, 1989, chap. 10.
12. **Mazat, J. P. and Mazat, F.,** Double-site enzyme and squatting: where one regulatory ligand is also a substrate of the reaction, *J. Theor. Biol.,* 121, 89, 1986.
13. **Laurent, M. and Yon, J. M.,** Yeast phosphofructokinase, in *Allosteric Enzymes,* Hervé, G., Ed., CRC Press, Boca Raton, FL, 1989, chap. 11.
14. **Johnson, L. N., Hajdu, J., Acharya, K. R., Stuart, D. I., McLaughlin, P. J., Oikonomakos, N. G., and Barford, D.,** Glycogen phosphorylase *B,* in *Allosteric Enzymes,* Hervé, G., Ed., CRC Press, Boca Raton, FL, 1989, chap. 4.
15. **Mazat, J. P., Langla, J., and Mazat, F.,** Modèle de squatting; evolution possible du mode de régulation des aspartokinases, in *L'Evolution des Protéines,* Hervé, G., Ed., Masson, Paris, 1983, 67.
16. **Rhee, S. G., Boon Chock, P., and Stadtman, E. R.,** Regulation of *Escherichia coli* glutamine synthetase, in *Allosteric Enzymes,* Hervé, G., Ed., CRC Press, Boca Raton, FL, 1989, chap. 5.
17. **Hervé, G.,** Aspartate transcarbamylase from *Escherichia coli,* in *Allosteric Enzymes,* Hervé, G., Ed., CRC Press, Boca Raton, FL, 1989, chap. 3.
18. **Weitzman, P. D. J.,** Citrate synthase, in *Allosteric Enzymes,* Hervé, G., Ed., CRC Press, Boca Raton, FL, 1989, chap. 7.
19. **Ladjimi, M. M. and Kantrowitz, E. R.,** A possible model for the concerted allosteric transition as deduced from site-directed mutagenesis studies, *Biochemistry,* 27, 276, 1988.
20. **Hervé, G.,** Evolution des protéines et évolution des enzymologistes, in *L'Evolution des Protéines,* Hervé, G., Ed., Masson, Paris, 1983, 1.
21. **Hervé, G.,** Is the association of concerted homotropic cooperative interactions and local heterotropic effects a general basic feature of regulatory enzymes?, *Biochimie,* 63, 103, 1981.
22. **Lad, P. M. and Hammes, G. G.,** Physical and chemical properties of rabbit muscle phosphofructokinase cross-linked with dimethyl suberimidate, *Biochemistry,* 13, 4530, 1974.
23. **Wang, J. H. and Tu, J. I.,** Allosteric properties of glutaraldehyde-modified glycogen phosphorylase B. Selective desentization of homotropic cooperativity, *J. Biol. Chem.,* 245, 176, 1970.

24. **Mitchell, C. G. and Weitzman, P. D. J.,** Reversible effects of cross-linking on the regulatory cooperativity of acinetobacter citrate synthase, *FEBS Lett.,* 151, 260, 1983.

25. **Tat-Kwang Lau, F. and Fersht, A. R.,** Conversion of allosteric inhibition to activation in phosphofructokinase by protein engineering, *Nature (London),* 326, 811, 1987.

26. **Karplus, M.,** Structure and function of hemoglobin: the cooperative mechanism, in *Allosteric Enzymes,* Hervé, G., Ed., CRC Press, Boca Raton, FL, 1989, chap. 2.

27. **Foote, F. and Lipscomb, W. N.,** Kinetics of aspartate transcarbamylase from *Escherichia coli* for the reverse direction of reaction, *J. Biol. Chem.,* 256, 11428, 1981.

28. **Whitehead, E. P. and Rossi, M.,** Allosteric interactions in deoxycytidylate aminohydrolase, in *Allosteric Enzymes,* Hervé, G., Ed., CRC Press, Boca Raton, FL, 1989, chap. 9.

29. **Tauc, P. and Hervé, G.,** unpublished observation.

INDEX